# Discontinuous Deformation Analysis in Rock Mechanics Practice

ISRM Book Series
Series editor: Xia-Ting Feng
*Institute of Rock and Soil Mechanics, Chinese Academy of Sciences, Wuhan, China*

ISSN : 2326-6872
eISSN: 2326-778X

Volume 5

International Society for Rock Mechanics

ISRM

# Discontinuous Deformation Analysis in Rock Mechanics Practice

## Yossef H. Hatzor
*Ben-Gurion University of the Negev, Beer-Sheva, Israel*

## Guowei Ma
*The University of Western Australia, Crawley, WA, Australia*

## Gen-hua Shi
*DDA Company, Belmont, CA, USA*

**CRC Press**
Taylor & Francis Group
Boca Raton  London  New York

CRC Press is an imprint of the
Taylor & Francis Group, an **informa** business

A BALKEMA BOOK

Published by:
CRC Press/Balkema
P.O. Box 447, 2300 AK Leiden, The Netherlands
e-mail: Pub.NL@taylorandfrancis.com
www.crcpress.com – www.taylorandfrancis.com

First issued in paperback 2021

© 2018 by Taylor & Francis Group, LLC
CRC Press/Balkema is an imprint of Taylor & Francis Group, an informa business

No claim to original U.S. Government works

ISBN-13: 978-1-03-209665-0 (pbk)
ISBN-13: 978-1-138-02768-8 (hbk)

Typeset by MPS Limited, Chennai, India

**Visit the Taylor & Francis Web site at**
**http://www.taylorandfrancis.com**

**and the CRC Press Web site at**
**http://www.crcpress.com**

*Library of Congress Cataloging-in-Publication Data*

# Dedications

Yossef dedicates this book to his beloved wife Orna who left behind a promising career in the arts in Israel and followed him to Berkeley for his PhD studies. Her support during those difficult times and ever since has made this work possible and has made him a much better person.

"שלי ושלכם שלה הוא... (נדרים, נא)".

Guowei wishes to express his deep love and appreciation to his wife and two kids for their patience, love, and support in every aspect in life. They are the greatest motivation for his career pursuit including the works in this book.

# Table of contents

# About the authors

**Yossef H. Hatzor**
Lemkin Professor of Rock Mechanics, Department of Geological and Environmental
Sciences, Ben-Gurion University of the Negev, P.O.B. 653, Beer-Sheva, 84105, Israel
Tel: +972-8-6472621
Fax: +972-8-6428717
E-mail: hatzor@bgu.ac.il

 Yossef H. Hatzor is a graduate of the geotechnical engineering program of the civil engineering department at U. C. Berkeley, 1992. He did his PhD research in rock mechanics and geological engineering under the supervision of Professor Richard E. Goodman. Prof. Hatzor currently functions as chair of the department of geological and environmental sciences, the director of the engineering geology program and of the rock mechanics laboratory at BGU.

Professor Hatzor focuses on the development, verification, validation, and application of Block Theory, the numerical Discontinuous Deformation Analysis, and the numerical Manifold Method, by developing analytical solutions that can be used as a basis for code verification, conducting laboratory experiments that can be used for code validation, and monitoring rock mass deformation in the field for addressing key rock mechanics issues. Professor Hatzor is involved in major geotechnical engineering projects in Israel, including dynamic stability analysis and reinforcement design of Masada world heritage site.

Professor Hatzor is the founding president of Israel Rock Mechanics Association, an ISRM national group, and served as IRMA president between 2003 and 2014. In 2007 Professor Hatzor was appointed Chair Professor in Rock Mechanics by BGU Senate. In 2011 Professor Hatzor won the competitive visiting professorship appointment for leading international scientists by the Chinese Academy of Sciences and he holds this position at the Rock and Soil Mechanics Institute of the CAS in Wuhan, where he collaborates with Professor Xia-Ting Feng and colleagues from the National Key Laboratory of Geomechanics and Geotechnical Engineering on projects involving deep tunnels in various geological conditions in China. Professor Hatzor is founder and co–chair of the ISRM DDA Commission. He is the recipient of several national and international research awards and is a member of the editorial boards of both the *International Journal of Rock Mechanics and Mining Sciences*, and *Rock Mechanics and Rock Engineering*.

**Guowei Ma** (马国伟教授)
Professor, School of Civil, Environmental and Mining Engineering
Faculty of Engineering, Computing and Mathematics
The University of Western Australia
35 Stirling Highway, Crawley WA6009, Australia
Tel: 61-8-6488-3102
Fax: 61-8-6488-1018
Email: ma@civil.uwa.edu.au

Prof. Guowei Ma obtained his BSc from Beijing University in 1989, his MEng from Xi'an Jiaotong University in 1992, and his PhD from Nanyang Technological University, Singapore in 2000. He subsequently worked at Xi'an Jiaotong University/China, Iwate University/Japan, University of Delaware/USA, Nanyang Technological University (NTU)/Singapore and the University of Western Australia (UWA)/Australia. He had been the Secretary General of the Society of Rock Mechanics & Engineering Geology (Singapore) and the Secretary General of the Association of Computational Mechanics (Singapore). Prof. Ma is currently the President of the Western Australia Chinese Scientists Association (WACSA).

Prof. Ma's research interests include rock dynamics, analysis of discontinuous deformation, dynamic constitutive models of materials and protective structures. He chaired two research programs, i.e. the Underground Technology and Rock Engineering (UTRE) research program and the Jurong Rock Cavern project in NTU.

Prof. Ma is also the Associate Editor of *International Journal of Protective Structures*, and an Editorial Board Member of 6 other international journals. He co-authored 2 books published by Springer, edited 2 conference proceedings and 3 special issues of international journals. He authored over 300 peer reviewed international journal and conference papers. Prof. Ma delivered about 30 keynote lectures including at the 12th Congress of International Society of Rock Mechanics, Beijing, China (2011), the 6th Asian Rock Mechanics Symposium, New Delhi, India (2010), and the past five International Conferences on Analysis of Discontinuous Deformation (Beijing 2007, Singapore 2009, Hawaii 2011, Fukuoka 2013, Wuhan 2015). In 2010, he was invited to be the Kwang-Hua Chair Professor of Tongji University, China. He was awarded the Collaborative Research Fund with Overseas, Hong Kong and Macau Scholars (previous Overseas Outstanding Young Researcher's Fund) of China in 2011. In late 2011, he received the National Natural Science Award of China. From November 2011, Prof. Ma has been the co-president of the inaugural Discontinuous Deformation Analysis (DDA) Commission of the International Society for Rock Mechanics. In 2012, Prof. Ma was selected by the Thousand Talents Program of China and he has been an adjunct professor in the College of Architecture and Civil Engineering, Beijing University of Technology since then.

**Gen-hua Shi** (石根华教授，著名华人科学家，岩石力学与工程专家，DDA 创始人)
Consulting Engineer in Rock Engineering and Structural Engineering
President, DDA Company. 1746 Terrace Drive, Belmont CA 94002 USA
Professor, University of Chinese Academy of Sciences, Beijing, China
Tel: 650-631-1804; Cell: 650-867-6248, Email: sghua@aol.com

**Dr. Gen-hua Shi,** acknowledged as a professional consultant in rock and structural engineering, is the chairman of the DDA Company in California, USA, and the chief scientist of the Discontinuous Deformation Analysis Laboratory, Yangtze River Scientific Research Institute. He obtained his BSc and MEng from Beijing University, China respectively in 1963 and 1968, and his PhD from University of California at Berkeley, USA in 1988. He has put forward the Key Block theory and Discontinuous Deformation Analysis method, now widely studied and applied in rock mechanics and rock engineering fields worldwide. He is also the inventor of the Numerical Manifold Method, which is a novel method for the analysis of both continuous and discontinuous material behaviors. He has been actively involved in many world-famous projects related to rock mechanics in-situ tests, nuclear waste storage, blasting design of rock engineering, stability analysis of rock slopes and rock foundations, underground excavation support design and construction, and dam design and dam foundation analysis. His papers have appeared in profound journals, and in significant conferences such as the series of North American Rock Mechanics Symposiums, and the series of Conferences on Analysis of Discontinuous Deformation. Dr. Shi is the recipient of the China Natural Science Award and other international awards. He resides in Belmont, California, USA.

# Acknowledgments

Many people should be thanked for the creation of this book. First and foremost, we would like to thank Professor Xia-Ting Feng, the Director of the National Key Laboratory of Geomechanics and Geotechnical Engineering, Institute of Rock and Soil Mechanics, Chinese Academy of Sciences, Wuhan, China. Prof. Feng suggested writing this book within the context of the *ISRM Book Series* he has instituted while he was serving as President of the International Society for Rock Mechanics, as the main product of the ISRM DDA Commission, which the three authors jointly chair. His encouragement and support along the way has assisted us in achieving this goal.

Yossef Hatzor would like to express his sincere gratitude to the Chinese Academy of Sciences for the Visiting Professorship grant for Senior International Scientists (No.2011T2G29) awarded to him in late 2011. The frequent visits to the Institute in Wuhan, twice a year for the duration of the fellowship period, have provided a perfect setting for the quiet time and concentration required for the completion of this book. Prof. Feng is deeply thanked for being such a gracious host during that time. The assistance of colleagues from the lab in Wuhan is greatly appreciated, particularly Dr. Yan Guo and Mr. Hang Ruan who provided all the technical and administrative assistance willingly and flawlessly. Y. Hatzor also wishes to thank his PhD and Post Doc students who have done a lot of the actual research, the results of which are summarized in chapters 4–7 of this book: Drs. Michael Tsesarsky, Ronnie Kamai, Dagan Bakun-Mazor, Gony Yagoda-Biran, Huirong Bao and Benguo He. Dr. He is also thanked for assisting in proof reading and for compiling the subject index. Finally, Alistair Bright and the editorial team at Taylor and Francis are thanked for their excellent work during the production process of this book.

Guowei Ma wishes to thank Dr. Gen-hua Shi for leading him into the DDA field and mentoring his team in DDA and NMM research. Guowei has been enjoying the experience in initiating and co-chairing the ISRM DDA Commission with Yossef. Special thanks are dedicated to Prof. Jian Zhao at Monash University, Australia and Dr. Yingxin Zhou at the Defense Science and Technology Agency, Singapore. With their support, Guowei had the honor to chair the Underground Technology and Rock Engineering (UTRE) research program and the Jurong Rock Cavern (JRC) research project at Nanyang Technological University (NTU), which greatly substantiated and enhanced the DDA research at NTU. His appreciation and gratitude also goes to his team members in DDA research at both Nanyang Technological University, Singapore and the University of Western Australia, Australia including Drs. Xinmei An, Lei He, Youjun Ning, Huihua Zhang, Lifeng Fan, Guoyang Fu, Xiaolei Qu, and Feng Ren.

Gen-hua Shi thanks Professor Zuyu Chen from the China Institute of Water Resources and Hydropower and Professor Ying Wang from the University of Chinese Academy of Sciences for their guidance and support during the last several years of research on his new contact theory. Dr. Shi also thanks Drs. Xu Li, Yuxing Ben, Wei Wu, Xinchao Lin, Yanqiang Wu, and PhD candidates Mengsu Hu, Xiaolong Cheng, Yunfan Xiao and Fei Zheng for their proofreading and figure presentations. Dr. Shi is also thankful for many helpful discussions and comments given by the participants attending the workshop on 3D contact theory, which was held in Dalian China from March 28 to April 4, 2016 and was supported by China Basic Research Program Grant No. 2014CB047100.

Finally, we all thank our mentor, Professor Richard Goodman, who has given us the motivation, curiosity and means to advance, albeit slightly, the state of the art in the mechanics of discontinuous rock masses.

# Foreword

This book brings the extraordinary power of DDA (Discontinuous Deformation Analysis) to the tool-baskets of engineers who are responsible for excavations and foundations in jointed and faulted rock masses. The mathematical basis for DDA originated in the brilliant PhD Dissertation and subsequent publications of Gen-hua Shi, and has been furthered by many applications in recent years. These include Professor Hatzor's resourceful analyses of works needed to protect ancient surficial and underground structures in Israel. Important developments have been published in the proceedings of the International Assoc. for Computer Methods and Advances in Geomechanics, and in other recent engineering literature. With this book, the potential applications of DDA will be further recognized and applied for the benefit of the entire engineering community.

Professor Richard E. Goodman

# Chapter 1

# Introduction

## 1.1 WHO SHOULD READ THIS BOOK?

We write this book with several sorts of readership in mind, all of whom are assumed to have at least undergraduate level training in Geomechanics. First and foremost we think of the committed Geological Engineering practitioners that need to solve a rock mechanics problem in a real rock mass. The emphasis here is on the word "real". Although, as you will see when you become familiar with this book, discontinuous deformation analysis (DDA) is highly technical and rests on a solid theoretical foundation, the rationale behind it comes from deep recognition of the importance structural geology has in determining the layout and means for solution of complex, yet real, rock engineering problems. Next, we have in mind the diligent code developer, who may not go to the field so often but is blessed with a strong mathematical background and computational skills which enable him or her to improve the state of the art in numerical modelling as applied to rock mechanics. The rock mechanics community gains from such people if they set their mind to work on rock engineering problems when they could just as well use their skills to solve other, more general, engineering problems. Finally, we have in mind the research community, graduate students and faculty members alike, in both civil engineering and the earth sciences. On one hand, DDA is a vibrant field of study with many challenging problems still unresolved. On the other DDA has matured to the level where it can be used as a powerful tool for studying earth science and civil engineering problems that require a discrete element approach. Therefore the research community can benefit from this book by becoming more familiar with this novel numerical method, identifying limitations, and by employing the method solve complex research problems in a rigorous yet efficient manner.

## 1.2 HOW TO USE THIS BOOK?

This book is written in a self-consistent manner so that previous knowledge of the DDA method by the reader is not assumed. Readers who are not familiar at all with the method will find it useful to first read Chapters 2 and 3 where the basic principles of 2D and 3D DDA are reviewed. We tried to provide in these chapters all the necessary background material needed to understand the theoretical foundation of the method. Some readers may find it useful to refer to the fundamentals covered in Chapters 2 and 3 to better understand the main developments of the method during the past two decades as reviewed in Chapter 1.

Readers who are familiar with DDA but are trained more in numerical modelling with less fieldwork experience, are encouraged to study Chapter 4 that provides guidelines for how input parameters should be obtained from the field in order to conduct a meaningful DDA simulation that will provide realistic results. Benchmark tests for DDA are reviewed in Chapter 5. The set of benchmark tests will be useful for code developers so that they can test the accuracy and performance of new enhancements proposed. In addition, users of other numerical discrete element methods may find this chapter useful for comparison between approaches. Chapters 6 and 7 review useful case studies where DDA has been applied to solve problems in underground and slope engineering, respectively. It will be useful particularly to practitioners as it will delineate to some extent the scope of the method and the sort of applications that may be attempted with it. Finally in Chapter 8 we provide the complete new contact theory developed by DDA author Dr. Gen-hua Shi, for 2D and 3D DDA, including proofs of all mathematical postulations and illustrated examples. This last chapter defines the state of the art currently, and it remains for future generations to implement Shi's new contact theory in improved numerical codes.

## 1.3 CONTINUOUS VS. DISCONTINUOUS DEFORMATION

Rock is an inherently discontinuous material because it typically consists of more than one mineral, the crystal boundaries of which form discontinuities at the micro scale (Fig. 1.1). Also at the micro scale voids and fissure are abundantly present in what we normally call "intact rock" (Fig. 1.2). At the meso scale, or hand specimen size, features that distort the continuity of the rock such as voids are visible to the naked eye (Fig. 1.3). At the outcrop scale these planes of discontinuity combine to form a discontinuous rock mass structure (Fig. 1.4).

For these reasons, modeling rock mass deformation using approaches that assume continuity is overly simplified and may lead to erroneous results. As argued by Hoek and Bray (1981), assuming material continuity rock slopes with average material strength of say 100 MPa and typical unit weight of say 25 kN/m$^3$ should be expected to extend to heights of some 4000 meters above ground, yet such rock slope heights are never encountered in nature. Consider for example the famous El Capitan cliff at Yosemite national park in California (Fig. 1.5). The height of El Capitan is "only" 884 m even though the strength of the granitic rock is greater than 100 MPa. Still, this sub-vertical cliff is way "off the chart" in the famous empirical collection of worldwide natural rock slope stability as expressed in slope height – slope angle space by Hoek and Bray (1981).

The reason rock slopes in nature do not extend to the heights that would have been predicted assuming continuity, is because rock masses are typically transected by sets of discontinuities, each with characteristic geometrical and mechanical properties. The intersections of the discontinuities in the rock mass result in a block system, the geometrical characteristics of which are determined by the geometrical characteristics of each set of discontinuities. Hence for example the average size of the blocks in the

*Figure 1.1*   Photomicrograph of a thin section (27 × 46 mm) of a pyroxenite rock under crossed polarizers (XPL) from Valle Sesia of the Italian Alps. The pyroxenite is composed mostly of clinopyroxene and orthopyroxene with minor amounts of olivine, plagioclase and phlogopite. It is composed of equal sized crystals with straight grain boundaries inter-secting at 120°, forming a mosaic equigranular texture (photo by Bar Elisha, caption by Prof. Yaron Katzir, Ben-Gurion University of the Negev).

*Figure 1.2*   Photomicrograph of a thin section of a porphyritic rhyolite dyke from the Timna igneous complex, southern Israel. The photograph was taken under a petrographic microscope using plane polarized light. The porphyritic rock includes phenocrysts of quartz (clear white) and alkali feldspar (cloudy, grey-brown) set in a fine grained matrix. The rock is fractured; fractures highlighted by copper mineralization including green malachite overgrowing earlier copper sulfide (opaque). Photo by Bar Elisha, caption by Prof. Yaron Katzir, Ben-Gurion University of the Negev.

*Figure 1.3* Hand specimen of a fine-grained olivine basalt from northern Israel with voids. Core diameter 54 mm.

*Figure 1.4* Intersection of three joint sets giving rise to a discontinuous rock mass structure as exposed in a rock slope of the Yalong river, Sichuan province, China. Horizontal dimension of figure approximately 10 m.

*Figure 1.5* The sub-vertical El Capitan cliff at Yosemite national Park, California.

system will be controlled by the mean spacing of the discontinuities in each set, and the shear strength of the block system will be controlled by the shear strength characteristics of the surfaces of the individual sets of discontinuities. Since on average the shear strength of discontinuities is much smaller than the shear strength of intact rock material, most of the deformation will take place by sliding along pre-existing planes of discontinuities rather than by shearing through the intact rock material itself. Therefore, the total height a natural rock slope can withstand is controlled much more by the orientation and shear strength of the discontinuities than the strength of the intact rock material. This is typical for rocks; in soils slopes typically fail by shear through the continuous material which generally cannot support discontinuities in the first place due to its negligible tensile strength, and this is where rock mechanics and soil mechanics approaches to solve slope stability problems drastically differ.

A case in point is the Snake path cliff at Masada national park in Israel (Fig. 1.6(A)) where the slope height is "only" 250 meters above ground surface but the material

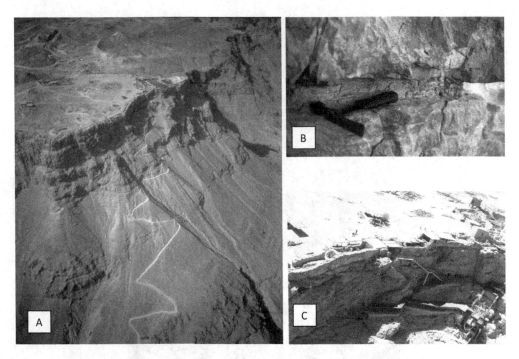

*Figure 1.6* A) The Snake path cliff at Masada national park, Israel, B) A characteristic bedding plane in Masada dolomites, C) A rock block that rests on a bedding plane in the Snake path cliff at Masada.

strength of the dolomite is greater than 300 MPa. Clearly, with such a high uniaxial compressive strength the rock slope would have been safe against failure in compression at the base even if it extended several kilometres high. But more importantly, failure by shear through the intact rock material is also very unlikely: a series of triaxial tests performed on solid cylinders from this dolomite (Hatzor and Palchik, 1997) provide a cohesion of 30 MPa and an internal friction angle of 45 degrees.

When testing a bedding plane interface in this material (Fig. 1.6(B)) in triaxial shear however, the obtained cohesion is zero and the residual friction angle only 23 degrees (Hatzor, 2003). This means that the block shown in the right corner of Fig. 1.6(C) which rests on a bedding plane that is dipping 19 degrees, is in fact in danger of sliding with a very low static factor of safety, even though the intact rock material is completely safe from failure, either by compression or shear.

The same rationale is also applicable for the design of underground excavations in rocks. When designing caverns for storage of hazardous materials for example the level of discontinuity of the rock mass will actually determine the safe dimensions of the chambers beyond which active reinforcement will have to be applied. This can be appreciated from a comparison between two rock masses in which historic caverns were excavated in the past, one in a weak but relatively continuous rock mass and the other in a stronger but highly discontinuous rock mass. Consider the bell shaped

*Figure 1.7*  Control of discontinuities on stability of underground openings in rocks. A) The 1000-year-old bell shaped caverns at Bet Guvrin, Israel, B) The 3000-year-old underground water storage system at Tel Beer Sehva, Israel.

cavern shown in Fig. 1.7A that has been excavated at Bet Guvrin national park in a soft but relatively continuous chalk. Although the compressive strength of the intact rock material is very low, between 3–9 MPa, (Hatzor *et al.*, 2002) the cavern has been standing unsupported with a free span of 24 meters since its excavation approximately

1000 years ago. This is in stark contrast to the underground water storage system that was discovered at Tel Beer Sheva national park that was excavated approximately 3000 years ago in a chalk with unconfined compressive strength of 30 MPa. Even though the strength of the material is much higher, the underground chambers span cannot exceed 7 meters, and when the ancient engineers attempted to increase the span they have experienced immediate collapse (Hatzor and Benary, 1998). This is because the rock mass is transected by a very dense network of discontinuities with average spacing of 10–20 cm only (Fig. 1.7B) giving rise to a rock mass structure that consists of many small blocks.

The numerical discontinuous deformation analysis (DDA) method has been developed for enabling robust numerical modelling of deformation in discontinuous media with geometric and mechanical characteristics of natural rock masses specifically in mind.

## 1.4  DDA HISTORY

Since the development of the joint element by Goodman, Taylor, and Brekke (Goodman *et al.*, 1968) to enable discontinuous deformation within the framework of the finite element method (FEM), numerical computations with geological discontinuities developed rapidly and have been applied to rock engineering extensively. Shortly after, Cundall (1971) introduced the distinct element method (DEM), a force method employing fictitious forces to regulate sliding, to prevent block overlapping, and to reach equilibrium. Indeed, Cundall's DEM is now widely used in rock mechanics practice. A decade later Shi and Goodman presented a novel method for computing the strains and displacements of a block system (Shi and Goodman, 1984, 1985) that best explains a set of displacement and strain observations made at a sufficient number of points. They called this approach discontinuous deformation analysis (DDA). Knowing only the measured displacements, at individual points or directions, the method computes the best least square fit over all displacements, deformations, and strains of each block as well as the opening, closing and sliding of all block interfaces. The error minimization is constrained by a robust kinematic analysis and for the first time a correction procedure that prevents block penetrations is introduced. Since in the original approach the displacements were observed and the block strains and movements were inferred, constitutive equations were not needed and thus were not introduced in the original DDA. This backward modelling approach has later been expanded by Shi (1988) to accommodate constitutive relations and loading for the modelled rock mass, thereby permitting general problems in the mechanics of block systems to be solved in either a forward or a backward mode. Given the geometry, loading, and the material constants of each block as well as the friction angle, cohesion and damping mechanism at the contacts between blocks, the forward DDA computes the blocks stresses, strains, sliding forces, contact forces, and movements.

DDA uses the displacements as unknowns and solves the equilibrium equations in the same manner as the matrix analysis of structures in the finite element method. Although it is intended primarily for discontinuous block systems, DDA is based on strict adherence to the rules of classical mechanics. The development of DDA required the introduction of a complete kinematic theory that enables one to obtain large deformation solutions for numerous blocks, without penetration. DDA's kinematic theory recognizes the connections of joints around the perimeters of blocks so that an optimal correction-logic can be applied.

Although it is theoretically possible to incorporate large numbers of blocks in a finite element mesh by introducing large numbers of joint elements or slip lines, in reality such an approach will inevitably be ill-conditioned. This is because the corrections required to satisfy the kinematical constraints of non-closing on one joint may be in conflict with corrections appropriate to another joint, and so on. A robust kinematic theory required to establish the optimum correction steps was not included, in fact, in any other previously proposed approach for modelling discontinuous deformation.

## 1.5 THREE DECADES OF DDA RESEARCH AND DEVELOPMENT

Since DDA was published in the 1980s its applicability to rock mechanics practice has been studied extensively by the international civil engineering research community and the results have been published in peer reviewed journal papers as well as in proceedings of a series of international symposia known as ICADD (International Conference on Analysis of Discontinuous Deformation). The first ICADD meeting was held in Taiwan in 1995 (ICADD1, 1995) followed by a DDA "Forum" that was held only once at Berkeley (Salami and Banks, 1996). The second ICADD meeting was held in Tokyo (ICADD2, 1997), then in Vail (ICADD3, 1999), Glasgow (ICADD4, 2001), Wuhan (ICADD5, 2002), Trondheim (ICADD6, 2003), Honolulu (ICADD7, 2005), Beijing (ICADD8, 2007), Singapore (ICADD9, 2009), Honolulu (ICADD10, 2011), Fukuoka (ICADD11, 2013), and Wuhan (ICADD12, 2015). In these series of proceedings improvements, verifications, and applications of the method have been published but not all of these studies were published as journal papers and therefore their results are less accessible. Below we summarize some important DDA R&D results that were published over the past three decades in the series of ICADD proceedings.

## 1.6 DDA VS. FEM AND DEM

Although it may seem DDA's forward analysis of discontinuous deformations for block systems resembles the distinct element method, in fact DDA is more closely related to the finite element method (FEM), due to the following shared attributes between the two methods:

1 Minimization of the total potential energy to establish the equilibrium equations
2 The displacements are the unknowns of the simultaneous equilibrium equations
3 Stiffness, mass, and loading submatrices are added to the coefficient matrix of the simultaneous equations. In DDA however the block stiffness matrices are simpler than the element stiffness matrices in the FEM.
4 DDA uses displacement locking of contacting blocks which resembles adding bar elements in FEM.

For these reasons, programming forward and backward DDA models will be natural for programmers familiar with the FEM.

There are some notable advantages of DDA over FEM when considering block systems. Firstly, DDA does not assume continuity at block boundaries, namely it is fundamentally a discontinuous approach. Secondly, the blocks forming the "elements" of the "mesh" can be convex or non-convex with any number of edges, and can even contain holes. Moreover, block meshes do not require the vertex of one block to be in contact with the vertex of another block. Thirdly, modelling only few block elements with FEM would require a much larger mesh than is required in DDA.

DDA is different by nature from the widely used distinct element method (Cundall, 1971). DDA is a displacement method, where the unknowns in the equilibrium equations are displacements, whereas DEM is a force method which attempts to adjust the contact forces to be constants using damping. DDA is an implicit method, characterized by a rigorous energy consumption approach whereas DEM is an explicit method.

## 1.7  MAIN FEATURES OF DDA

Several features are characteristic of DDA:

1  Complete kinematic theory and its numerical realization
2  Perfect first order displacement approximation
3  Strict postulate of equilibrium
4  Correct energy consumption
5  Large deformation

The reliability of the method is attributed to the fact that mathematical and numerical analyses in DDA attempt to address the mechanical phenomena associated with block movements as close as possible to the physical reality. Some of the main features in DDA are briefly discussed below following Shi (1988).

### 1.7.1  Block system kinematics

Block kinematics has to be dealt with when large displacements and large deformations are involved. Block system kinematics is different from existing particle kinematics and single rigid body kinematics which can be described by simple equations. Block system kinematics has two constraints: there can be no-penetration and no-tension between blocks. These constraints are described in DDA by inequalities. The condition of no-penetration is easy to check from an output drawing. But if the no-tension constraint is not fulfilled, the result may be incorrect and then even the no-interpenetration condition may not be guaranteed. DDA imposes the constraint inequalities to the linear equation system by "penalty submatrices" and in that sense it can be said to be a "penalty" method. As mentioned earlier, it should be pointed out that prior to the development of DDA theory, no-tension and no-penetration constraints could not be fulfilled rigorously with existing procedures, even for modelling a single joint or crack.

## 1.7.2  Complete first order displacement approximation

The displacement function for each block is equivalent to the following complete first order approximations of displacements:

$$u = a_1x + a_2y + a_3; \quad v = b_1x + b_2y + b_3$$

This displacement function can be generalized in DDA to a higher order approximation. Each block has independent displacement functions, therefore the best approximation can be chosen independently for each block. For example, the block having the complete third order approximation has the same number of unknowns as an eight noded finite element. In the finite element scheme, an eight noded element however, will usually lack terms $x^3$ and $y^3$.

## 1.7.3  Equilibrium, dynamics and energy consumption

The equilibrium equations in DDA are established by minimizing the total potential energy and are solved directly. The first order approximation requires six equilibrium equations per block. The external forces and internal stresses $\sigma_x, \sigma_y, \tau_{xy}$ also reach equilibrium. Forward DDA allows implicit dynamic computations, like that of the finite element method, except that the forward analysis of a discontinuous block system also considers the acceleration of strains:

$$\frac{\partial^2 \varepsilon_x(t)}{\partial t^2}; \quad \frac{\partial^2 \varepsilon_y(t)}{\partial t^2}; \quad \frac{\partial^2 \gamma_{xy}(t)}{\partial t^2}$$

In DDA Coulomb's friction law is applied for block contacts each with a characteristic friction coefficient and cohesion. Indeed, frictional sliding is the main source of energy consumption in DDA.

Prior to the introduction of DDA a damping coefficient has typically been introduced at block contacts so as to quickly suppress vibrations. It may be argued that this damping term may not be realistic and by using such a fast energy consumption approach both vibration of fictitious forces as well as real vibrations may be eliminated. This of course may seriously compromise the accuracy of dynamic simulations of discontinuous deformation.

## 1.7.4  Large Deformation

Forward DDA considers both static and dynamic deformation. For dynamic problems an implicit algorithm is used. Time steps are used in both static and dynamic analyses. The only difference is that in static computation the velocity at the beginning of each time step is assumed to be zero whereas dynamic computation inherits the velocity from the previous time step. When the block system undergoes large deformation each step starts with the deformed block shape and position as obtained in the previous step, and the equilibrium equations are written and solved for the updated block geometry.

Large displacement and large deformation are important in discontinuous deformation analysis. As the blocks move or deform, the updated block shape and position will produce different block contacts and different interactive forces, which will change the whole block structure and affect the modes of failure. This process, which is characteristic of discontinuous deformation, affects the accuracy of the solution which is very sensitive to it, much more than when solving a continuum mechanics problem.

### 1.7.5 Efficiency

Discontinuous deformation analysis uses a selective equation system to decide on the opening, sliding and closing of the contacts between blocks. On average, in each time step five time selections are required, therefore the computation time is about five times greater than that of finite element computation with the same number of unknowns. In the special case where the number of blocks equals the number of elements, the discontinuous deformation method has about four times the degrees of freedom of that of the finite element method for a continuum. These efficiency considerations with respect to computing time and memory requirements apply to both statics and dynamics of block systems.

## 1.8 SOME LIMITATIONS OF THE ORIGINAL DDA

In a recent review paper Lin (2013) discusses some key DDA issues that call for further attention and some of his comments are briefly reviewed here. First, since the original DDA was developed to solve an inverse problem, this determined the choice of the governing parameters. In each discrete block, which is considered a constant strain element, there are six unknowns: the $x$ and $y$ displacements, the rotation, the $x$ and $y$ strains, and the shear strain. Although this form is convenient when some of the strains are measurable in an inverse problem setting and where the underlying kinematics of the block system must be determined, it does have some restrictions for forward modelling mainly because the unknowns are not independent. Lin (2013) proposes a better choice for the original displacement field (Equation 1.1) that will also retain the number of unknowns in the system of equations at six as originally suggested by Lin and Lee (1996):

$$u_{x,y} = a_0 x + a_1 y + a_2; \quad v_{x,y} = b_0 x + b_1 y + b_2 \tag{1.1}$$

This formulation would also have the advantage that finite rotations and finite strains could easily be introduced.

Regarding rotation, as we shall see in Chapter 2, in the original DDA the first order displacement $(u, v)$ at any point $(x, y)$ within a block is:

$$\begin{Bmatrix} u \\ v \end{Bmatrix} = \begin{bmatrix} 1 & 0 & -(y - y_0) & (x - x_0) & 0 & (y - y_0)/2 \\ 0 & 1 & (x - x_0) & 0 & (y - y_0) & (x - x_0)/2 \end{bmatrix} \begin{Bmatrix} u_0 \\ v_0 \\ r_0 \\ \varepsilon_x \\ \varepsilon_y \\ r_{xy} \end{Bmatrix} \tag{1.2}$$

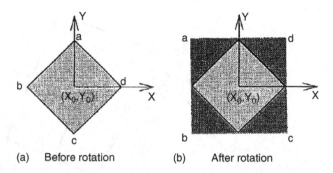

Figure 1.8  Rigid block rotation by 60° with DDA first order approximation would result in expansion of block area (Ohnishi et al., 1995).

where $(u_0, v_0)$ are the rigid displacements at the center point $(x_0, y_0)$ within the block; $r_0$ is the angle of rigid body rotation about the center of the block $(x_0, y_0)$; $\varepsilon_x, \varepsilon_y$, and $\varepsilon_{xy}$ are the average elastic strains of the block. Since Equation 1.2 is only the first order approximation of displacement, the rigid rotation $r_0$ is defined as:

$$r_0 = \frac{1}{2}\left(\frac{\partial v}{\partial x} - \frac{\partial u}{\partial y}\right) \tag{1.3}$$

with $x = x_0$ and $y = y_0$. But Equation 1.3 is accurate only for very small deformation and therefore it cannot be expected to be valid for large rotations. As clearly demonstrated by Ohnishi and co-workers (Ohnishi et al., 1995), suppose the displacement of the block only involves rigid rotation, then Equation 1.3 becomes:

$$\begin{aligned} u &= -(y - y_0)r_0 \\ v &= (x - x_0)r_0 \end{aligned} \tag{1.4}$$

Now consider the block in Fig. 1.8a. If it is rotated by 60 degrees counter clockwise, according to Equation 1.4 the vertices will move to the new positions shown in Fig. 1.8b so that the block area will actually increase. In fact, the correct displacement for this case should be calculated as:

$$\begin{aligned} u &= (x - x_0)(\cos r_0 - 1) - (y - y_0)\sin r_0 \\ v &= (x - x_0)\sin r_0 + (y - y_0)(\cos r_0 - 1) \end{aligned} \tag{1.5}$$

It is readily apparent that Equation 1.4 is indeed a first order approximation of Equation 1.5 that provides the exact solution. When the rotation angle $r_0$ is large enough, the term $(\cos r_0 - 1)$ can no longer be approximated as zero and similarly $\sin r_0$ can no longer be approximated as $r_0$. But the non-linearity of Equation 1.5 makes it difficult to be used directly in the original DDA linear system of equations, whereas performing the computation with the original form of Equation 1.2 is fast and efficient and the error is tolerable, provided that the rotations are not too large.

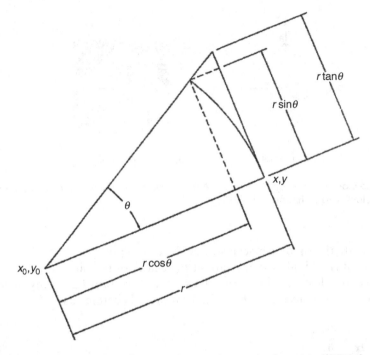

*Figure 1.9* The exact solution for the rotation follows the arc. In the original DDA formulation the displacement is along the tangent with magnitude $r\theta$ which falls between $r\sin\theta$ and $r\tan\theta$ (MacLaughlin and Sitar, 1996).

Ke (1996) has proposed an "ad hoc" solution to this problem, a post adjustment procedure which allows keeping the system of linear equations as in the original DDA. MacLaughlin and Sitar (1996) demonstrated the discrepancy between the exact and original DDA for rigid rotation (Fig. 1.9). They proposed a more accurate displacement function using the Taylor's series polynomial approximations for $\sin\theta$ and $\cos\theta$ instead of the trigonometric functions, and developed a solution which is internally consistent with the potential energy terms. One trivial solution to the rotation problem would simply be to shorten the time step to ensure the rotations in each time step are sufficiently small. Alternatively, the order of the polynomial can be increased (e.g. Koo and Chern, 1996; Ma *et al.*, 1996).

Another difficulty with the original DDA formulation is the implementation of stiff springs once contacts have been detected automatically. The high stiffness evoked by the penalty parameter means that the time step size must be reduced to avoid uncontrolled block bouncing. The advantage of the penalty method, however, is that it is straightforward and easy to be implemented. The augmented Lagrange multiplier method represents a good compromise (e.g. Amadei *et al.*, 1996) as it does not require an increase in the size of the system of equations in the original DDA and yet the simplicity of the penalty method is retained. This would allow also introduction of soft springs at the contacts which is particularly advantageous in dynamic simulations.

Indeed, the way damping is modelled in DDA is an issue that also requires some attention. In the original DDA damping may be modelled as "kinetic". DDA resets the velocity to zero at the beginning of a time step in a "static" analysis and completely transfers it to the next time step in a "dynamic" analysis. Any ratio between zero and 1.0 of this velocity transfer will represent a measure of "kinetic damping" that is imposed by the user in a dynamic simulation. The physical meaning of this procedure is ambiguous however. If accurate dynamic displacement in the time domain is sought it would inevitably lead to erroneous results when less than 100% of the velocity is transferred to the next time step.

In the discrete element method three damping schemes have typically been employed: 1) element level damping, 2) global damping, and 3) contact damping. For instance, in FLAC (ITASCA, 1993) a local adaptive damping has been designed to converge its solution scheme, but for the implicit DDA the forces should all be in equilibrium at the end of each time step, and therefore such a scheme would not apply. A viscous element damping can readily be introduced; Ohnishi with co-workers (Ohnishi *et al.*, 2012) have shown how a viscous contact damping can be implemented at contacts, and at the boundary, so as to obtain non-reflecting boundaries. Note that the "kinetic damping" discussed here should not be confused with the algorithmic damping associated with DDA's time integration scheme which is discussed in some detail by Wang *et al.* (1996) and Doolin and Sitar (2004).

## 1.9 BLOCK DISCRETIZATION

In the original DDA blocks are assumed to be stiff, of infinite strength, and simply deformable. While this assumption is valid for the range of stresses considered in most civil engineering applications where intact rock strength is typically much higher than the *in situ* stress level, these assumptions may cease to be valid when modelling rock deformation in high *in-situ* stress environments or extreme load conditions.

To enhance stress distribution within DDA's simply deformable blocks, Ke (1995) proposed an artificial joint concept which he has further developed to enable DDA modeling of fracture propagation in a column subjected to uniaxial tension (Ke, 1997) and showed how both the strength of the solid elements as well as the texture of the modeled mass affect the fracturing behavior. Ke's artificial joint concept consists of cutting blocks into sub-blocks and binding the sub-blocks within a block with "cement" to provide an incipient joint with tensile strength.

Amadei with coworkers (Amadei *et al.*, 1996; Lin *et al.*, 1995) proposed sub-blocking for DDA which was further developed to enable block fracturing. Koo and Chern (1997) presented a new approach for modeling progressive fracture in jointed rock masses considering both tensile (mode 1) and shear (mode 2) failure modes. They used the Coulomb–Mohr failure criterion with tension cut off to model strength of intact rock elements. These initial developments lead to coupling between DDA and FEM as discussed in a further sub section.

Ma *et al.* (2007) proposed a Moving Least Squares (MLS) approximation where interpolative nodes are scattered within the blocks or along their borders, in a manner similar to coupling DDA with Element Free Galerkin (EFG) method thus inheriting all the advantages of Mesh Free Methods directly in DDA such as an ability to model

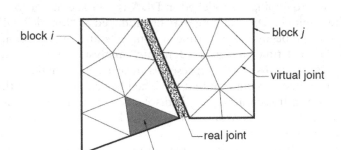

*Figure 1.10* Illustration of a nodal-based DDA model (Bao and Zhao, 2010b).

crack propagation within blocks, blocks fracture, etc. Zhu *et al.* (2007) proposed a sub-block meso-damage evolution model where each block is divided into smaller sub-blocks and two kinds of contacts are implemented between the sub-blocks using the Augmented Lagrangian Method, one continuous and the other discontinuous. The continuity condition can change along the sub-block contacts and the continuous contacts can convert into discontinuous ones using the Mohr-Coulomb criterion.

Grayeli and Mortazavi (2007) proposed an internal discretization scheme based on the Delaunay mesh generation scheme and the DDA formulation was obtained based on a FEM discretization. An elasto-plastic yield criterion was implemented in the modified code and its performance was compared with analytical results and commercial codes (FLAC, 1998). A block fragmentation model for DDA utilizing the FEM adaptive mesh generation technique was proposed by Jiao and Zhang (2012). Randomly distributed mechanical parameters statistically satisfying Weibull's distribution are assigned to the blocks to simulate the heterogeneity of rock mass and the artificial joints provide the potential paths along which the cracks generate and propagate.

These initial improvements paved the way to a new branch of current DDA research trying to incorporate fracture mechanics into the original DDA method so that internal block failure processes can be modeled. Bao and Zhao (2010) adopted a Nodal-based DDA method to conduct fracture analysis (Fig. 1.10), and this direction is explored by other researchers as well (e.g. Ben *et al.*, 2013; Jiao *et al.*, 2014; Zhao *et al.*, 2013b).

The Mohr-Coulomb criterion is commonly used for block fracturing. There are three parameters involved including the friction angle $\varphi$, cohesion $c$ and tensile strength $T_0$. The three-parameter Mohr-Coulomb criterion is expressed in terms of major and minor principal stresses $\sigma_3$ and $\sigma_1$ (with $\sigma_3 \geq \sigma_1$) as follows:

$$\sigma_1 = -C_0 + \sigma_3 \tan^2\left(\frac{\pi}{2} + \frac{\varphi}{2}\right) \tag{1.6}$$

when $\sigma_1 < \sigma_{1c}$ and

$$\sigma_3 = T_0 \tag{1.7}$$

when $\sigma_1 \geq \sigma_{1c}$. $C_0$ is the unconfined compressive strength of the block material, which is related to the cohesion $c$ and friction angle $\varphi$ as follows:

$$C_0 = 2c \tan\left(\frac{\pi}{2} + \frac{\varphi}{2}\right) \tag{1.8}$$

and $\sigma_{1c}$ is a critical transitional stress between the shear and tensile fracture modes:

$$\sigma_{1c} = -C_0 + T_0 \tan^2\left(\frac{\pi}{2} + \frac{\varphi}{2}\right) \tag{1.9}$$

To conclude this section we would like to point out that we believe the numerical manifold method (Shi, 1995), the presentation of which is beyond the scope of this book, provides the best platform to proceed in this research direction. Indeed, much of the more recent developments in this field utilize the numerical manifold method.

## 1.10  HIGHER ORDER DISPLACEMENT FUNCTION

The original DDA uses first order approximation of the displacement field in its governing equations and we have already shown above errors that may be encountered particularly with respect to rigid block rotation. A second order displacement function for DDA has been proposed very early on by Koo *et al.* (1995) who have provided a very comprehensive solution and implemented it in a modified DDA code which was verified using classical benchmark tests. They have extended the first order function by adding the complete second order term:

$$\begin{aligned} u &= a_1 + a_2 x + a_3 y + a_4 x^2 + a_5 xy + a_6 y^2 \\ v &= b_1 + b_2 x + b_3 y + b_4 x^2 + b_5 xy + b_6 y^2 \end{aligned} \tag{1.10}$$

The following unknowns of an individual block are chosen:

$$\begin{pmatrix} u_0 & v_0 & r_0^c & \varepsilon_x^c & \varepsilon_y^c & r_{xy}^c & \varepsilon_{x,x} & \varepsilon_{x,y} & \varepsilon_{y,x} & \varepsilon_{y,y} & r_{xy,x} & r_{xy,y} \end{pmatrix}$$

where $u_0$ and $v_0$ are rigid body translation of the block, $r_0^c$, $\varepsilon_x^c$, $\varepsilon_y^c$ and $r_{xy}^c$ are the constant terms of the rotation, normal and shear strains, $\varepsilon_{x,x}$, $\varepsilon_{x,y}$, $\varepsilon_{y,x}$, $\varepsilon_{y,y}$, $r_{xy,x}$, and $r_{xy,y}$ are the gradients of the normal and shear strains. The displacement functions can be written in matrix form as:

$$\begin{pmatrix} u \\ v \end{pmatrix}_{2 \times 1} = T_{2 \times 12} d_{12 \times 1} \tag{1.11}$$

in which $T$ is expressed as follows:

$$T = \begin{pmatrix} T_1 & T_2 \end{pmatrix}$$

$$T_1 = \begin{pmatrix} 1 & 0 & -(y-y_0) & (x-x_0) & 0 & \frac{1}{2}(y-y_0) \\ 0 & 1 & (x-x_0) & 0 & (y-y_0) & \frac{1}{2}(x-x_0) \end{pmatrix}$$

$$T_2 = \begin{pmatrix} \frac{1}{2}(x^2-x_0^2) & (xy-x_0y_0) & -\frac{1}{2}(y^2-y_0^2) & 0 & 0 & \frac{1}{2}(y^2-y_0^2) \\ 0 & -\frac{1}{2}(x^2-x_0^2) & (xy-x_0y_0) & -\frac{1}{2}(y^2-y_0^2) & -\frac{1}{2}(x^2-x_0^2) & 0 \end{pmatrix}$$

and $d = (u_0 \quad v_0 \quad r_0^c \quad \varepsilon_x^c \quad \varepsilon_y^c \quad r_{xy}^c \quad \varepsilon_{x,x} \quad \varepsilon_{x,y} \quad \varepsilon_{y,x} \quad \varepsilon_{y,y} \quad r_{xy,x} \quad r_{xy,y})^T$.

A second-order displacement function incorporating a six-node triangular mesh was derived by Grayeli and Mortazavi (2005) to enhance DDA's capabilities for use in practical applications. The stress distribution around a tunnel obtained with the modified DDA was compared with FEM and the analytical Kirsch solution (Kirsch, 1898) showing good agreement.

A third order displacement function for DDA was developed by Koo and Chern (1996). The coefficients of the displacement functions were used as the displacement variables. The complete third order displacement functions have the following form:

$$u = d_1 + d_3x + d_5y + d_7x^2 + d_9xy + d_{11}y^2 + d_{13}x^3 + d_{15}x^2y + d_{17}xy^2 + d_{19}y^3$$
$$v = d_2 + d_4x + d_6y + d_8x^2 + d_{10}xy + d_{12}y^2 + d_{14}x^3 + d_{16}x^2y + d_{18}xy^2 + d_{20}y^3$$

$$(1.12)$$

Writing in a matrix form, the displacement field can be expressed as follows:

$$\begin{pmatrix} u \\ v \end{pmatrix}_{2\times1} = T_{2\times20}d_{20\times1} \tag{1.13}$$

where $T$ is

$$T = \begin{pmatrix} 1 & 0 & x & 0 & y & 0 & x^2 & 0 & xy & 0 & y^2 & 0 & x^3 & 0 & x^2y & 0 & xy^2 & 0 & y^3 & 0 \\ 0 & 1 & 0 & x & 0 & y & 0 & x^2 & 0 & xy & 0 & y^2 & 0 & x^3 & 0 & x^2y & 0 & xy^2 & 0 & y^3 \end{pmatrix}$$

and $d^T = \begin{pmatrix} d_1^T & d_2^T \end{pmatrix}$

$$d_1^T = \begin{pmatrix} d_1 & d_2 & d_3 & d_4 & d_5 & d_6 & d_7 & d_8 & d_9 & d_{10} \end{pmatrix}$$

$$d_2^T = \begin{pmatrix} d_{11} & d_{12} & d_{13} & d_{14} & d_{15} & d_{16} & d_{17} & d_{18} & d_{19} & d_{20} \end{pmatrix}$$

A third order displacement function has also been introduced into the original DDA (Huang et al., 2010) exhibiting very accurate bending deformation of a cantilever beam. A method for approximating the continuous displacement function with complete high order polynomials has been proposed for DDA (Wu et al., 2012) and it is also proved to be highly accurate when solving a single block cantilever problem.

Finally, Shi (1988) generalized the displacement function in a series form as follows:

$$u = \sum_{j=1}^{m} a_j f_j(x, y)$$

$$v = \sum_{j=1}^{m} b_j f_j(x, y)$$

(1.14)

Written in matrix form:

$$\begin{pmatrix} u \\ v \end{pmatrix} = \begin{pmatrix} f_1 & 0 & f_2 & 0 & \cdots & \cdots & f_m & 0 \\ 0 & f_1 & 0 & f_2 & \cdots & \cdots & 0 & f_m \end{pmatrix} \begin{pmatrix} a_1 \\ b_1 \\ a_2 \\ b_2 \\ \vdots \\ \vdots \\ a_m \\ b_m \end{pmatrix}$$

(1.15)

For high order displacement functions, the sub-matrices and equilibrium equations can be obtained accordingly based on the minimization of the total potential energy similar to the original DDA.

One issue that is important to realize with regard to the implementation of higher order displacement functions in a discontinuous framework is that under large deformations the updated shapes of the blocks will differ drastically from the original shapes because of the large internal deformation the blocks will be permitted to undergo, and therefore new contacts will form and new interactive forces will be introduced which will change the whole structure and may affect the modes of failure more drastically than in a continuum mechanics problem. The difficulties associated with implementation of higher order in DDA and how they may be addressed are discussed by Wang *et al.* (2007).

## 1.11 COUPLING DDA WITH OTHER NUMERICAL METHODS

### 1.11.1 Coupling DDA with FEM

Coupling DDA with the FEM has been attempted by many using FEM meshes inside the DDA blocks in order to obtain a more accurate description of the deformation, taking advantage of the fact that both DDA and FEM use the principal of total potential energy minimization to obtain the solution equations for system equilibrium. Both DDA blocks and sub block triangular elements can therefore be treated initially as DDA blocks, using the standard DDA formulation (e.g. Clatworthy and Scheele, 1999). Combination of DDA and FEM has also been utilized to model dynamic wave propagation for geophysical applications (e.g. Shyu *et al.*, 1999). Coupling between 3D-DDA

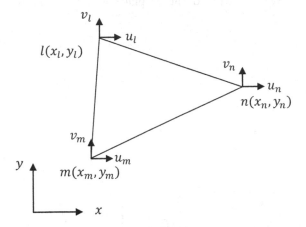

*Figure 1.11* A typical triangular element.

and FEM had been achieved by Cao *et al.* (2007) who obtained the displacements and strains by proper internal discretization of deformable blocks using finite element meshes while the contacts between the blocks were modeled by DDA. Based on the variational principle of minimum potential energy the simultaneous equilibrium equations of the coupled method were then established. They demonstrated the validity and advantages of their enhancement in computing the deformation of a multilayer asphalt concrete pavement subjected to moving vehicle loads.

In order to refine the stress distribution inside the blocks and improve the flexibility of block boundaries, a nodal based DDA was developed by Shyu (1993) by introducing the finite element mesh into the DDA blocks. Both triangular and quadrilateral elements were studied. For triangular elements, each element has three nodes $(l, m, n)$ and six displacement variables in two orthogonal directions $(u_l, v_l, u_m, v_m, u_n, v_n)$ as shown in Fig. 1.11. The displacement field of a triangular element with constant strain is given by:

$$\begin{pmatrix} u \\ v \end{pmatrix} = \begin{pmatrix} 1 & 0 & x & 0 & y & 0 \\ 0 & 1 & 0 & x & 0 & y \end{pmatrix} \begin{pmatrix} 1 & 0 & x_l & 0 & y_l & 0 \\ 0 & 1 & 0 & x_l & 0 & y_l \\ 1 & 0 & x_m & 0 & y_m & 0 \\ 0 & 1 & 0 & x_m & 0 & y_m \\ 1 & 0 & x_n & 0 & y_n & 0 \\ 0 & 1 & 0 & x_n & 0 & y_n \end{pmatrix} \begin{pmatrix} u_l \\ v_l \\ u_m \\ v_m \\ u_n \\ v_n \end{pmatrix} \qquad (1.16)$$

For quadrilateral elements, each element has four nodes $(k, l, m, n)$ and eight displacement variables in two orthogonal directions $(u_k, v_k, u_l, v_l, u_m, v_m, u_n, v_n)$ as shown in Fig. 1.12.

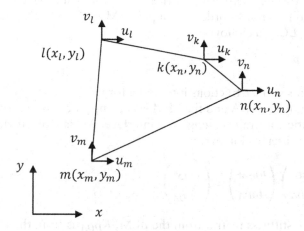

*Figure 1.12* Typical quadrilateral element.

The displacement field of a quadrilateral element with constant strain is given by:

$$\begin{pmatrix} u \\ v \end{pmatrix} = \begin{pmatrix} 1 & 0 & x & 0 & y & 0 & xy & 0 \\ 0 & 1 & 0 & x & 0 & y & 0 & xy \end{pmatrix} \begin{pmatrix} n_{11} & 0 & n_{12} & 0 & n_{13} & 0 & n_{14} & 0 \\ 0 & n_{11} & 0 & n_{12} & 0 & n_{13} & 0 & n_{14} \\ n_{21} & 0 & n_{22} & 0 & n_{23} & 0 & n_{24} & 0 \\ 0 & n_{21} & 0 & n_{22} & 0 & n_{23} & 0 & n_{24} \\ n_{31} & 0 & n_{32} & 0 & n_{33} & 0 & n_{34} & 0 \\ 0 & n_{31} & 0 & n_{32} & 0 & n_{33} & 0 & n_{34} \\ n_{41} & 0 & n_{42} & 0 & n_{43} & 0 & n_{44} & 0 \\ 0 & n_{41} & 0 & n_{42} & 0 & n_{43} & 0 & n_{44} \end{pmatrix}$$

$$(1.17)$$

where

$$N = \begin{pmatrix} n_{11} & n_{12} & n_{13} & n_{14} \\ n_{21} & n_{22} & n_{23} & n_{24} \\ n_{31} & n_{32} & n_{33} & n_{34} \\ n_{41} & n_{42} & n_{43} & n_{44} \end{pmatrix} = \begin{pmatrix} 1 & x_k & y_k & x_k y_k \\ 1 & x_l & y_l & x_l y_l \\ 1 & x_m & y_m & x_m y_m \\ 1 & x_n & y_n & x_n y_n \end{pmatrix}^{-1}$$

$$(1.18)$$

## 1.11.2 Coupling DDA with BEM

The boundary element method (BEM) is an attractive alternative for modelling semi-infinite or infinite domain problems. Lin and Al-Zahrani (2001) coupled the BEM and DDA so that the BEM is used to model the continuum far field while the DDA is used to simulate the discrete near field.

The static BEM formulation is considered in the coupled analysis. The BEM governing equation in a matrix form is as follows:

$$HU = GP$$

$$(1.19)$$

where $H$ and $G$ are the coefficient matrices, $U$ is the displacement vector and $P$ is the surface traction vector. In order to couple BEM with the DDA, Equation 1.19 is pre-multiplied by $LG^{-1}$ as follows:

$$LG^{-1}HU = LP \tag{1.20}$$

where $L$ transforms surface tractions into nodal forces.

The coupling of the DDA and the BEM is accomplished by enforcing the equilibrium and kinematic constraints along their interface. An assembled global discretized equation can be written as follows:

$$\begin{pmatrix} K_{BEM} & K_{BD} \\ K_{BD} & K_{DDA} \end{pmatrix} \begin{pmatrix} u_{BEM} \\ a_{DDA} \end{pmatrix} = \begin{pmatrix} F^I_{BEM} \\ F^I_{DDA} \end{pmatrix} \tag{1.21}$$

where $K_{BEM}$ is the stiffness matrix from the BEM, $K_{DDA}$ is from the DDA, $K_{BD}$ is the interaction between the DDA and the BEM, $u_{BEM}$ is the BEM interface displacement vector, $a_{DDA}$ is the coefficient vector of DDA. Both $F^I_{BEM}$ and $F^I_{DDA}$ are modified by the interaction terms and thus have the superscript $I$.

The open-close iteration based contact algorithm in the DDA has also been implemented into the dual reciprocity boundary element method (DRBEM) by Fu (2014), Fu, Ma and Qu (2015, 2016). The governing equation is written as follows:

$$HU - GP + M\ddot{U} + Mg - E^d\sigma^0 - \widehat{G}F_c = 0 \tag{1.22}$$

where $M$ is the equivalent mass matrix, $g$ is the gravitational acceleration, $E^d$ is the coefficient matrix for initial stress, $\sigma^0$ is the initial stress, $\widehat{G}$ is the coefficient matrix for concentrated forces and $F_C$ is the concentrated force vector.

All the contact forces are treated as concentrated forces and handled by the following equation to be added in the governing equations:

$$F_p = \widehat{G}_p^{P_q} F \tag{1.23}$$

where $F_p$ is the resultant force vector corresponding to block $p$ $(p=i,j)$, $F$ is the concentrated force and it could be the normal contact force $(F_n)$, shear contact force $(F_s)$ or frictional force $(F_f)$, $\widehat{G}_p^{P_q}$ is the coefficient matrix for the concentrated force $F$ applied at point $P_q$ $(q=1,2,3)$ in block $p$.

### 1.11.3  Coupling DDA with NMM

In the original DDA method the blocks are "simply deformable", namely, each single block has constant stress throughout the block. It is therefore difficult to consider local displacements and stress conditions with DDA. On the other hand, the Numerical Manifold Method (Shi, 1997) can simulate both continuous and discontinuous deformation of the blocky systems. Even so, rigid body rotation of blocks cannot be treated properly with NMM because NMM does not solve rigid body rotation in an explicit form. In order to make full use of the advantages of both methods, Miki *et al.* (2010) presented the formulation of the coupled DDA and NMM.

The total potential energy $\Pi_{sys}$ of the blocky system, which includes DDA blocks and NMM elements, can be expressed as:

$$\Pi_{sys} = \Pi_{sys}^{NMM} + \Pi_{sys}^{DDA} + \sum_{B,i} \sum_{E,j} \Pi_{i,j} \tag{1.24}$$

The last term in Equation 1.24 represents the potential energy for the contact between DDA block $i$ and NMM element $j$. The corresponding matrices and vectors in the kinematic equations can also be derived by minimizing the potential energy.

## 1.12 IMPROVED CONTACT ALGORITHMS

DDA contact algorithm is based on a penalty method. In the penalty method, stiff springs are set in normal and/or shear directions between blocks to transfer the inequality problem of contact constraint into equality problem of computing contact displacements and contact forces. The penalty method proved to be effective in many numerical areas and has been widely applied. It's relatively simple implementation and it does not increase the dimension of the system of equilibrium equations. The main shortcoming of the penalty method is that it can only fulfil the contact constraint approximately, and the contact treatment precision is affected by selected penalty number. With a reasonably large penalty number the penalty method can treat the contact of blocks well, however if the penalty number is too large the system of equilibrium equations may become ill-conditioned resulting in un-acceptable errors. There is no simple way to guess a priory the best value of the penalty method in the original DDA and this does present an obstacle when high accuracy is sought. To amend this problem many researchers have proposed different remedies since the original DDA has been proposed.

Amadei *et al.* (1996) modified the original DDA contact using the Augmented Lagrangian Method (ALM) so as to retain the simplicity of the penalty method and yet to minimize the disadvantages of the penalty method and the classical Lagrange Multiplier Method, the implementation of which would require an increase in the system of the governing equations. The essential concept behind the ALM is to use both a penalty number, $p$ representing the stiffness of the contact spring, and a Lagrange multiplier $\lambda^*$ representing the contact force $\lambda$, for each block to block contact to iteratively calculate the contact force. An iterative method is used to calculate the Lagrange multiplier until the distance, $d$, of penetration of one block into the other and the residual force between block contacts are both below minimum specified tolerances. The final exact contact forces can always be obtained by the iterative method even with a small initial value of the penalty parameter, although if the penalty parameter is too small many iterations may be required and consequently some efficiency may be lost.

To better understand the ALM modification to the original DDA contact algorithm consider Fig. 1.13. In the original two dimensional DDA all contacts between blocks can be transformed to contacts between an angle and an edge. $P_1$ in Fig. 1.13 is a vertex of block $i$, $P_2P_3$ is an edge of block $j$. After a step displacement $P_1$ moves to $P_0$. Assume $(x_k, y_k)$ and $(u_k, v_k)$ are the coordinates and displacements of $P_k(k = 0 \sim 3)$ respectively.

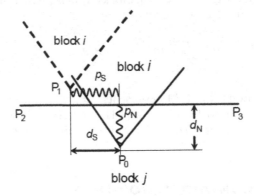

Figure 1.13 Angle–edge contact in DDA (Ning et al., 2010).

The normal and shear contact displacements $d_n$ and $d_s$ of $P_1$ to $P_2P_3$ in a time step can be expressed as (Ning et al., 2010):

$$d_n = \frac{S_0}{l} + ED_i + GD_j, \quad d_s = \frac{S_0'}{l} + E'D_i + G'D_j \tag{1.25}$$

where $D_i$ and $D_j$ are unknown displacement vectors of block $i$ and $j$ respectively,

$$S_0 = \begin{vmatrix} 1 & x_0 & y_0 \\ 1 & x_2 & y_2 \\ 1 & x_3 & y_3 \end{vmatrix}, \quad S_0' = [x_3 - x_2 y_3 - y_2] \begin{Bmatrix} x_1 - x_0 \\ y_1 - y_0 \end{Bmatrix},$$

$$l = \sqrt{(x_2 - x_3)^2 + (y_2 - y_3)^2}$$

$E, G, E'$, and $G'$ are $1 \times 6$ matrices $(r = 1, \ldots, 6)$, and

$$e_r = \frac{1}{l}[(y_2 - y_3)t_{1r}(x_0, y_0) + (x_3 - x_2)t_{2r}(x_0, y_0)]$$

$$g_r = \frac{1}{l}[(y_3 - y_0)t_{1r}(x_2, y_2) + (x_0 - x_3)t_{2r}(x_2, y_2)]$$

$$\quad + \frac{1}{l}[(y_0 - y_2)t_{1r}(x_3, y_3) + (x_2 - x_0)t_{2r}(x_3, y_3)]$$

$$e_r' = \frac{1}{l}[(y_2 - y_3)t_{1r}(x_1, y_1) + (x_3 - x_2)t_{2r}(x_1, y_1)]$$

$$g_r' = \frac{1}{l}[(x_2 - x_3)t_{1r}(x_0, y_0) + (y_2 - y_3)t_{2r}(x_0, y_0)]$$

in which $t_{1r}(x, y)$ and $t_{2r}(x, y)$ are displacement functions of block $i$ and $j$ respectively.

In the penalty method, two springs with stiffness of $p_n$ and $p_s$ are used in the normal and shear directions to constrain the contact displacements to zero. Then the strain energy of normal and shear springs can be respectively expressed as:

$$\Pi_n = \frac{1}{2}p_n d_n^2, \quad \Pi_s = \frac{1}{2}p_s d_s^2 \tag{1.26}$$

In the Augmented Lagrangian Method a penalty number $p$ and a Lagrangian multiplier $\lambda^*$ are used to compute the contact force iteratively as:

$$\lambda_{m+1}^* = \lambda_m^* + pd \tag{1.27}$$

where $\lambda_m^*$ and $\lambda_{m+1}^*$ are the Lagrangian multipliers of time step $m$ and $m + 1$, respectively, and $d$ is the contact displacement. Then at the $m^{\text{th}}$ time step the contact strain energy can be expressed as:

$$\Pi_s = \Pi_\lambda + \Pi_p = \lambda_m^* d + \frac{1}{2}pd^2 \tag{1.28}$$

where $\Pi_p = \frac{1}{2}pd^2$ is the same as the spring strain energy in the penalty method, so only $\Pi_\lambda = \lambda_m^* d$ needs to be taken into account additionally. $\Pi_\lambda$ in the normal and shear directions can be deduced as:

$$\Pi_\lambda = \lambda_m^* d_n = \lambda_m^* \left( \frac{S_0}{l} + ED_i + GD_j \right)$$
$$\Pi_\lambda' = \lambda_m^{*\prime} d_s = \lambda_m^{*\prime} \left( \frac{S_0'}{l} + E'D_i + G'D_j \right) \tag{1.29}$$

According to the minimum potential energy principle, four $6 \times 1$ sub-matrices are added to the system sub-matrices $F_i$ or $F_j$. As the normal contact force may become different from that in the penalty method, the friction force sub-matrices should also be adjusted. Some verification of the ALM for DDA are presented by Ning *et al.* (2010) and by Bao *et al.* (2014).

Among the different types of contacts, the modelling of the vertex-to-vertex contact is the most challenging one as the contact reference edges in the vertex-to-vertex contact are not unique, which may lead to an indeterminate state in the numerical analysis. The original DDA method employs the penetration distance to determine the contact edge, which may not work for the cases where two vertices are detected in contact without a penetration. An enhanced algorithm was proposed by Bao and Zhao (2010a) by choosing the initial contact edge of the vertex-to-vertex contact edge in the DDA. A temporary contact spring, which has the same stiffness as a normal contact spring, is applied between the two vertices when they are detected to collide in the next time step. At the end of the first open-close iteration, the relative movements between the contact vertices will appear because the special vertex-to-vertex contact spring only works like a weak pin joint, which allows small movement. After the proper contact edge is obtained, the temporary vertex-to-vertex contact spring is removed and the open-close iteration will continue.

In the open-close iteration of the DDA method, each contact pattern converges at time step $n$ when the contact pattern at iteration $i$ is the same as the one at previous iteration $i - 1$. When the number of iteration equals to 6 and the contact pattern does not converge, the time step size will be reduced and a new iteration cycle will start. At iteration 1 of the new cycle, the previous pattern is defined as that at iteration 5 of the previous cycle, which may be different from the contact pattern at the end of step $n - 1$. This will cause significant and unexpected change in time step size. Wu and Lin (2013) improved the open-close iteration by reloading the contact information at the end of previous step $n - 1$ as the previous iteration when the iteration is 1 in the new cycle. The computational results indicate that the new DDA program results in correct stress evaluation.

A new contact theory was proposed by Shi (2013), in which entrance blocks are used for contacts between two dimensional and three dimensional blocks. The boundary of an entrance block is the entrance surface, which is a contact cover system. Each contact cover could be contact vectors, contact edges, contact angles or contact polygons. Each contact cover defines a contact point and all closed contact points together define the movements, rotations and deformations of all blocks. Given a reference point, the concept of entrance blocks simplifies the contact computations. The essentials of Shi's new contact theory are provided in Chapter 8.

## 1.13 INCORPORATION OF VISCOUS DAMPING

In the original DDA code written by Shi viscous damping was ignored, although the corresponding viscous damping matrix does appear in the simultaneous equations of equilibrium. The reason for not implementing viscous damping in the original DDA code was related to uncertainties regarding the viscous damping coefficient that should be used. Therefore, in the original DDA code no damping other than "kinetic damping" exists. The lack of viscous damping however precluded accurate modeling of dynamic impact problems such as rock falls, but DDA is well suited for modeling such problems and therefore several research groups have attempted very early on to incorporate viscous damping in the DDA code. The critical point here is to understand the energy dissipation mechanism during rock fall and then to simulate it correctly by introducing viscous damping which has a different physical meaning than the friction coefficient or coefficient of restitution. In particular the energy of the impact is clearly expected to be influenced strongly by the viscosity coefficient. It is important to realize, however, that there is no independent way to establish the viscosity coefficient and therefore it must be tweaked based on field observations of the particular case at hand. Shinji *et al.* (1997) inserted a viscosity coefficient into the original DDA equations as an "equivalent friction coefficient" and, based on field observations, suggested some empirical relationship between the viscosity coefficient and the slope conditions, primarily talus and vegetation cover. This representation of viscosity however is clearly a gross approximation. A formal implementation of viscous damping into DDA contacts has been proposed by Sasaki *et al.* (2005) and was used to model rock fall problems in dangerous slopes in Japan that are frequently subjected to earthquake excitation. The case of the 1994 Niigata prefecture earthquake was used as a case study where Voigt-type viscous damping elements were implemented between blocks and

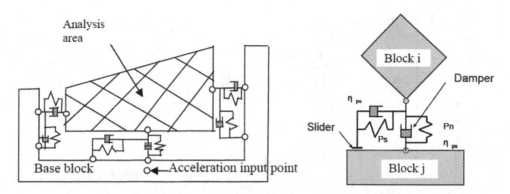

*Figure 1.14* Boundary condition of earthquake response analysis (Left) and Voigt-type viscous damping of friction (Right) implemented in DDA to model earthquake induced rock falls in Niigata prefecture, Japan (Sasaki *et al.*, 2005).

non-reflective boundaries were implemented on the boundaries of the jointed domain (Fig. 1.14).

## 1.14 IMPROVED FRICTION LAW FOR DISCONTINUITIES

In the original DDA the "no tension – no penetration" criterion is used to impose kinematical constraints on the deformation of the constant springs. If the normal spring undergoes tension it is removed so that no tension can exist between initially mating surfaces. Therefore, the normal spring is set only for compression between blocks. Because zero penetration is not feasible when calculating the spring force, a very small amount of penetration $\delta_m$ is allowed in practical computations.

The Coulomb–Mohr law is applied for modeling interface frictional resistance in DDA. If the shear force calculated by the program is higher than the available shear strength of the interface $\tau_m$ the tangential spring is removed and Coulomb frictional resistance replaces the role of the tangential spring. The shear resistance $\tau_m$ in DDA is given by:

$$\tau_m = C + f_n \tan \phi \qquad (1.30)$$

where $C$ and $\phi$ are the cohesion and friction angle of the discontinuity, both of which are assumed to be constants throughout the analysis, and $f_n$ is the compressive force of the normal spring which is updated in every time step according to the dynamic deformation of the modelled block system.

When studying dynamic deformation in geological materials which is typically concentrated along pre-existing discontinuities, the frictional resistance can hardly be assumed to be constant as assumed in the original DDA. This has already been recognized by Goodman and Seed (1966) in their classic paper which provides the analytical solution for dynamic block displacement on frictional interfaces due to cyclic loading. In contrast to the so called "Newmark's method" for slope stability

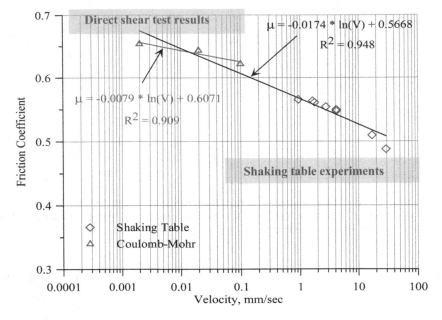

*Figure 1.15* Friction degradation in concrete surfaces as a function of sliding velocity (Bakun-Mazor *et al.*, 2012). Open triangles are results from pseudo-static direct shear experiments, open diamonds are back analyzed results from shaking table experiments.

(Newmark, 1965), Goodman and Sid, allowed in their solution incremental frictional degradation between displacement cycles. Clearly if the friction coefficient is maintained constant throughout the dynamic simulation the forward modeling results may be un-conservative because cyclic motions damage the asperities and deteriorate the frictional resistance the initially rough rock discontinuity surface may offer. Velocity weakening of concrete wedge surfaces during cyclic displacement has been measured experimentally using shaking table experiments of dynamic sliding of tetrahedral wedges (Bakun-Mazor *et al.*, 2012) and the results were compared to velocity degradation obtained during slow direct shear tests (Fig. 1.15). Similar results have been reported for rock discontinuities as well.

Velocity weakening is but part of a more complex constitutive model for rock joint friction that takes into account both the *rate* as well as the *state* of sliding, largely known today as "rate and state friction" (Dieterich, 1972). The original rate and state friction law has been modified and improved over the years (e.g. Dieterich, 1978, 1979; Dieterich, 1981; Dieterich and Kilgore, 1994, 1996; Kilgore *et al.*, 1993; Ruina, 1983) and has been used to model earthquake mechanisms (e.g. Scholz, 2002). One of the more standard representations of the rate and state constitutive law is (Dieterich, 1979; Ruina, 1983):

$$\mu = \mu^* + A\ln\left(\frac{V}{V^*} + 1\right) + B\left(\frac{\theta V^*}{D_c} + 1\right)$$
(1.31)

where $A$ and $B$ are dimensionless empirical fitting parameters, $D_c$ is a characteristic sliding distance from one steady state to another, $\theta$ is a state variable, $V^*$ is the reference velocity, and $\mu^*$ is the coefficient of friction when the contact surface slips under a constant slip rate $V^*$. Various evolution laws have been proposed in the geophysical and seismological literature for the state variable. The two most commonly used state evolution laws are the 'Dieterich law' (Dieterich, 1979):

$$\frac{d\theta}{dt} = -\frac{V\theta}{D_c} \tag{1.32}$$

and the 'Ruina law' (Ruina, 1983):

$$\frac{d\theta}{dt} = -\frac{V\theta}{D_c}\ln\left(\frac{V\theta}{D_c}\right) \tag{1.33}$$

When dynamic frictional sliding takes place at steady state conditions $d\theta/dt = 0\left(\frac{V\theta}{D_c} = 1\right)$ and Equation 1.31 can be written as:

$$\mu_{ss} = \mu^* + (A - B)\ln\left(\frac{V_{ss}}{V^*} + 1\right) \tag{1.34}$$

Osada and Tanityama (2005) realized the significance of incorporating dynamic friction into DDA and implemented the formulation of the law as proposed by Ruina (1983) into the code. Dong and Osada (2007) demonstrated the significance of rate and state friction by checking the validation of block response in DDA to cyclic frictional sliding (following Kamai and Hatzor, 2005; 2008). Other researchers have proposed different methods to incorporate displacement or velocity weakening of the sliding interface in DDA (Wang et al., 2013; Wu, 2010).

Another issue is the assumed linearity of the normal and tangential springs in the original DDA. In reality the normal deformation is not necessarily linear when joints are compressed, and when sliding initiates shear resistance in reality does not necessarily drop to zero. A nonlinear model for discontinuities in DDA has been proposed by Chen and Ohnishi (1999) who have also considered the tensile strength for the interface. While the exact details of the mechanics of the normal and shear response may vary, this issue is certainly important as it affects the contact algorithm and thus the mechanical response of the entire block system.

## 1.15  GRAVITY TURN ON AND SEQUENTIAL EXCAVATION

DDA is a fully dynamic method that uses inertia forces to determine the motion of block systems. The inertia forces are updated after each time step and are dependent on the block displacement in the previous time step. Consequently, sudden loading at the beginning of the analysis, whether by gravity or any other applied load, produces unwanted artificial accelerations and displacements that propagate throughout the time steps of the analysis. MacLaughlin and Sitar (1999) identified this problem and were the first to propose a static "gravity turn-on" phase which precedes the dynamic analysis, and added two extra iterations in each time step in order to guarantee accurate determination of the contact forces between blocks. They demonstrated the enhanced accuracy of this approach using the classic block on an inclined plane problem.

The gravity turn on concept has already been discussed in the context of the FEM (Desai and Abel, 1972) and its significance in geotechnical engineering has been recognized by early development of analytical (e.g. Goodman and Brown, 1963) and numerical (e.g. Dunlop et al., 1970) solutions for rock slope stability. MacLaughlin and Sitar (1996) explain this issue using the case of an open pit mine: in reality the removal of the overburden will result in upward displacement of the floor, however if the post excavation geometry is modeled without regard to the initial stress conditions the floor of the pit will displace downward as the material deforms elastically when gravity load is introduced.

In deep underground excavations this is particularly important. When modeling a high *in situ* stress environment it may take many time steps for the model to actually equilibrate under the imposed *in situ* stresses and gravitational load, a process that has been attributed to "seating" of the contacts (MacLaughlin and Sitar, 1999) which actually is a correct way for describing this problem. Therefore, if the underground opening is removed before the stresses stabilize everywhere in the block system the results may be overly conservative because the real available frictional resistance of the joints hasn't been realized yet in the model as normal stresses have not reached the correct value (see Fig. 1.16). To address this problem MacLaughlin and Sitar (1999) proposed a preliminary pseudo-static phase during which the unwanted displacements are dissipated by setting the artificial velocities at the beginning of every time step in this stage to zero.

An alternative approach would be to actually model the excavation sequence where first the imposed *in situ* stresses are allowed to fully develop everywhere in the mesh (the number of time steps this stage will require increases with the number of blocks in the mesh because indeed all contact springs need to be equilibrated under the loads), and only then the excavation is removed and true displacements are realized.

Sequence excavation modeling in high *in situ* stress environments has been demonstrated with the NMM (Tal et al., 2014) and DDA (Hatzor et al., 2015). In the sequence excavation code developed by Tal et al. (2014) instead of using the additional iterations suggested by MacLaughlin and Sitar (1999) a special provision is made in the code to run the simulation in static conditions (dynamic parameter = 0) before the excavation is removed and when the excavation material is removed to switch the analysis to dynamic (dynamic parameter = 1). One added advantage of the preliminary stage in sequence excavation modeling is that it allows tweaking the best value for the contact springs; the penalty value at which the steady state stresses in that stage are as close as possible to the imposed stresses (before the opening is introduced) is in all likelihood the optimal contact spring value for the specific problem at hand.

## 1.16 DYNAMIC WAVE PROPAGATION AND BLASTING

DDA is a dynamic method and as such it is suitable for modeling dynamic problems such as seismic site response or shock wave propagation through discontinuous media. Several modifications have been proposed to the original DDA in order to enhance modeling capabilities of dynamic wave propagation.

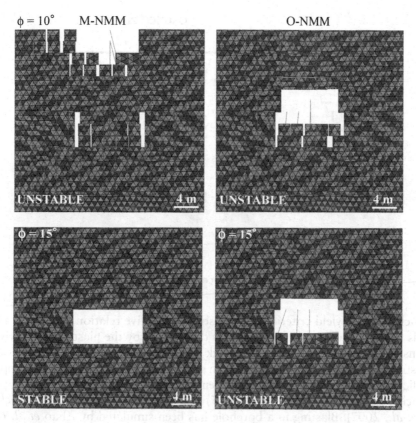

*Figure 1.16* The significance of modeling the excavation sequence. On the right panels (Original NMM) the opening exists from the first time step whereas on the left panels (Modified NMM) the opening is removed only after all stresses have stabilized (gravity turn on). Note that with gravity turn on a friction angle of 15 degrees on all joints is sufficient for stability (Tal et al., 2014).

Tsai and Wang (1995) modified DDA to include a sliding boundary that transmits the frictional forces from the ground to the superstructure. Wang *et al.* (1995) conducted a pioneering study on shock wave propagation through an elastic bar discussing the interrelationship between the contact spring stiffness and the time step and proposing new criteria to select the best choice of time step and contact spring stiffness for dynamic contact problems. The influence of various time integration scheme on the accuracy of the dynamic solution is discussed by Wang *et al.* (1996). To improve rock blasting simulation capability a plastic constitutive relation has been added to the original 2D-DDA code and energy dissipation was taken into account by Yang and Ning (2005). They argued that frictional sliding along pre-existing discontinuities cannot be assumed to be the main energy consumption mechanism in blasting simulations, as is assumed in the original DDA. They implemented a Drucker-Prager plastic yield criterion for the solid block elements so that linear strain-hardening was added to the original 2D-DDA code. While a block reaches plastic deformation according to

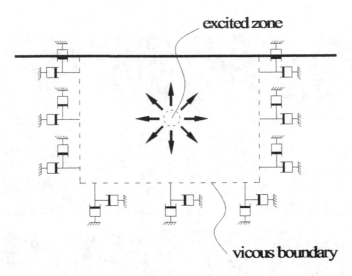

*Figure 1.17* The concept of a viscous boundary (Bao *et al.*, 2012).

the Drucker-Prager yield criterion, the plastic constitutive relation is used. If a block expands beyond a certain threshold that is determined by the plastic constitutive law it is considered to be fractured and a crack is recorded. For this fractured block the tensile strength is zeroed and the increase in strain is ignored. They used this approach to predict the fragment size that will be generated and demonstrated their approach for bench blasting (Yang and Ning, 2005) and demolition blasting of a brick wall (Ning *et al.*, 2007). Blasting in a borehole has been simulated by Zhao *et al.* (2007) with provisions made in the original DDA code for fracture initiation and propagation assuming a Weibull type tensile strength distribution in the joints. This approach has been extended for modeling full face blasting during tunneling (Zhao *et al.*, 2010).

To enable accurate modeling and fast computation of blasting phenomena with DDA the boundaries of the modeled domain should be relatively close to the blasting point. If however the boundaries of the modeled domain are positioned too close to the basting point, inevitable reflections from the boundaries may obscure the computation. To avoid this problem non-reflective boundaries must be introduced. Two types of non-reflective boundaries for DDA are typically considered (Ning and Zhao, 2012): viscous boundary condition (VBC) and superposition boundary condition (SBC).

Jiao *et al.* (2007) were the first to improve the existing fixed boundary condition in the original DDA by introducing a viscous boundary condition into the original DDA code. They adopted the VBC proposed by Lysmer and Kuhlemeyer (1969) which is based on the use of independent dashpots in the normal and shear directions of specific boundaries (Fig. 1.17). Their proposed VBC was verified using analytical solution for $P$ wave propagation through an elastic bar and attenuation of blasting wave. Bao *et al.* (2012) introduced a new viscous boundary submatrix with high absorbing efficiency which was developed specifically for DDA, based on the viscous boundary condition originally introduced by Lysmer and Kuhlemeyer (1969). In their derivation, the analytical velocity of a dashpot is employed instead of using the finite

*Figure 1.18* Observed finite displacement in an old masonry structure used to constrain paleo peak ground acceleration.

difference method to obtain the velocity. Their VBC is verified using analytical solutions for $P$ wave propagation through an elastic bar and $S$ wave propagation through a stack of horizontal layers. A recent comprehensive review and original developments of boundary settings for DDA is provided by Fu *et al.* (2015).

## 1.17 MASONRY STRUCTURES

The robust simplex integration scheme in DDA and complete dynamic formulation makes it a very attractive candidate for studying deformation of masonry structures. This promising research direction has been pioneered by Professor Bicanic from the university of Glasgow (Bicanic and Stirling, 2001; Bicanic *et al.*, 2003) who has originally used the semicircular masonry arch under self-weight as a benchmark test for DDA to show how a departure from the classical interface conditions (no sliding, no crushing, no tension) influences masonry arch stability. Observed finite block displacements in historic masonry structures (Fig. 1.18) have been used to constrain paleo peak ground acceleration (Kamai and Hatzor, 2008; Yagoda-Biran and Hatzor, 2010) by inserting sinusoidal input accelerations to the modeled structure and finding and frequency and acceleration at which the damage obtained numerically best fits the damage mapped in the field.

DDA can also be utilized to consider alternative reinforcement schemes for ancient masonry structures. This has been demonstrated in the case of Angkor, one of the most significant monuments of the Khmer culture in Cambodia (Hayashi *et al.*, 2012; Koyama *et al.*, 2013; Sasaki *et al.*, 2011; Tian *et al.*, 2012).

## 1.18 IMPROVED ROCKBOLT ELEMENT

The original DDA includes rockbolt submatrices where the bolt is modeled as a fixed line on two ends with a given stiffness and length. The block system is then allowed to deform and the bolts to extend, but the original bolt elements are linear, of infinite stiffness, and a perfect coupling between the bolt element and the rock mass is assumed. The location of the end point of the bolt and its stiffness are defined by the user. During each time step the resistance provided by the bolt stiffness is incorporated into the global stiffness matrix by minimizing the potential energy. The bolt stiffness is then mobilized with the occurrence of relative displacement of the two end points. The applicability of this simple rockbolt element has been demonstrated for both underground (Bin Shi *et al.*, 2010; Hatzor *et al.*, 2015; Yeung, 1991, 1993) and rock slope (Hatzor *et al.*, 2004; Shi, 2012; Tsesarsky and Hatzor, 2009) engineering.

The original DDA bolt is capable of simulating mechanically anchored bolt systems, the usage of which has become less popular over time. Clapp and MacLaughlin (2003) further pointed out several deficiencies of the bolt element in the original DDA:

1   The original bolt element keeps deforming infinitely throughout the analysis as it is assumed to have linear elastic stiffness and infinite strength.
2   No option for measurement points along the bolt trace are provided in the original code to check the shear stress and displacement distribution along the bolt length.
3   The original bolt element does not interact realistically with the rock mass.
4   The single segment spring bolt allows forces to be distributed uniformly along the length of the bolt instead of dispersing the load gradually away from the load point.

A more representative bolting system which takes into consideration the bond and its interaction with the bolt element on one side and with the rock mass on the other is therefore required. Moosavi and Grayeli (2006) implemented bond strength characteristics into the original DDA bolt element for simulation of the performance of fully grouted cable bolts during deformation of a discontinuous rock mass where most of the deformation is assumed to concentrate along the interface between the rock and the cable bolt.

A unified rock bolt element for 2D-DDA has been developed by Professor Zhao Zhiye and his research group in Singapore (Zhao *et al.*, 2013a). Different types of rockbolts are modeled in a unified framework and different failure modes are possible in the proposed analysis procedure. The details of the coupling between the rock bolts and rock blocks and its implementation in 2D-DDA are discussed by Nie *et al.* (2014).

## 1.19   GRANULAR MATERIALS

### 1.19.1   Historical overview

Thanks to the powerful simplex integration technique used in DDA (Shi, 1988) and NMM (Shi, 1996), blocks in the original DDA can have any shape, concave or convex, with or without holes, but they must still be closed polygons. This may cause numerical problems when attempting to model with the original DDA interactions between circular blocks since they would have to be modeled as polygons with many edges. Many researchers wanted to transfer the strengths of DDA as a powerful, dynamic, discrete element method into the world of granular materials and for that purpose embarked on efforts to develop circular or elliptical disc elements for DDA. The first efforts in this direction were presented by Ohnishi and his group (Ohnishi *et al.*, 1995; Ohnishi and Miki, 1996) who have also provided the mathematical derivation and implemented this into the original DDA with the elliptical elements enabling internal deformation of the initially circular discs. This research direction was followed by many workers in the soil mechanics and geotechnical engineering fields (e.g. O'Sullivan and Bray, 2001; Thomas *et al.*, 1996).

The deterioration of railroad track material (ballast) was studied by applying cyclic loading to a stack of coarse granular fragments and measuring the obtained plastic deformation with DDA (Ishikawa *et al.*, 1997). To better represent the high friction angle of such materials polygon elements generated by Voronoi tessellation were employed and DDA results with polygon and circular disc elements were compared. They concluded that the contact mechanism in the original DDA must be improved in order to properly model coarse granular deformation. They have later extended their analysis to study time dependent deformation of granular materials with DDA (Ishikawa and Ohnish, 2001) and further incorporated also elliptic disc elements in the analysis (Ohnishi *et al.*, 2005).

DDA has been used to model ball milling in mining industry context with more numerical efficient contact sorting and contact detection algorithms by simulating rigid, non-fracturing, 2D ore particles and implementing viscos damping both in the normal and shear directions (Balden *et al.*, 2001).

Early in 2001, an algorithmic breakthrough was achieved allowing 3D transient dynamic modelling of particulate systems of real-shaped particles (Munjiza and Latham, 2002). Whereas with spheres it is relatively simple to establish whether particles are in contact from the position of their centers and their radii, and to establish forces and trajectory paths associated with collisions, considerable algorithmic sophistication is required for collisions of rock fragment-shaped particles. Guo and Lin (2007) used 2D DDA with polygonal particle shapes to study, numerically, the mechanical behavior of coarse granular media and obtained surprisingly good agreement with laboratory experiments.

Some researchers have tried extending DDA to problems which are typically considered as soil mechanics problems. For example, the effect of soil expansion in response to changes in water on slope stability has been studied with DDA by introducing time step wise linear material property degradation in response to increasing water content (Lin and Qiu, 2010).

## 1.19.2  Modelling particulate media with DDA

Ke and Bray (1995), Ohnishi *et al.* (1995), and Ohnishi and Miki (1996) developed DDA codes for circular and elliptical disks. For circular disks, each disk is assumed to be rigid, except for small displacements along its boundary represented by the compression of the contact springs. The displacement vector of any point $P(x, y)$ within disk $i$ is represented by the displacement variables consisting of two rigid body translations and one rotation $(u_0, v_0, \gamma_0)$ For pure translation:

$$\begin{pmatrix} u \\ v \end{pmatrix} = \begin{pmatrix} 1 & 0 \\ 0 & 1 \end{pmatrix} \begin{pmatrix} u_0 \\ v_0 \end{pmatrix} \tag{1.35}$$

For pure rigid-body rotation, a linear approximation is as follows:

$$\begin{pmatrix} u \\ v \end{pmatrix} = \begin{pmatrix} -(y - y_0) \\ x - x_0 \end{pmatrix} \gamma_0 \tag{1.36}$$

Combining Equations 1.35 and 1.36 yields:

$$\begin{pmatrix} u \\ v \end{pmatrix} = \begin{pmatrix} 1 & 0 & -(y - y_0) \\ 0 & 1 & x - x_0 \end{pmatrix} \begin{pmatrix} u_0 \\ v_0 \\ \gamma_0 \end{pmatrix} \tag{1.37}$$

After solving $(u_0, v_0, \gamma_0)$, Equation 1.37 may not compute $(u, v)$ of a specified point $P(x, y)$ within disk $i$ if $\gamma_0$ is very small, as the radii is lengthened. To avoid such an error, Equation 1.36 is used to calculate the components of $(u, v)$ due to $\gamma_0$.

For elliptical disks (Ohnishi and Miki, 1996), assuming that $(x_0, y_0)$ is the centre of gravity of an ellipse, an ellipse is represented as:

$$\begin{aligned} x &= \frac{ab}{\sqrt{b^2 \cos^2 s + a^2 \sin^2 s}} \cos s + x_0 \\ & \hspace{6em} 0 \leq s \leq 2\pi \\ y &= \frac{ab}{\sqrt{b^2 \cos^2 s + a^2 \sin^2 s}} \sin s + y_0 \end{aligned} \tag{1.38}$$

where $a$ and $b$ are the major and minor axes, respectively and $s$ is an angular parameter. When only translations $(u_0, v_0)$ are involved, the displacements $(u, v)$ at any point in the block are given by:

$$\begin{pmatrix} u \\ v \end{pmatrix} = \begin{pmatrix} 1 & 0 \\ 0 & 1 \end{pmatrix} \begin{pmatrix} u_0 \\ v_0 \end{pmatrix} \tag{1.39}$$

Assuming that only rotation $(r_0)$ of the ellipse around $(x_0, y_0)$ is involved, the shape of the ellipse can exactly be represented by:

$$\begin{aligned} x &= \frac{ab}{\sqrt{b^2 \cos^2 s + a^2 \sin^2 s}} \cos(s + r_0) + x_0 \\ & \hspace{6em} 0 \leq s \leq 2\pi \\ y &= \frac{ab}{\sqrt{b^2 \cos^2 s + a^2 \sin^2 s}} \sin(s + r_0) + y_0 \end{aligned} \tag{1.40}$$

When the rotation angle $r_0$ is small, the displacements are simplified to the following equation:

$$\begin{pmatrix} u \\ v \end{pmatrix} = \begin{pmatrix} -(y - y_0) \\ x - x_0 \end{pmatrix} \gamma_0 \tag{1.41}$$

The displacements of the ellipse due to normal strains $(\varepsilon_x, \varepsilon_y)$ are:

$$\begin{pmatrix} u \\ v \end{pmatrix} = \begin{pmatrix} x - x_0 & 0 \\ 0 & y - y_0 \end{pmatrix} \begin{pmatrix} \varepsilon_x \\ \varepsilon_y \end{pmatrix} \tag{1.42}$$

When only shear strain $r_{xy}$ is involved, the shape of an ellipse is represented by:

$$x = \frac{ab}{\sqrt{b^2 \cos^2 s + a^2 \sin^2 s}} \cos\left(s - \frac{r_{xy}}{2}\right) + x_0$$

$$y = \frac{ab}{\sqrt{b^2 \cos^2 s + a^2 \sin^2 s}} \sin\left(s + \frac{r_{xy}}{2}\right) + y_0 \tag{1.43}$$

Assuming small shear strain $r_{xy}$, the displacements are simplified as follows:

$$\begin{pmatrix} u \\ v \end{pmatrix} = \begin{pmatrix} \dfrac{y - y_0}{2} \\ \dfrac{x - x_0}{2} \end{pmatrix} \gamma_{xy} \tag{1.44}$$

Summing Equations 1.39, 1.41, 1.42, and 1.44, we obtain the displacements for any point within the elliptical disk, which is identical to that of a polygon in the original DDA:

$$\left\{ \begin{matrix} u \\ v \end{matrix} \right\} = \begin{bmatrix} 1 & 0 & -(y - y_0) & (x - x_0) & 0 & (y - y_0)/2 \\ 0 & 1 & (x - x_0) & 0 & (y - y_0) & (x - x_0)/2 \end{bmatrix} \begin{pmatrix} u_0 \\ v_0 \\ r_0 \\ \varepsilon_x \\ \varepsilon_y \\ r_{xy} \end{pmatrix} = T_i d_i \tag{1.45}$$

The remaining formulation of particle-based DDA is the same as the original DDA.

## 1.20 PORE PRESSURE AND FLUID FLOW

### 1.20.1 Historical overview

In the original DDA no provisions were made to incorporate water pressures in the joints, but rarely in rock mechanics are joints actually dry and is the rock mass drained, particularly when studying rock mechanics problems well below the ground water table as in deep mining, tunneling, and petroleum engineering applications. Moreover, water pressures in the joints have a profound influence on rock mass deformation in general and on the stability of rock blocks in particular.

Jing *et al.* (2001) derived, for the first time, explicit expressions for contributions from fluid pressure to the global stiffness matrix and load vectors of the discrete block systems for rigid blocks, triangle and quadrilateral elements (used for internal block discretization) by closed form integration. The flow algorithm used a residual flow method for locating the free surfaces in unconfined flow problems, which was then coupled to the mechanical equations of motion for the coupled hydro-mechanical analysis. A laboratory experiment of fluid flow in a 2-D rectangular network was simulated to verify the residual flow algorithm in their modified DDA code, and an acceptable agreement between the measured and calculated results was presented.

A coupled hydro-mechanical DDA model (HYDRO-DDA) has also been proposed by Rouainia et al. (2001) where the discontinuous medium is represented with DDA and is interfaced with a continuum formulation for flow through porous media. The two frameworks communicate via mapping of an equivalent porosity field from the solid to the fluid phase and an inverse mapping of the pressure field back to DDA. The fluid system, by means of Dracy's law using a FEM mesh, responds to pressure or flux boundary conditions and to porosity changes caused by changes in the discontinuity patterns. DDA is used to model the blocky discontinuous solid phase, responding to force or displacement boundary conditions and to the state of effective stress. The procedure operates by establishing an initial fluid pressure distribution in a static rock framework and passing this information to DDA which produces a deformed solid framework shape. This change in geometry is passed back to the hydro part of the code for refinement of fluid pressure distribution and returned back to DDA until the results converge according to some tolerance. The motivation behind their study came from the petroleum engineering industry where the fracturing of a mudrock sealing cap of an over pressured sandstone reservoir was of interest, but of course this approach can be further utilized to model other petroleum engineering related issues such as, for example, borehole breakouts and hydraulic fracturing. Indeed, a different fluid-solid coupling scheme for DDA was proposed by Ben et al. (Ben et al., 2012) for single phase compressible fluid, and it was later utilized for modeling hydraulic fracturing with DDA (Ben et al., 2013).

Challenging geotechnical engineering problems, which would typically fall into the realm of classical soil mechanics, have been attempted to be analyzed with DDA. Oh et al. (2002) modeled wave induced sea bed response to the dynamic impact of waves incorporating into DDA effective stresses and Biot's equations, perhaps for the first time, arguing that an effective, rather than total, stress approach is needed in such cases. Moreover, the progressive soil deformation will become discontinuous over time under these unique loading conditions and thus the justification for attempting a discontinuous approach, on contrast to methods that have been applied until then to address these problems numerically. Koyama et al. (2012) modeled triggering large landslides due to torrential rainfall by introducing pore pressures into the unsaturated zone using FEM and modeling the dynamic deformation with DDA. The validity of their numerical approach was tested using large scale physical experiments. Chen et al. (2013) introduced seepage forces into DDA to analyze the stability of coastal breakwater structures against tsunami damage. Hydro-mechanical coupling with DDA has also been applied to investigate the seepage rate into oil – storage caverns under different in situ stress conditions after the excavation is formed (Zhao et al., 2013b).

A three-dimensional fluid-structure coupling between Smoothed Particles Hydrodynamics (SPH) and 3D-DDA for modelling rock-fluid interactions has recently been presented by Mikola and Sitar (2013). The Navier-Stokes equation is simulated using the SPH method and the motions of the blocks are tracked by a Lagrangian algorithm based on a newly developed, explicit, 3D-DDA formulation. The coupled model is employed to investigate the water entry of a sliding block and the resulting wave(s) providing a new computational tool for coastal and offshore engineering.

## 1.20.2  Coupling DDA and fluid flow

As evident from the historical overview above, coupling between fluid flow and stress/deformation in fractured rock masses has drawn the attention of many researchers. Flow and mechanical deformation analysis should be coupled because fluid flow along joints and through porous media interacts with the stress/deformation of the rock matrix. The existence of liquids causes a change in apertures of the discontinuities while the alteration of apertures changes the hydraulic pressure in the discontinuities and rock matrix in turn (Rutqvist and Stephansson, 2003). Therefore, it is critical to take into account the hydro-mechanical coupling process when simulating the behaviors of the rock masses.

In some of the approaches reviewed above (Ben *et al.*, 2013; Jing *et al.*, 2001; Rouainia *et al.*, 2001) DDA is used to model the blocky discontinuous media and Darcy's law is modelled using a fixed finite element mesh with a finer mesh along discontinuities for fluid flow analysis. Non-negativeness of the dissipation and isothermal conditions are assumed, and the fluid flow through porous media is governed by Darcy's law. The specific discharge vector $q$ takes the following form:

$$q = -\frac{k}{\mu_r}(\nabla h + \rho_r \nabla z)$$

(1.46)

where $k$ is the hydraulic conductivity tensor and $h$ is the hydraulic head. $\mu_r$ and $\rho_r$ are the relative viscosity and relative density of the fluid respectively and $z$ is the elevation. The finite element approximation (Smith and Griffiths, 1998) is used to solve the hydraulic head, with suitable boundary conditions and fluid material properties.

The fluid pressure acting on the vertex of every DDA block is determined from the fluid flow analysis. Where a solid block vertex lies within a finite element, the pressure at the centroid of the element is calculated and applied as force components to the vertex of the block. The force components on the vertices of a block are written as follows:

$$F_1 = \frac{L}{2}\left(p_{1m} + \frac{1}{3}(p_{2m} - p_{1m})\right)$$
$$F_2 = \frac{L}{2}\left(p_{1m} + \frac{2}{3}(p_{2m} - p_{1m})\right)$$

(1.47)

where $L = \sqrt{(x_2 - x_1)^2 + (y_2 - y_1)^2}$ and $p_{1m}$ and $p_{2m}$ are the pressure at the centroid of the corresponding finite elements in the mesh where the two block vertices lie, respectively. The horizontal and vertical components of the force components $F_1$ and $F_2$ are determined and added to the external load and given as boundary conditions in the DDA. This produces a changed element geometry according to block system kinematics and consequently leads to a change in the fluid pressure distribution.

## 1.21 CURRENT DEVELOPMENT OF 3-D DDA

3-D DDA is being extensively studied currently, mainly its basic theory and contact algorithm. Shi (2001a, b, c) provided basic formulations for matrices such as mass matrix, stiffness matrix, point load matrix, body load matrix, initial stress matrix and fixed point matrix for different potential terms. Liu *et al.* (2004) and Yeung *et al.* (2003; 2004) highlighted the application of 3-D DDA for dealing with 3-D mechanical interactions of blocks, using the 'common-plane' technique. Jiang and Yeung (2004) developed a point-to-face model for 3-D DDA, but this model did not consider the vertex-to-vertex, vertex-to-edge and edge-to-edge contact modes. Moosavi *et al.* (2005) investigated dynamic 3-D DDA using analytical solutions, but they did not investigate the effect of the numerical parameters on the results. Beyabanaki and Jafari (2005) presented modified point-to-face frictionless contact constraints for 3-D DDA. Wu *et al.* (2005) developed a new contact-searching algorithm for frictionless vertex-to-face contact problems, and presented the 3-D DDA formulation for normal contact force. Yeung *et al.* (2007) and Wu (2008) presented algorithms for edge-to-edge contacts. Beyabanaki *et al.* (2008) presented a new algorithm to search and calculate geometrical contacts in 3-D DDA. Beyabanaki *et al.* (2009a, b) implemented trilinear and serendipity hexahedron FEM meshes into 3-D DDA. Beyabanaki *et al.* (2009b, c; 2010) further formulated 3-D DDA with high-order displacement functions.

In terms of geometric algorithms, a polyhedron is discretized into geometric elements including vertexes, edges and facets. The 3-D contact detection of polyhedral blocks includes vertex-to-vertex, vertex-to-edge, vertex-to-face, edge-to-edge, edge-to-face and face-to-face contacts (Ahn *et al.*, 2011). Intuitively, all other contact patterns, such as vertex-to-vertex, vertex-to-edge, face-to-face, edge-to-face and edge-to-edge, respectively, can be converted into vertex-to-face contacts (Wu *et al.*, 2005). Jiang *et al.* (2004) proposed a vertex-to-face contact algorithm based on geometric analysis for 3-D DDA. Keneti and Jafari (2008) developed a new contact detection algorithm considering main planes and dominant contacts to identify contact points and types. Beyabanaki and Mikola (2008) offered an approach to identify the contact pattern between two blocks using a closest point searching algorithm as well as other 3-D contact algorithms for improved efficiency. Wu *et al.* (2014) presented an efficient and robust spatial contact detection algorithm using a novel multi-shell cover system and decomposition of geometrical sub-units, which greatly reduced the contact detection volume and iterations. Zhang *et al.* (2016) identified contact types between polyhedral blocks using an extended hierarchy territory algorithm and a new loop search procedure.

Jiao *et al.* (2015) presented a new 3D spherical DDA model for rock failure (SDDARF3D) for simulating the whole process of rock failure. An integrated system coupling 3D binocular photogrammetry and DDA for stability analysis of tunnels in blocky rock mass was proposed by Zhu *et al.* (2015) recently.

These studies demonstrate the extensive research currently being pursued on contact definition, identification and implementation in DDA. In Chapter 8, we present the complete new contact theory developed by Dr. Gen-hua Shi which is general and therefore applicable to both two and three dimensional DDA. It remains for the research community to find ways to implement these new algorithms in the future, in efficient computational platforms.

# Chapter 2

# Theory of the discontinuous deformation analysis (DDA)

## 2.1 GOVERNING EQUATIONS AND DISPLACEMENT APPROXIMATION

As stated in Chapter 1, the DDA method possesses unique features in terms of system variables, single block matrix formulation, simplex integration, contact algorithm and time integration, etc.

The traditional DDA method adopts three displacement variables $(u_0, v_0, r_0)$ and three strain variables $(\varepsilon_x, \varepsilon_y, r_{xy})$ as unknowns for an individual block. The equilibrium equations of the DDA method are derived from the minimization of the total potential energy. The Coulomb's law of friction controls the contact modes between blocks and an iterative process called open-close iterations is enforced at each time step to satisfy the no-tension and no-penetration criteria and ensure the contact accuracy. The contacts between blocks are identified and continuously updated during the entire analysis process.

### 2.1.1 Governing equations

The DDA method uses an incremental solution procedure, the equations of motion are solved at each time step, and the incremental change in energy is determined at each time step as the system attempts to reach equilibrium.

For any virtual displacement, the sum of both internal and external work for the whole system should be zero, which can be represented as follows:

$$\delta(U + W) = \delta(\Pi) = 0 \tag{2.1}$$

where $U$ is the elastic potential energy of the whole system, $W$ is the work done by the external loads including these among the blocks, and $\Pi$ is the total potential energy of the whole system.

Based on the minimized potential energy, the system of equations can be represented as follows:

$$K d + C \dot{d} + M \ddot{d} = F \tag{2.2}$$

where $K$ is the stiffness matrix, $M$ is the mass matrix, $C$ is the viscosity matrix, $F$ is the external force vector, $d, \dot{d}$ and $\ddot{d}$ are the displacement, velocity and acceleration vector respectively.

The kinematic equation of motion Equation 2.2 is solved by the Newmark–$\beta$ scheme with two parameters with $\beta = 0.5$ and $\gamma = 1.0$ respectively, assuming no damping,

$$\left(K + \frac{2}{\Delta t^2} M\right) d(t + \Delta t) = F(t + h) + \frac{2}{\Delta t} M \dot{d}(t) \tag{2.3}$$

The sub-matrices from each block and the contact sub-matrices among the blocks are assembled into the global matrices according to the degrees-of-freedom of each block. The global equilibrium equations can be written in the following form,

$$\begin{bmatrix} K_{11} & K_{12} & K_{13} & \cdots & K_{1n} \\ K_{21} & K_{22} & K_{23} & \cdots & K_{2n} \\ K_{31} & K_{32} & K_{33} & \cdots & K_{3n} \\ \vdots & \vdots & \vdots & \ddots & \vdots \\ K_{n1} & K_{n2} & K_{n3} & \cdots & K_{nn} \end{bmatrix} \begin{Bmatrix} d_1 \\ d_2 \\ d_3 \\ \vdots \\ d_n \end{Bmatrix} = \begin{Bmatrix} F_1 \\ F_2 \\ F_3 \\ \vdots \\ F_n \end{Bmatrix} \tag{2.4}$$

where sub-matrices $K_{ii}$ depend on the material properties of block $i$ and $j$ and sub-matrices $K_{ij}$ are defined by the contacts between blocks $i$ and $j$. $d_i$ and $F_i$ are $6 \times 1$ sub-matrices, and $F_i$ is the loading distributed to the six unknown variables of block $i$. The matrix $K_{ij}$ $(i, j = 1, 2, \ldots, n)$ is derived by the differentiation:

$$\frac{\partial^2 \Pi}{\partial d_{ri} \partial d_{sj}} \quad (r, s = 1, 2, \ldots, 6) \tag{2.5}$$

and $F_i$ $(i = 1, 2, \ldots, n)$ is obtained from the differentiation:

$$-\frac{\partial \Pi}{\partial d_{ri}} \bigg|_{d_{ri} = 0} \quad (r = 1, 2, \ldots, 6) \tag{2.6}$$

The details of the derivation for the matrix $K_{ij}$ and force vector $F_i$ can be referred to in Sections 2.2 and 2.3.

### 2.1.2  Displacement approximation of a single block

The complete first order approximation of block displacements has been used in the DDA:

$$\begin{aligned} u &= a_1 + a_2 x + a_3 y \\ v &= b_1 + b_2 x + b_3 y \end{aligned} \tag{2.7}$$

where $(u, v)$ are the displacements at point $(x, y)$. At any point in the block $(x_0, y_0)$, the displacements $(u_0, v_0)$ are determined as follows:

$$\begin{aligned} u_0 &= a_1 + a_2 x_0 + a_3 y_0 \\ v_0 &= b_1 + b_2 x_0 + b_3 y_0 \end{aligned} \tag{2.8}$$

Subtracting Equation 2.8 from Equation 2.7,

$$u = a_2(x - x_0) + a_3(y - y_0) + u_0$$
$$v = b_2(x - x_0) + b_3(y - y_0) + v_0$$

$$(2.9)$$

The parameters $a_2$, $a_3$, $b_2$ and $b_3$ are obtained by the following equations:

$$\varepsilon_x = \frac{\partial u}{\partial x} = a_2 \rightarrow a_2 = \varepsilon_x$$

$$\varepsilon_y = \frac{\partial u}{\partial y} = b_3 \rightarrow b_3 = \varepsilon_y$$

$$(2.10)$$

$$\left. \begin{array}{l} \dfrac{1}{2}\gamma_{xy} = \dfrac{1}{2}\left(\dfrac{\partial v}{\partial x} + \dfrac{\partial u}{\partial y}\right) = \dfrac{1}{2}(b_2 + a_3) \\[3mm] r_0 = \dfrac{1}{2}\left(\dfrac{\partial v}{\partial x} - \dfrac{\partial u}{\partial y}\right) = \dfrac{1}{2}(b_2 - a_3) \end{array} \right\} \rightarrow \left\{ \begin{array}{l} a_3 = \dfrac{1}{2}\gamma_{xy} - r_0 \\[3mm] b_2 = \dfrac{1}{2}\gamma_{xy} + r_0 \end{array} \right.$$

Substituting the parameter derived from Equation 2.10 into Equation 2.9 and we get:

$$u = \varepsilon_x(x - x_0) + \left(\frac{1}{2}\gamma_{xy} - r_0\right)(y - y_0) + u_0$$

$$v = \left(\frac{1}{2}\gamma_{xy} + r_0\right)(x - x_0) + \varepsilon_y(y - y_0) + v_0$$

$$(2.11)$$

Written in a matrix form, the complete first order approximation of the displacements of any point within the block can be given as follows:

$$\begin{Bmatrix} u \\ v \end{Bmatrix} = \begin{bmatrix} 1 & 0 & -(y - y_0) & (x - x_0) & 0 & (y - y_0)/2 \\ 0 & 1 & (x - x_0) & 0 & (y - y_0) & (x - x_0)/2 \end{bmatrix} \begin{pmatrix} u_0 \\ v_0 \\ r_0 \\ \varepsilon_x \\ \varepsilon_y \\ r_{xy} \end{pmatrix} = T_i d_i \quad (2.12)$$

where $(x, y)$ are the coordinates of any point within the block, $(x_0, y_0)$ are the coordinates of a point within the block (normally the centroid of the block is selected). $u_0$ and $v_0$ are the rigid body translations at the point $(x_0, y_0)$ along $x$ and $y$ directions respectively, $r_0$ is the rotation angle in radian around the point $(x_0, y_0)$, $\varepsilon_x$, $\varepsilon_y$ and $r_{xy}$ are the normal and shear strains of the block. $d_i$ is the vector form of the unknowns $(u_0, v_0, r_0, \varepsilon_x, \varepsilon_y, r_{xy})$ as shown in Fig. 2.1 for block $i$ and $T_i$ is the matrix form of the first order displacement function.

Due to the adoption of the linear displacement function term, the angular rotation will produce a strain term, which may be very large if the angular rotation within a time step is large, leading to unexpected block expansion and to distortion of blocks. A number of techniques have been proposed by researchers to solve this problem as mentioned in Section 1.8.

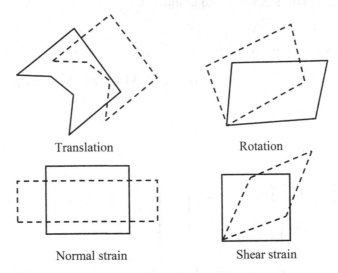

Translation          Rotation

Normal strain          Shear strain

*Figure 2.1* Displacement variables of DDA (Shi, 1988).

Based on Equations 1.4 and 1.5 suggested by Ohnishi and co-workers (Ohnishi *et al.*, 1995), MacLaughlin and Sitar (1996) proposed to use the second order term to account for the effects of angular rotation:

$$\cos r_0 = 1 - \frac{r_0^2}{2} \tag{2.13}$$

Then the displacements can be expressed as follows:

$$
\begin{aligned}
u &= u_0 - (x - x_0)\frac{r_0^2}{2} - (y - y_0)r_0 \\
v &= v_0 - (x - x_0)r_0 - (y - y_0)\frac{r_0^2}{2}
\end{aligned}
\tag{2.14}
$$

Ke (1996) proposed to use a post-adjustment with a maximum allowable rotation of 0.1 radian within a time step. The displacements of point $(x, y)$ include the contributions of $r_0$-induced and $r_0$-unrelated terms, as shown below:

$$
\begin{aligned}
u &= (x - x_0)(\cos r_0 - 1) - (y - y_0)\sin r_0 + \sum_{j=1}^{M} t_{1j}d_j \quad (j \neq 3) \\
v &= (x - x_0)\sin r_0 + (y - y_0)(\cos r_0 - 1) + \sum_{j=1}^{M} t_{2j}d_j \quad (j \neq 3)
\end{aligned}
\tag{2.15}
$$

where $M$ is the number of block unknowns, $d^i$ and $t_{ij}$ is the elements of the displacement function $t^i$.

An iterative method was developed by Cheng and Zhang (2000) which can reduce the block distortion to a minimum.

The exact solution for the $x$ and $y$ components $(u, v)$ of the displacement of an arbitrary point $(x, y)$ within a deformable block are given by:

$$u = u_0 + (x - x_0)(\cos r_0 - 1) - (y - y_0)\sin r_0 + (x - x_0)\varepsilon_x + (y - y_0)\gamma_{xy}/2$$
$$v = v_0 + (x - x_0)\sin r_0 + (y - y_0)(\cos r_0 - 1) + (y - y_0)\varepsilon_y + (x - x_0)\gamma_{xy}/2 \tag{2.16}$$

Consider that,

$$\cos r_0 - 1 = -\sin^2 r_0/(1 + \cos r_0) \tag{2.17}$$

Substituting Equation 2.17 into Equation 2.16, we have:

$$u = u_0 + f_1 \sin r_0 + (x - x_0)\varepsilon_x + (y - y_0)\gamma_{xy}/2$$
$$v = v_0 + f_2 \sin r_0 + (y - y_0)\varepsilon_y + (x - x_0)\gamma_{xy}/2 \tag{2.18}$$

where:

$$f_1 = -(x - x_0)\frac{\sin r_0}{1 + \cos r_0} - (y - y_0)$$

$$f_2 = -(y - y_0)\frac{\sin r_0}{1 + \cos r_0} - (x - x_0)$$

In the treatment of the terms $f_1$ and $f_2$ in iteration $i$ within each time step, the value of $r_0$ can be taken as the value of $r_0$ in step $i = 1$. This process goes along with the global iteration analysis within each time step.

## 2.2 FORMULATION OF MATRICES FOR EACH SINGLE BLOCK

The equilibrium equations of the DDA method are derived by minimizing the total potential energy of the block system. For single blocks, the total potential energy may include elastic strains, initial stress, point loading, line loading, body force, bolting connection, inertia force, displacement constraints at a point or in a direction, etc.

### 2.2.1 Sub-matrices of elastic strains

The strain energy $\Pi_e$ produced by the elastic stresses of block $i$ is:

$$\Pi_e = \iint \frac{1}{2}(\varepsilon_x \sigma_x + \varepsilon_y \sigma_y + r_{xy}\tau_{xy})dxdy \tag{2.19}$$

where the integration is over the entire area of block $i$. For each displacement step, the blocks are assumed to be linearly elastic.

The relationship between stress and strain is expressed as follows:

$$\sigma_i = E_i e_i \tag{2.20}$$

For plane stress problems:

$$\begin{pmatrix} \sigma_x \\ \sigma_y \\ \tau_{xy} \end{pmatrix} = \frac{E}{1-v^2} \begin{pmatrix} 1 & v & 0 \\ v & 1 & 0 \\ 0 & 0 & \frac{1-v}{2} \end{pmatrix} \begin{pmatrix} \varepsilon_x \\ \varepsilon_y \\ \gamma_{xy} \end{pmatrix} \tag{2.21}$$

For plane strain problems:

$$\begin{pmatrix} \sigma_x \\ \sigma_y \\ \tau_{xy} \end{pmatrix} = \frac{E}{(1+v)(1-2v)} \begin{pmatrix} 1-v & v & 0 \\ v & 1-v & 0 \\ 0 & 0 & \frac{1-2v}{2} \end{pmatrix} \begin{pmatrix} \varepsilon_x \\ \varepsilon_y \\ \gamma_{xy} \end{pmatrix} \tag{2.22}$$

For block $i$, the rigid body motion terms do not induce strain, so $\varepsilon_i$ can be replaced by $d_i$, and matrix $E_i$ can be expanded to a $6 \times 6$ matrix.

Substituting Equation 2.22 into Equation 2.19, the strain energy is represented in terms of the block deformation parameters:

$$\Pi_e = \iint \frac{1}{2} (\varepsilon_x \quad \varepsilon_y \quad r_{xy}) \begin{pmatrix} \sigma_x \\ \sigma_y \\ \tau_{xy} \end{pmatrix} dxdy = \iint \frac{1}{2} (\varepsilon_x \quad \varepsilon_y \quad r_{xy}) E \begin{pmatrix} \varepsilon_x \\ \varepsilon_y \\ \gamma_{xy} \end{pmatrix} dxdy$$

$$= \frac{1}{2} \iint d_i^T E_i d_i dxdy = \frac{S}{2} d_i^T E_i d_i \tag{2.23}$$

where $S$ is the area of block $i$ as the integration is over the entire area of the block.

The derivatives are calculated to minimize the strain energy $\Pi_e$:

$$k_{rs} = \frac{\partial^2 \Pi_e}{\partial d_{ri} \partial d_{si}} = \frac{S}{2} \frac{\partial^2}{\partial d_{ri} \partial d_{si}} d_i^T E_i d_i = SE_i \quad (r,s=1,\ldots,6) \quad \rightarrow \quad K_{ii} \tag{2.24}$$

which is added to the sub-matrix $K_{ii}$ in the global Equation 2.4.

## 2.2.2 Sub-matrices of initial stress

In the original DDA (Shi, 1988), block $i$ is assumed to have initial constant stresses $(\sigma_x^0 \; \sigma_y^0 \; \tau_{xy}^0)$. The potential energy can be represented as follows:

$$\Pi_\sigma = -\iint (\varepsilon_x \sigma_x^0 + \varepsilon_y \sigma_y^0 + r_{xy} \tau_{xy}^0) dxdy = -S(\varepsilon_x \sigma_x^0 + \varepsilon_y \sigma_y^0 + r_{xy} \tau_{xy}^0)$$

$$= -Sd_i^T \begin{pmatrix} 0 \\ 0 \\ 0 \\ \sigma_x^0 \\ \sigma_y^0 \\ \tau_{xy}^0 \end{pmatrix} = -Sd_i^T \sigma_0 \tag{2.25}$$

The potential energy $\Pi_\sigma$ is minimized by taking the derivatives:

$$f_r = -\frac{\partial \Pi_\sigma}{\partial d_{ri}} = S\sigma_0 \quad (r = 1, \ldots, 6) \quad \rightarrow \quad F_i \tag{2.26}$$

$f_r$ is a $6 \times 1$ sub-matrix and it is added to $F_i$ in the global Equation 2.4.

## 2.2.3 Sub-matrices of point loading

Assuming the point loading force acting on a point $(x, y)$ of block $i$ is $(F_x, F_y)$, the potential energy of the point loading is simply expressed as:

$$\Pi_p = -(F_x u + F_y v) = -(u \quad v) \begin{pmatrix} F_x \\ F_y \end{pmatrix} = -d_i^T T_i^T \begin{pmatrix} F_x \\ F_y \end{pmatrix} \tag{2.27}$$

To minimize $\Pi_p$, the derivatives are calculated as follows:

$$f_r = -\frac{\partial \Pi_p(0)}{\partial d_{ri}} = \frac{\partial}{\partial d_{ri}} d_i^T T_i^T \begin{pmatrix} F_x \\ F_y \end{pmatrix} = \begin{pmatrix} t_{11} & t_{21} \\ t_{12} & t_{22} \\ t_{13} & t_{23} \\ t_{14} & t_{24} \\ t_{15} & t_{25} \\ t_{16} & t_{26} \end{pmatrix} \begin{pmatrix} F_x \\ F_y \end{pmatrix} \quad \rightarrow \quad F_i \tag{2.28}$$

$f_r$ $(r = 1, \ldots, 6)$ forms a $6 \times 1$ sub-matrix and it is added to $F_i$ in the global Equation 2.4.

## 2.2.4 Sub-matrices of line loading

Assume the loading is distributed on a straight line from point $(x_1, y_1)$ to point $(x_2, y_2)$ as shown in Figure 2.2. The equation of the loading line is:

$$\begin{aligned} x &= (x_2 - x_1)t + x_1 \\ y &= (y_2 - y_1)t + y_1 \end{aligned} \quad 0 \leq t \leq 1 \tag{2.29}$$

The length of this line segment is

$$l = \sqrt{(x_2 - x_1)^2 + (y_2 - y_1)^2} \tag{2.30}$$

The loading is a variant along the loading line and represented as follows:

$$\begin{aligned} F_x &= F_x(t) \\ F_y &= F_y(t) \end{aligned} \quad 0 \leq t \leq 1 \tag{2.31}$$

The potential energy of the line loading $(F_x(t), F_y(t))$ is:

$$\Pi_l = -\int_0^1 (u \quad v) \begin{pmatrix} F_x(t) \\ F_y(t) \end{pmatrix} l\,dt = -d_i^T \int_0^1 T_i^T \begin{pmatrix} F_x(t) \\ F_y(t) \end{pmatrix} l\,dt \tag{2.32}$$

The derivatives of $\Pi_l$ are calculated to minimize the potential energy $\Pi_l$:

$$f_r = -\frac{\partial \Pi_l}{\partial d_{ri}} = \frac{\partial}{\partial d_{ri}} d_i^T \int_0^1 T_i^T \begin{pmatrix} F_x(t) \\ F_y(t) \end{pmatrix} l dt = \int_0^1 T_i^T \begin{pmatrix} F_x(t) \\ F_y(t) \end{pmatrix} l dt \quad \rightarrow \quad F_i \qquad (2.33)$$

$f_r$ $(r = 1, \ldots, 6)$ is added to $F_i$ in the global Equation 2.4.

When the line loading $(F_x(t), F_y(t)) = (F_x, F_y)$ is constant, the matrix integration of Equation 2.33 has an analytical formula, which is derived as follows:

$$\int_0^1 T_i^T \begin{pmatrix} F_x(t) \\ F_y(t) \end{pmatrix} l dt = \left( \int_0^1 T_i^T l dt \right) \begin{pmatrix} F_x(t) \\ F_y(t) \end{pmatrix} \qquad (2.34)$$

The integrations of the elements have to be calculated in order to compute the integration of matrix $T_i$:

$$\int_0^1 ((x_2 - x_1)t + (x_1 - x_0)) l dt = \frac{l}{2}(x_2 + x_1 - 2x_0)$$

$$\int_0^1 ((y_2 - y_1)t + (y_1 - y_0)) l dt = \frac{l}{2}(y_2 + y_1 - 2y_0) \qquad (2.35)$$

Therefore,

$$\left( \int_0^1 T_i^T l dt \right) \begin{pmatrix} F_x \\ F_y \end{pmatrix} = l \begin{pmatrix} F_x \\ F_y \\ -\frac{1}{2}(y_2 + y_1 - 2y_0)F_x + \frac{1}{2}(x_2 + x_1 - 2x_0)F_y \\ \frac{1}{2}(x_2 + x_1 - 2x_0)F_x \\ \frac{1}{2}(y_2 + y_1 - 2y_0)F_y \\ \frac{1}{4}(y_2 + y_1 - 2y_0)F_x + \frac{1}{4}(x_2 + x_1 - 2x_0)F_y \end{pmatrix} \rightarrow F_i \qquad (2.36)$$

This $6 \times 1$ sub-matrix is added to $F_i$ in the global Equation 2.4.

## 2.2.5 Sub-matrices of body force

If the body force $(f_x, f_y)$ is constant acting uniformly on the volume of block $i$, the potential energy is:

$$\Pi_v = -\iint (f_x u + f_y v) dx dy = -d_i^T \iint T_i^T dx dy \begin{pmatrix} f_x \\ f_y \end{pmatrix} \qquad (2.37)$$

$$\iint T_i^T \, dxdy = \begin{bmatrix} S & 0 \\ 0 & S \\ -S_y + y_0 S & S_x - x_0 S \\ S_x - x_0 S & 0 \\ 0 & S_y - y_0 S \\ (S_y - y_0 S)/2 & (S_x - x_0 S)/2 \end{bmatrix} = \begin{bmatrix} S & 0 \\ 0 & S \\ 0 & 0 \\ 0 & 0 \\ 0 & 0 \\ 0 & 0 \end{bmatrix} \tag{2.38}$$

The derivatives of $\Pi_v$ are calculated to minimize the potential energy $\Pi_v$:

$$f_r = -\frac{\partial \Pi_V(0)}{\partial d_{ri}} = \begin{pmatrix} f_x S \\ f_y S \\ 0 \\ 0 \\ 0 \\ 0 \end{pmatrix} \quad (r = 1, \ldots, 6) \quad \rightarrow \quad F_i \tag{2.39}$$

$f_r$ forms a $6 \times 1$ sub-matrix and is added to $F_i$ in the global Equation 2.4.

## 2.2.6  Sub-matrices of bolting connection

Consider a bolt connecting a point $(x_1, y_1)$ in block $i$ and a point $(x_2, y_2)$ in block $j$ as shown in Figure 2.3. Both points are not necessarily the vertices of the two blocks. The length of the bolt is:

$$l = \sqrt{(x_1 - x_2)^2 + (y_1 - y_2)^2} \tag{2.40}$$

$$\begin{aligned} dl &= \frac{1}{l}[(x_1 - x_2)(dx_1 - dx_2) + (y_1 - y_2)(dy_1 - dy_2)] \\ &= \frac{1}{l}[(x_1 - x_2)(u_1 - u_2) + (y_1 - y_2)(v_1 - v_2)] \\ &= (u_1 \quad v_1) \begin{pmatrix} l_x \\ l_y \end{pmatrix} - (u_2 \quad v_2) \begin{pmatrix} l_x \\ l_y \end{pmatrix} = d_i^T T_i^T \begin{pmatrix} l_x \\ l_y \end{pmatrix} - d_j^T T_j^T \begin{pmatrix} l_x \\ l_y \end{pmatrix} \end{aligned} \tag{2.41}$$

where $(l_x, l_y)$ are the direction cosines of the bolt:

$$\begin{aligned} l_x &= \frac{1}{l}(x_1 - x_2) \\ l_y &= \frac{1}{l}(y_1 - y_2) \end{aligned} \tag{2.42}$$

Assuming the stiffness of the bolt is $s$, the bolt force is:

$$f = -s\frac{dl}{l} \tag{2.43}$$

The strain energy of the bolt is:

$$\Pi_b = -\frac{1}{2}fdl = \frac{s}{2l}dl^2 = \frac{s}{2l}\left(d_i^T T_i^T \begin{pmatrix} l_x \\ l_y \end{pmatrix} - d_j^T T_j^T \begin{pmatrix} l_x \\ l_y \end{pmatrix}\right)^2$$

$$= \frac{s}{2l}d_i^T T_i^T \begin{pmatrix} l_x \\ l_y \end{pmatrix}(l_x \quad l_y)T_i d_i - \frac{s}{l}d_i^T T_i^T \begin{pmatrix} l_x \\ l_y \end{pmatrix}(l_x \quad l_y)T_j d_j$$

$$+ \frac{s}{2l}d_j^T T_j^T \begin{pmatrix} l_x \\ l_y \end{pmatrix}(l_x \quad l_y)T_j d_j$$

$$= \frac{s}{2l}d_i^T E_i E_i^T d_i - \frac{s}{l}d_i^T E_i G_j^T d_j + \frac{s}{2l}d_j^T G_j G_j^T d_j \qquad (2.44)$$

where

$$E_i = T_i^T \begin{pmatrix} l_x \\ l_y \end{pmatrix}$$

$$G_j = T_j^T \begin{pmatrix} l_x \\ l_y \end{pmatrix}$$

The derivative of $\Pi_b$ is:

$$k_{rs} = \frac{\partial^2 \Pi_b}{\partial d_{ri} \partial d_{si}} = \frac{s}{2l}\frac{\partial^2}{\partial d_{ri} \partial d_{si}}d_i^T E_i E_i^T d_i = \frac{s}{l}E_i E_i^T \quad \rightarrow \quad K_{ii} \qquad (2.45)$$

which is added to the sub-matrix $K_{ii}$ in the global Equation 2.4.

$$k_{rs} = \frac{\partial^2 \Pi_b}{\partial d_{ri} \partial d_{sj}} = -\frac{s}{l}\frac{\partial^2}{\partial d_{ri} \partial d_{sj}}d_i^T E_i G_j^T d_j = -\frac{s}{l}E_i G_j^T \quad \rightarrow \quad K_{ij} \qquad (2.46)$$

which is added to the sub-matrix $K_{ij}$ in the global Equation 2.4.

$$k_{rs} = \frac{\partial^2 \Pi_b}{\partial d_{rj} \partial d_{si}} = -\frac{s}{l}\frac{\partial^2}{\partial d_{rj} \partial d_{si}}d_i^T E_i G_j^T d_j = -\frac{s}{l}G_j E_i^T \quad \rightarrow \quad K_{ji} \qquad (2.47)$$

which is added to the sub-matrix $K_{ji}$ in the global Equation 2.4.

$$k_{rs} = \frac{\partial^2 \Pi_b}{\partial d_{rj} \partial d_{sj}} = \frac{s}{2l}\frac{\partial^2}{\partial d_{rj} \partial d_{sj}}d_j^T G_j G_j^T d_j = \frac{s}{l}G_j G_j^T \quad \rightarrow \quad K_{jj} \qquad (2.48)$$

which is added to the sub-matrix $K_{jj}$ in the global Equation 2.4.

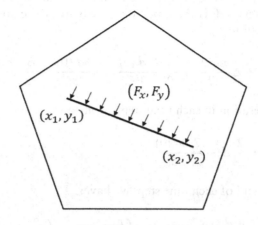

*Figure 2.2* Line loading.

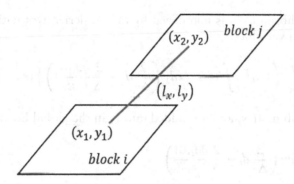

*Figure 2.3* Bolting connection.

## 2.2.7 Sub-matrices of inertia force

Using $(u(t), v(t))$ to denote the time dependent displacement of any point $(x, y)$ of block $i$ and $M$ to represent the mass per unit area, the inertia force of block $i$ is as follows:

$$\begin{pmatrix} f_x \\ f_y \end{pmatrix} = -M \begin{pmatrix} \dfrac{\partial^2 u(t)}{\partial t^2} \\ \dfrac{\partial^2 v(t)}{\partial t^2} \end{pmatrix} = -M T_i \frac{\partial^2 d_i(t)}{\partial t^2} \tag{2.49}$$

The potential energy of the inertial force of block $i$ is written as follows:

$$\Pi_i = -\iint (u \quad v) \begin{pmatrix} f_x \\ f_y \end{pmatrix} dx\, dy = M \iint d_i^T T_i^T T_i \frac{\partial^2 d(t)}{\partial t^2} dx\, dy \tag{2.50}$$

Assume $d_i(0) = 0$ is the block displacement at the beginning of the time step, $\Delta$ is the time step, and $d_i(\Delta) = d_i$ is the block displacement at the end of the time step. Using the time integration:

$$d_i = d_i(\Delta) = d_i(0) + \Delta \frac{\partial d_i(0)}{\partial t} + \frac{\Delta^2}{2} \frac{\partial^2 d_i(0)}{\partial t^2} = \Delta \frac{\partial d_i(0)}{\partial t} + \frac{\Delta^2}{2} \frac{\partial^2 d_i(0)}{\partial t^2} \qquad (2.51)$$

Assume the acceleration in each time step is constant:

$$\frac{\partial^2 d(t)}{\partial t^2} = \frac{\partial^2 d(0)}{\partial t^2} = \frac{2}{\Delta^2} d_i - \frac{2}{\Delta} \frac{\partial d_i(0)}{\partial t} \qquad (2.52)$$

Therefore, at the end of each time step, we have:

$$\Pi_i = M \iint d_i^T T_i^T T_i \frac{\partial^2 d(t)}{\partial t^2} dxdy = M d_i^T \iint T_i^T T_i dxdy \left( \frac{2}{\Delta^2} d_i - \frac{2}{\Delta} \frac{\partial d_i(0)}{\partial t} \right) \qquad (2.53)$$

To reach equilibrium, $\Pi_i$ is minimized by taking derivatives with respect to the block displacement variables:

$$f_r = -\frac{\partial \Pi_i}{\partial d_{ri}} = -\frac{\partial}{\partial d_{ri}} \left( M d_i^T \iint T_i^T T_i dxdy \left( \frac{2}{\Delta^2} d_i - \frac{2}{\Delta} \frac{\partial d_i(0)}{\partial t} \right) \right) \quad (r = 1, \ldots, 6) \quad (2.54)$$

$f_r$ forms a $6 \times 1$ sub-matrix, and it is added into $F_i$ in the global Equation 2.4:

$$M \iint T_i^T T_i dxdy \left( \frac{2}{\Delta^2} d_i - \frac{2}{\Delta} \frac{\partial d_i(0)}{\partial t} \right) \; \rightarrow \; F_i \qquad (2.55)$$

As there is an unknown $d_i$ in the Equation 2.55, this equation should be transformed into two as follows:

$$\frac{2M}{\Delta^2} \iint T_i^T T_i dxdy \; \rightarrow \; K_{ii} \qquad (2.56)$$

$$\frac{2M}{\Delta} \left( \iint T_i^T T_i dxdy \right) v_0 \; \rightarrow \; F_i \qquad (2.57)$$

## 2.2.8 Sub-matrices of displacement constraints at a point

When the displacements $(u, v)$ of a point $(x, y)$ are set to be constant values $(u_m, v_m)$, two very stiff springs in $x$ and $y$ directions respectively are used to force the displacements $(u, v)$ to be $(u_m, v_m)$, as shown in Figure 2.4.

The stiffness of the springs is the same as $p$ and the spring forces are:

$$\binom{f_x}{f_y} = -p \binom{u - u_m}{v - v_m} \qquad (2.58)$$

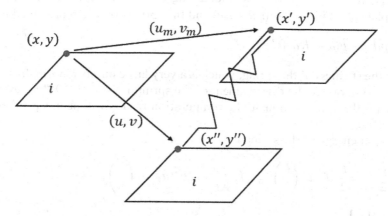

Figure 2.4 Fixed point.

The strain energy of the springs is:

$$\Pi_m = \frac{p}{2}(u - u_m \quad v - v_m)\begin{pmatrix} u - u_m \\ v - v_m \end{pmatrix}$$

$$= \frac{p}{2}(u \quad v)\begin{pmatrix} u \\ v \end{pmatrix} - p(u \quad v)\begin{pmatrix} u_m \\ v_m \end{pmatrix} + \frac{p}{2}(u_m \quad v_m)\begin{pmatrix} u_m \\ v_m \end{pmatrix}$$

$$= \frac{p}{2}d_i^T T_i^T T_i d_i - p d_i^T T_i^T \begin{pmatrix} u_m \\ v_m \end{pmatrix} + \frac{p}{2}(u_m \quad v_m)\begin{pmatrix} u_m \\ v_m \end{pmatrix} \qquad (2.59)$$

The derivatives are calculated to minimize the spring strain energy $\Pi_m$:

$$k_{rs} = \frac{\partial^2 \Pi_m}{\partial d_{ri} \partial d_{si}} = p T_i^T T_i \quad \rightarrow \quad K_{ii} \qquad (2.60)$$

$k_{rs}$ $(r, s = 1, \ldots, 6)$ forms a $6 \times 6$ matrix and it is added to the sub-matrix $K_{ii}$ in the global Equation 2.4.

And the following vector can be added to the global force vector:

$$f_r = -\frac{\partial \Pi_m}{\partial d_{ri}} = p T_i^T \begin{pmatrix} u_m \\ v_m \end{pmatrix} \quad \rightarrow \quad F_i$$

$f_r$ $(r = 1, \ldots, 12)$ is added to $F_i$ in the global Equation 2.4.

## 2.2.9 Sub-matrices of displacement constraints in a direction

In some engineering problems, some blocks are fixed at specific points in specified directions. More generally, the displacement $\delta$ along a direction at a point could be a known value. The measured displacements can be applied to the blocky system by using a very stiff spring which has the measured displacement as pre-tension distance.

Denote $(l_x, l_y)(l_x^2 + l_y^2 = 1)$ as the direction along which the displacement $\delta$ exists. The spring displacement is $d = \delta - (l_x u + l_y v)$ and the spring force is represented by:

$$f = -pd = -p(\delta - (l_x u + l_y v)) \tag{2.61}$$

where $p$ is the stiffness of the spring, which is a very large number, normally from $10E$ to $1000E$, to guarantee the displacement of the spring is $10^{-3}$ to $10^{-4}$ times the total displacement. If $p$ is large enough, the computation results will be independent of the selections of $p$.

The strain energy of the spring is:

$$\Pi_c = \frac{p}{2}d^2 = \frac{p}{2}(u \quad v)\begin{pmatrix} l_x \\ l_y \end{pmatrix}(l_x \quad l_y)\begin{pmatrix} u \\ v \end{pmatrix} - p\delta(u \quad v)\begin{pmatrix} l_x \\ l_y \end{pmatrix} + \frac{p}{2}\delta^2 \tag{2.62}$$

Let $L = \begin{Bmatrix} l_x \\ l_y \end{Bmatrix}$ and $c = T^T L$

Then

$$(u \quad v)\begin{pmatrix} l_x \\ l_y \end{pmatrix} = d_i^T T_i^T L = d_i^T c_i$$

$$(l_x \quad l_y)\begin{pmatrix} u \\ v \end{pmatrix} = L_i^T T_i^T d_i = c_i^T d_i$$

Therefore, we have

$$\Pi_c = \frac{p}{2}d_i^T c_i c_i^T d_i - p\delta d_i^T c_i + \frac{p}{2}\delta^2 \tag{2.63}$$

A $6 \times 6$ matrix is obtained by taking the derivatives of the strain energy of the stiff spring:

$$k_{rs} = \frac{\partial^2 \Pi_c}{\partial d_{ri} \partial d_{si}} = \frac{p}{2}\frac{\partial^2}{\partial d_{ri} \partial d_{si}}d_i^T c_i c_i^T d_i = p c_i c_i^T \quad \rightarrow \quad K_{ii} \tag{2.64}$$

$k_{rs}$ $(r, s = 1, \ldots, 6)$ forms a $6 \times 6$ matrix and it is added to the sub-matrix $K_{ii}$ in the global Equation 2.4.

We minimize the spring strain energy $\Pi_c$ by taking the derivatives of $\Pi_c$ at $d = 0$:

$$f_r = -\frac{\partial \Pi_c(0)}{\partial d_{ri}} = p\delta \frac{\partial}{\partial d_{ri}}d_i^T c_i = p\delta c_i \quad \rightarrow \quad F_i \tag{2.65}$$

$f_r$ $(r = 1, \ldots, 6)$ is added to $F_i$ in the global Equation 2.4.

## 2.3 INTERACTIONS BETWEEN BLOCKS

The DDA blocks are interacting with neighboring blocks through contacting and separating, and their movement must obey the imposed contact criterion, i.e., 'no penetration, no tension' (Shi, 1988). The effect of the contact can be represented by applying two stiff contact springs in the normal and shear directions or frictional forces along the sliding edge. The normal and shear contact springs are added if the blocks are in contact and are not sliding relative to one another, and deleted if the blocks separate or the normal contact force is tensile. If the blocks are in contact and sliding relative to one another, a normal spring is added together with the frictional forces.

The solution of the analysis requires the exact number of contacts and their relevant information. However, the total number of contacts is unknown prior to the solution of the problem. This phenomenon is particularly serious in multi-body contact problems. For such a problem, a possible solution is to use a trial-and-error iteration procedure, which in DDA is called "open-close" iterations (Shi, 1988). Open-close iterations are applied to identify the contacts and to arrange the correct locations of contact springs for each time step. After determining the contact points and the associated contact forces, the normal contact and shear contact sub-matrices or friction force sub-matrices are formulated, calculated, and added to the global simultaneous equations.

## 2.3.1   Contact detection

Since blocks interact at their boundaries, for two-dimensional blocks only three types of contacts exist: vertex-to-vertex, vertex-to-edge (Fig. 2.5), and edge-to-edge (Fig. 2.6). Among these contacts, the vertex-to-edge contact and vertex-to-vertex contact are the two basic types; the edge-to-edge contact can be converted to combination of the two basic types. The edge-to-edge contact may exist in four forms (Fig. 2.6), and each form could be treated as different combinations of vertex-to-vertex and vertex-to-edge contacts.

The vertex-to-vertex contact and vertex-to-edge contact are then transformed to point-line crossing inequalities (Shi, 1988). A vertex-to-edge contact has the edge as the only entrance line. The vertex-to-edge contact occurs when the vertex passes the edge. The vertex-to-vertex contact can be further divided into two types: the one between a convex angle and a concave angle (Fig. 2.7a), and the one between two convex angles (Fig. 2.7b–d). Both types have two entrance lines. For the first type (the contact between a convex angle and a concave angle), the two edges of the concave angle are the entrance lines. For the contact between two convex angles, the criteria in Table 2.1 are used to determine the entrance lines. The vertex-to-vertex contact occurs when one of the entrance lines is passed by the corresponding vertex.

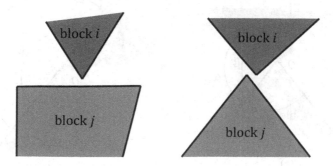

*Figure 2.5* Vertex-to-edge contact and vertex-to-vertex contact between two blocks.

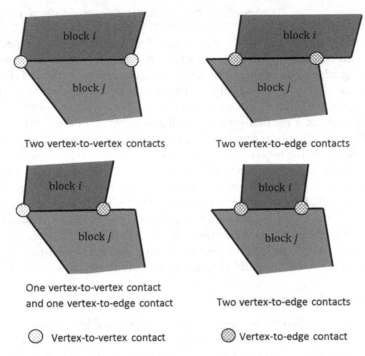

Two vertex-to-vertex contacts

Two vertex-to-edge contacts

One vertex-to-vertex contact
and one vertex-to-edge contact

Two vertex-to-edge contacts

⊙ Vertex-to-vertex contact    ⊛ Vertex-to-edge contact

*Figure 2.6* Edge-to-edge contacts between two blocks.

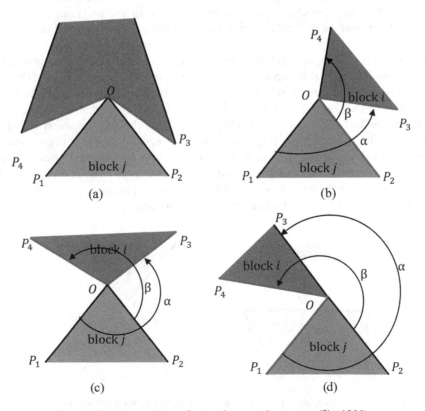

*Figure 2.7* Entrance lines of an angle-to-angle contact (Shi, 1988).

*Table 2.1*  Determination of the entrance lines for vertex-to-vertex contacts between two convex angles (Shi, 1988).

| Ranges of these two angles | | Entrance lines | |
| --- | --- | --- | --- |
| $\alpha \leq 180°$ | $\beta \leq 180°$ | $OP_3$ | $OP_2$ |
| $\alpha \leq 180°$ | $\beta > 180°$ | $OP_3$ | $OP_4$ |
| $\alpha > 180°$ | $\beta \leq 180°$ | $OP_1$ | $OP_2$ |
| $\alpha > 180°$ | $\beta > 180°$ | $OP_1$ | $OP_4$ |

Contacts are determined by calculating the penetration distance $d_n$ and the stiff springs are applied to ensure no-penetration. For the contact between two convex angles, a normal spring is used between the vertex and its corresponding entrance line, which has a smaller entrance distance. For the contact between a convex angle and a concave angle, a stiff spring is applied if only one of the two entrance lines is passed, and two stiff springs are applied if both entrance lines are passed. For the vertex-to-edge contact, only one stiff spring is used.

For each vertex-entrance line contact pair, there are three possible contact modes: open, sliding and locked. At the beginning of each time step, the contact modes for all contact pairs are assumed to be locked except those contact pairs inherited from the previous time step. For a locked contact pair, a normal spring is applied to push the vertex away from the entrance line in the normal direction and a shear spring is applied to avoid the tangential displacement between the vertex and the entrance line.

## 2.3.2  Contact constraint enforcement

After contact detection, the contact pairs are identified. Then a suitable strategy is required to deal with the contact constraints. A number of methods have been developed and used in the numerical codes handling contact problems. The most commonly used ones include the Penalty method, the Lagrange multiplier method and the augmented Lagrangian method.

### Penalty method

The penalty method, most widely used due to its simplicity, was originally adopted in the DDA method to enforce the contact constraints at block interfaces. This method does not increase the number of governing equations because of contacts, and the solution is easily obtained by simply adding contact sub-matrices to the stiffness matrix.

However, the penalty method is just an approximation of the actual contact constraints and it reaches the exact result when the penalty parameters $k_N$ and $k_T$ are approaching infinity. In practical applications, however, it is impossible to use very large penalty parameters, as this would lead to an ill-conditioned coefficient matrix.

The formulation for the contact contribution for the lock state is:

$$\Pi_c = \frac{1}{2}k_N(d_N)^2 + \frac{1}{2}k_T(d_T)^2 \tag{2.66}$$

where $d_N$ and $d_T$ are the normal and tangential gap functions.

### Lagrange multiplier method

The classical Lagrange multiplier method is one of the best methods to solve block contact problems. In this method, the strain energy of the contact force is calculated by multiplying the unknown contact force $\lambda$ by the penetration distance $d$. This method provides better and more stable results in solving contact problems. However, it increases the number of governing equations as an unknown contact force $\lambda$ is introduced, and extra computational effort is required.

The formulation of the Lagrange multiplier method is as follows:

$$\Pi_c = \lambda_N d_N + \lambda_T d_T \tag{2.67}$$

where $\lambda_N$ and $\lambda_T$ are the Lagrange multipliers.

### Augmented Lagrangian method

The augmented Lagrangian method is extended from the classical Lagrange multiplier method. It retains the simplicity of the penalty method and minimise the disadvantages of the penalty and the classical Lagrange multiplier methods. The essence of the augmented Lagrangian method is to use both a penalty number $p$ and a Lagrange multiplier $\lambda^* \approx \lambda$ at each bock contact (Landers and Taylor, 1986). To avoid the increase in the number of governing equations, an iterative method is used to calculate the Lagrange multiplier $\lambda^*$ until the distance of penetration $d$ is below a minimum tolerance. The advantage of this method is that the satisfaction of the no-penetration constraint can be improved even if the penalty parameters $k_N$ and $k_T$ are much smaller than those in the penalty method. The augmented Lagrangian multiplier method is described in detail in Pietrzak and Curnier (1999):

$$\Pi_N = \lambda_N d_N + \frac{1}{2} k_N (d_N)^2, \quad \Pi_T = \lambda_T d_T + \frac{1}{2} k_T (d_T)^2 \tag{2.68}$$

## 2.3.3 Open-close iterations

For each time step, at the beginning of the time interval, the locked directions of the closed contacts are identified. The contact springs are applied in each lock direction to prevent the penetration along the spring direction. The sub-matrices of contact springs are added to the simultaneous equilibrium equations prior to solving the equations. After solving the equations, if there is any penetration in a position where no contact spring is used, go back to the beginning of this time interval and apply a contact spring; if there is a contact spring having tension, go back to the beginning of this time interval and delete this contact spring. This procedure of lock selections is referred to as "open-close" iterations in DDA.

At each time step, open-close iterations are used to enforce no-penetration and no-tension conditions between contacting blocks before proceeding to the next time step. The open-close iterative procedure, previously introduced as ad-hoc rules by Shi (1988) and then discussed more systematically by Doolin and Sitar (2002), is summarized in Table 2.2. The vertex-to-vertex and vertex-to-edge contact types are denoted as V-V and V-E respectively. There are three contact states including open, sliding and locked. "Open" refers to blocks that have no interpenetrating vertices in

*Table 2.2* Rules for Open-Close Iteration following Doolin and Sitar (2002).

| Type | Previous state | Contact condition | New state | Operation |
|---|---|---|---|---|
| **V-V, V-E** | Open | $N > 0$ | Open | No change |
| **V-E** | Open | $N < 0$ & $\|S\| > \|f\|$ | Sliding | Apply a normal spring and a pair of friction forces |
| **V-V, V-E** | Open | $N < 0$ & $\|S\| < \|f\|$ | Locked | Apply normal and shear springs |
| **V-E** | Sliding | $N > 0$ | Open | Remove the normal spring and the pair of friction forces |
| **V-E** | Sliding | $N < 0$ & $\|S\| > \|f\|$ OR $N < 0$ & $\psi = -\text{sgn}\, f$ | Sliding | No change |
| **V-E** | Sliding | $N < 0$ & $\|S\| < \|f\|$ OR $N < 0$ & $\psi = \text{sgn}\, f$ | Locked | Remove the pair of friction forces and apply a shear spring |
| **V-V, V-E** | Locked | $N > 0$ | Open | Remove the normal and shear springs |
| **V-E** | Locked | $N < 0$ & $\|S\| > \|f\|$ | Sliding | Remove the shear spring and apply a pair of friction forces. |
| **V-V, V-E** | Locked | $N < 0$ & $\|S\| < \|f\|$ | Locked | No change |

**N** is the normal contact force, **S** is the shear contact force, $\psi$ is the direction of shear displacement, and **f** is the frictional force.

*Table 2.3* Open-close iterative procedure (Doolin and Sitar, 2002).

1. Initialize data
2. **For** each $t_i, i = 1, \ldots, n$ time steps **do**
3.    Identify/update contacts
4.    Assemble stiffness matrix
5.    Integrate
6.    **Repeat** (Open-close iteration)
7.      Add/subtract contact matrices
8.      Solve the unknowns **d**
9.      Resolve contact conditions
10.   **Until** no-tension and no penetration
11.   Update block geometry
12.   Update block stresses, velocities
13. **End for**

common. "Sliding" refers to a pair of blocks sharing a vertex-to-edge contact where the shearing force exceeds the frictional resistance. "Locked" denotes contacts that are interpenetrating and the shearing force is lower than the frictional resistance.

The complete DDA algorithm for open-close iteration is listed in Table 2.3.

In the DDA method, a locking state operator $\lambda_i$ $(i = 1, \ldots, n)$ for different contacts is used for adding and removing terms to the stiffness matrix and force vector. $\lambda_i = -1$ indicates that the contact condition between the previous and current open-close iteration has changed from open to closed, $\lambda_i = 0$ means that the contact condition is unchanged, and $\lambda_i = 1$ indicates that the contact condition has changed from closed to open. Both sliding and locked contacts are considered as closed.

### 2.3.4 Formulation of contact matrices

The contact displacement of the normal spring Figure 2.8 at the end of the current time step is:

$$d_n = \frac{\Delta}{l} \tag{2.69}$$

#### Normal contact

If $P_1$ passes through edge $\overrightarrow{P_2 P_3}$ (Figure 2.8), $d_n$ should be negative. $\Delta$ is the area of triangle $P_1 P_2 P_3$, and $l$ is the length of edge $\overrightarrow{P_2 P_3}$ at the end of the current time step:

$$l = \sqrt{(x_2 + u_2 - x_3 - u_3)^2 + (y_2 + v_2 - y_3 - v_3)^2}$$

$$\approx \sqrt{(x_2 - x_2)^2 + (y_2 - y_3)^2} \tag{2.70}$$

$$\Delta = \begin{vmatrix} 1 & x_1 + u_1 & y_1 + v_1 \\ 1 & x_2 + u_2 & y_2 + v_2 \\ 1 & x_3 + u_3 & y_3 + v_3 \end{vmatrix} \approx S_{n0} + \{y_2 - y_3 \quad x_3 - x_2\} \begin{Bmatrix} u_1 \\ v_1 \end{Bmatrix}$$

$$+ \{y_3 - y_1 \quad y_1 - y_3\} \begin{Bmatrix} u_2 \\ v_2 \end{Bmatrix} + \{y_1 - y_2 \quad y_2 - y_1\} \begin{Bmatrix} u_3 \\ v_3 \end{Bmatrix} \tag{2.71}$$

$$S_0 = \begin{vmatrix} 1 & x_1 & y_1 \\ 1 & x_2 & y_2 \\ 1 & x_3 & y_3 \end{vmatrix} \tag{2.72}$$

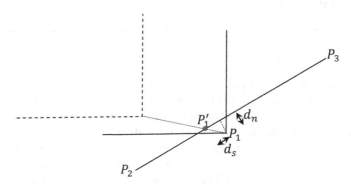

Figure 2.8 Normal and shear contact displacements.

So $d_n$ can be written as follows:

$$d_n = \frac{1}{l}\left( S_0 + \{y_2 - y_3 \quad x_3 - x_2\}\begin{Bmatrix} u_1 \\ v_1 \end{Bmatrix} + \{y_3 - y_1 \quad x_1 - x_3\}\begin{Bmatrix} u_2 \\ v_2 \end{Bmatrix}\right.$$

$$\left. + \{y_1 - y_2 \quad x_2 - x_1\}\begin{Bmatrix} u_3 \\ v_3 \end{Bmatrix}\right) \tag{2.73}$$

Since $P_1$ belongs to block $i$ while edge $\overrightarrow{P_2P_3}$ belongs to block $j$, the displacements of $P_1$, $P_2$ and $P_3$ can be represented as follows:

$$\begin{pmatrix} u_1 \\ v_1 \end{pmatrix} = T_i(x_1, y_1)d_i$$

$$\begin{pmatrix} u_2 \\ v_2 \end{pmatrix} = T_j(x_2, y_2)d_j \tag{2.74}$$

$$\begin{pmatrix} u_3 \\ v_3 \end{pmatrix} = T_j(x_3, y_3)d_j$$

Then

$$d_n = \frac{S_0}{l} + (e_1 \quad e_2 \quad e_3 \quad e_4 \quad e_5 \quad e_6)\begin{pmatrix} d_{1i} \\ d_{2i} \\ d_{3i} \\ d_{4i} \\ d_{5i} \\ d_{6i} \end{pmatrix} + (g_1 \quad g_2 \quad g_3 \quad g_4 \quad g_5 \quad g_6)\begin{pmatrix} d_{1j} \\ d_{2j} \\ d_{3j} \\ d_{4j} \\ d_{5j} \\ d_{6j} \end{pmatrix}$$

$$\tag{2.75}$$

where

$$e_r = \frac{1}{l}((y_2 - y_3)t_{1r}(x_1, y_1) + (x_3 - x_2)t_{2r}(x_1, y_1))$$

$$g_r = \frac{1}{l}((y_3 - y_1)t_{1r}(x_2, y_2) + (x_1 - x_3)t_{2r}(x_2, y_2))$$

$$+ \frac{1}{l}((y_1 - y_2)t_{1r}(x_3, y_3) + (x_2 - x_1)t_{2r}(x_3, y_3)) \tag{2.76}$$

The strain energy of the normal contact spring forces is:

$$\Pi_{nc} = \frac{p}{2}d_n^2 = \frac{p}{2}\left(\sum_{r=1}^{6} e_r d_{ri} + \sum_{r=1}^{6} g_r d_{rj} + \frac{S_0}{l}\right)^2$$

$$= \frac{p}{2}\left(\left(\sum_{r=1}^{6} e_r d_{ri}\right)^2 + \left(\sum_{r=1}^{6} g_r d_{rj}\right)^2\right.$$

$$+ 2\left(\sum_{r=1}^{6} e_r d_{ri}\right)\left(\sum_{r=1}^{6} g_r d_{rj}\right) + \frac{2S_0}{l}\left(\sum_{r=1}^{6} e_r d_{ri}\right)$$

$$\left. + \frac{2S_0}{l}\left(\sum_{r=1}^{6} g_r d_{rj}\right) + \frac{S_0^2}{l^2}\right) \tag{2.77}$$

Four $6 \times 6$ sub-matrices and two $6 \times 1$ sub-matrices are obtained and then added to $K_{ii}$, $K_{ij}$, $K_{ji}$, $K_{jj}$, $F_i$ and $F_j$ respectively, by minimizing the strain energy $\Pi_{nc}$. The derivative of $\Pi_{nc}$

$$k_{rs} = \frac{\partial^2 \Pi_{nc}}{\partial d_{ri} \partial d_{si}} = \frac{p}{2}\frac{\partial^2}{\partial d_{ri}\partial d_{si}}\left(\sum_{r=1}^{6} e_r d_{ri}\right)^2$$

$$= p\begin{pmatrix} e_1 \\ e_2 \\ e_3 \\ e_4 \\ e_5 \\ e_6 \end{pmatrix}(e_1 \quad e_2 \quad e_3 \quad e_4 \quad e_5 \quad e_6) \rightarrow K_{ii} \tag{2.78}$$

$k_{rs}$ $(r, s = 1, \ldots, 6)$ is added to $K_{ii}$ in the global Equation 2.4.
The derivative of $\Pi_{nc}$

$$k_{rs} = \frac{\partial^2 \Pi_{nc}}{\partial d_{ri} \partial d_{sj}} = \frac{p}{2}\frac{\partial^2}{\partial d_{ri}\partial d_{sj}}\left(\sum_{r=1}^{6} e_r d_{ri}\right)\left(\sum_{r=1}^{6} g_r d_{rj}\right)$$

$$= p\begin{pmatrix} e_1 \\ e_2 \\ e_3 \\ e_4 \\ e_5 \\ e_6 \end{pmatrix}(g_1 \quad g_2 \quad g_3 \quad g_4 \quad g_5 \quad g_6) \rightarrow K_{ij} \tag{2.79}$$

$k_{rs}$ $(r, s = 1, \ldots, 6)$ is added to $K_{ij}$ in the global Equation 2.4.

The derivative of $\Pi_{nc}$

$$k_{rs} = \frac{\partial^2 \Pi_{nc}}{\partial d_{rj} \partial d_{si}} = \frac{p}{2} \frac{\partial^2}{\partial d_{ri} \partial d_{si}} \left( \sum_{r=1}^{6} e_r d_{ri} \right) \left( \sum_{r=1}^{6} g_r d_{rj} \right)$$

$$= p \begin{pmatrix} g_1 \\ g_2 \\ g_3 \\ g_4 \\ g_5 \\ g_6 \end{pmatrix} (e_1 \quad e_2 \quad e_3 \quad e_4 \quad e_5 \quad e_6) \quad \to \quad K_{ij} \tag{2.80}$$

$k_{rs}$ $(r, s = 1, \ldots, 6)$ is added to $K_{ji}$ in the global Equation 2.4.
The derivative of $\Pi_{nc}$

$$k_{rs} = \frac{\partial^2 \Pi_{nc}}{\partial d_{rj} \partial d_{sj}} = \frac{p}{2} \frac{\partial^2}{\partial d_{rj} \partial d_{sj}} \left( \sum_{r=1}^{6} g_r d_{rj} \right)^2$$

$$= p \begin{pmatrix} g_1 \\ g_2 \\ g_3 \\ g_4 \\ g_5 \\ g_6 \end{pmatrix} (g_1 \quad g_2 \quad g_3 \quad g_4 \quad g_5 \quad g_6) \quad \to \quad K_{ij} \tag{2.81}$$

$k_{rs}$ $(r, s = 1, \ldots, 6)$ is added to $K_{jj}$ in the global Equation 2.4.
The derivative of $\Pi_{nc}$

$$f_r = -\frac{\partial \Pi_{nc}(0)}{\partial d_{ri}} = -\frac{pS_0}{l} \frac{\partial}{\partial d_{ri}} \left( \sum_{r=1}^{6} e_r d_{ri} \right) = -\frac{pS_0}{l} \begin{pmatrix} e_1 \\ e_2 \\ e_3 \\ e_4 \\ e_5 \\ e_6 \end{pmatrix} \quad \to \quad F_i \tag{2.82}$$

$f_r$ $(r = 1, \ldots, 6)$ is added to $F_i$ in the global Equation 2.4.
The derivative of $\Pi_{nc}$

$$f_r = -\frac{\partial \Pi_{nc}(0)}{\partial d_{rj}} = -\frac{pS_0}{l} \frac{\partial}{\partial d_{rj}} \left( \sum_{r=1}^{6} e_r d_{rj} \right) = -\frac{pS_0}{l} \begin{pmatrix} g_1 \\ g_2 \\ g_3 \\ g_4 \\ g_5 \\ g_6 \end{pmatrix} \quad \to \quad F_j \tag{2.83}$$

$f_r$ $(r = 1, \ldots, 6)$ is added to $F_j$ in the global Equation 2.4.

### Shear contact

Assume $P_0$ with coordinates $(x_0, y_0)$ is the projection of vertex $P_1$ on the edge $\overrightarrow{P_2P_3}$, and the coordinates of $P_0$ can be represented as follows:

$$x_0 = (1 - t)x_2 + x_3 t$$
$$y_0 = (1 - t)y_2 + y_3 t \tag{2.84}$$

The shear displacement is:

$$d_s = \frac{1}{l}\overrightarrow{P_0P_1} \cdot \overrightarrow{P_2P_3} = \frac{1}{l}\,(x_1 + u_1 - x_0 - u_0 \quad y_1 + v_1 - y_0 - v_0)\begin{Bmatrix} x_3 + u_3 - x_2 - u_2 \\ y_3 + v_3 - y_2 - v_2 \end{Bmatrix} \tag{2.85}$$

Ignoring the second-order infinitesimal small terms and rearranging Equation 2.85, gives,

$$d_s = \frac{S_0}{l} + \frac{1}{l}(x_3 - x_2 \quad y_3 - y_2)\begin{Bmatrix} u_1 - u_0 \\ v_1 - v_0 \end{Bmatrix}$$

$$= \frac{S_0}{l} + \frac{1}{l}\,(x_3 - x_2 \quad y_3 - y_2)\begin{Bmatrix} u_1 \\ v_1 \end{Bmatrix} + \frac{1}{l}(x_2 - x_3 \quad y_2 - y_3)\begin{Bmatrix} u_0 \\ v_0 \end{Bmatrix} \tag{2.86}$$

$$S_0 = (x_3 - x_2 \quad y_3 - y_2)\begin{Bmatrix} x_1 - x_0 \\ y_1 - y_0 \end{Bmatrix} \tag{2.87}$$

Since $P_1$ belongs to block $i$ while $P_0$ belongs to block $j$, the displacements of $P_1$ and $P_0$ can be represented as follows:

$$\begin{pmatrix} u_1 \\ v_1 \end{pmatrix} = T_i(x_1, y_1)d_i$$

$$\begin{pmatrix} u_0 \\ v_0 \end{pmatrix} = T_j(x_0, y_0)d_j \tag{2.88}$$

Then

$$d_s = \frac{S_0}{l} + (e_1 \quad e_2 \quad e_3 \quad e_4 \quad e_5 \quad e_6)\begin{pmatrix} d_{1i} \\ d_{2i} \\ d_{3i} \\ d_{4i} \\ d_{5i} \\ d_{6i} \end{pmatrix} + (g_1 \quad g_2 \quad g_3 \quad g_4 \quad g_5 \quad g_6)\begin{pmatrix} d_{1j} \\ d_{2j} \\ d_{3j} \\ d_{4j} \\ d_{5j} \\ d_{6j} \end{pmatrix} \tag{2.89}$$

where

$$e_r = \frac{1}{l}((y_3 - y_2)t_{1r}(x_1, y_1) + (x_3 - x_2)t_{2r}(x_1, y_1))$$

$$g_r = \frac{1}{l}((-x_1 + 2(1-t)x_2 - (1-2t)x_3)t_{1r}(x_2, y_2)$$

$$+ (-y_1 + 2(1-t)y_2 - (1-2t)y_3)t_{2r}(x_2, y_2)) \tag{2.90}$$

$$+ \frac{1}{l}((x_1 - (1-2t)x_2 - 2tx_3)t_{1r}(x_3, y_3)$$

$$+ (y_1 - (1-2t)y_2 - 2ty_3)t_{2r}(x_3, y_3))$$

The strain energy of the shear contact spring forces is:

$$\Pi_{sc} = \frac{p}{2}d_s^2 = \frac{p}{2}\left(\sum_{r=1}^{6} e_r d_{ri} + \sum_{r=1}^{6} g_r d_{rj} + \frac{S_0}{l}\right)^2$$

$$= \frac{p}{2}\left(\left(\sum_{r=1}^{6} e_r d_{ri}\right)^2 + \left(\sum_{r=1}^{6} g_r d_{rj}\right)^2\right.$$

$$+ 2\left(\sum_{r=1}^{6} e_r d_{ri}\right)\left(\sum_{r=1}^{6} g_r d_{rj}\right) + \frac{2S_0}{l}\left(\sum_{r=1}^{6} e_r d_{ri}\right)$$

$$\left. + \frac{2S_0}{l}\left(\sum_{r=1}^{6} g_r d_{rj}\right) + \frac{S_0^2}{l^2}\right) \tag{2.91}$$

Four $6 \times 6$ sub-matrices and two $6 \times 1$ sub-matrices are obtained and then added to $K_{ii}$, $K_{ij}$, $K_{ji}$, $K_{jj}$, $F_i$ and $F_j$ respectively, by minimizing the strain energy $\Pi_{sc}$. The derivative of $\Pi_{sc}$

$$k_{rs} = \frac{\partial^2 \Pi_{sc}}{\partial d_{ri} \partial d_{si}} = \frac{p}{2} \frac{\partial^2}{\partial d_{ri} \partial d_{si}}\left(\sum_{r=1}^{6} e_r d_{ri}\right)^2$$

$$= p \begin{pmatrix} e_1 \\ e_2 \\ e_3 \\ e_4 \\ e_5 \\ e_6 \end{pmatrix} (e_1 \quad e_2 \quad e_3 \quad e_4 \quad e_5 \quad e_6) \rightarrow K_{ii} \tag{2.92}$$

$k_{rs}$ $(r, s = 1, \ldots, 6)$ is added to $K_{ii}$ in the global Equation 2.4.

The derivative of $\Pi_{sc}$

$$k_{rs} = \frac{\partial^2 \Pi_{sc}}{\partial d_{ri} \partial d_{sj}} = \frac{p}{2} \frac{\partial^2}{\partial d_{ri} \partial d_{sj}} \left( \sum_{r=1}^{6} e_r d_{ri} \right) \left( \sum_{r=1}^{6} g_r d_{rj} \right)$$

$$= p \begin{pmatrix} e_1 \\ e_2 \\ e_3 \\ e_4 \\ e_5 \\ e_6 \end{pmatrix} (g_1 \quad g_2 \quad g_3 \quad g_4 \quad g_5 \quad g_6) \rightarrow K_{ij} \tag{2.93}$$

$k_{rs}$ $(r, s = 1, \ldots, 6)$ is added to $K_{ij}$ in the global Equation 2.4.
The derivative of $\Pi_{sc}$

$$k_{rs} = \frac{\partial^2 \Pi_{sc}}{\partial d_{rj} \partial d_{si}} = \frac{p}{2} \frac{\partial^2}{\partial d_{rj} \partial d_{si}} \left( \sum_{r=1}^{6} e_r d_{ri} \right) \left( \sum_{r=1}^{6} g_r d_{rj} \right)$$

$$= p \begin{pmatrix} g_1 \\ g_2 \\ g_3 \\ g_4 \\ g_5 \\ g_6 \end{pmatrix} (e_1 \quad e_2 \quad e_3 \quad e_4 \quad e_5 \quad e_6) \rightarrow K_{ji} \tag{2.94}$$

$k_{rs}$ $(r, s = 1, \ldots, 6)$ is added to $K_{ji}$ in the global Equation 2.4.
The derivative of $\Pi_{sc}$

$$k_{rs} = \frac{\partial^2 \Pi_{sc}}{\partial d_{rj} \partial d_{sj}} = \frac{p}{2} \frac{\partial^2}{\partial d_{rj} \partial d_{sj}} \left( \sum_{r=1}^{6} g_r d_{rj} \right)^2$$

$$= p \begin{pmatrix} g_1 \\ g_2 \\ g_3 \\ g_4 \\ g_5 \\ g_6 \end{pmatrix} (g_1 \quad g_2 \quad g_3 \quad g_4 \quad g_5 \quad g_6) \rightarrow K_{jj} \tag{2.95}$$

$k_{rs}$ $(r, s = 1, \ldots, 6)$ is added to $K_{jj}$ in the global Equation 2.4.
The derivative of $\Pi_{sc}$

$$f_r = -\frac{\partial \Pi_{sc}(0)}{\partial d_{ri}} = -\frac{pS_0}{l} \frac{\partial}{\partial d_{ri}} \left( \sum_{r=1}^{6} e_r d_{ri} \right) = -\frac{pS_0}{l} \begin{pmatrix} e_1 \\ e_2 \\ e_3 \\ e_4 \\ e_5 \\ e_6 \end{pmatrix} \rightarrow F_i \tag{2.96}$$

$f_r$ $(r = 1, \ldots, 6)$ is added to $F_i$ in the global Equation 2.4.

The derivative of $\Pi_{sc}$

$$f_r = -\frac{\partial \Pi_{sc}(0)}{\partial d_{rj}} = -\frac{pS_0}{l}\frac{\partial}{\partial d_{rj}}\left(\sum_{r=1}^{6} d_r d_{rj}\right) = -\frac{pS_0}{l}\begin{pmatrix} g_1 \\ g_2 \\ g_3 \\ g_4 \\ g_5 \\ g_6 \end{pmatrix} \rightarrow F_j \tag{2.97}$$

$f_r$ $(r = 1, \ldots, 6)$ is added to $F_j$ in the global Equation 2.4.

### Frictional force

For sliding mode, besides the normal spring, a pair of frictional forces instead of a shear spring should be added. The frictional force is calculated based on the normal contact compressive force from the former iteration:

$$F_f = p_n |d_n| \tan(\varphi) + c \tag{2.98}$$

where $p_n$ is the normal spring stiffness, $\varphi$ is the friction angle, $d_n$ is the normal penetration distance taken from the previous iteration, and $c$ is cohesion.

The friction force $F_f$ is along the direction

$$\frac{1}{l}(x_3 - x_2 \quad y_3 - y_2)$$

and

$$l = \sqrt{(x_3 - x_2)^2 + (y_3 - y_2)^2}$$

is the length of line $\overrightarrow{P_2 P_3}$.

The potential energy of the frictional force $F_f$ at $P_1$ on block $i$ is:

$$\Pi_f = \frac{F}{l}(u_1 \quad v_1)\begin{pmatrix} x_3 - x_2 \\ y_3 - y_2 \end{pmatrix} = \frac{F}{l}d_i^T T_i^T(x_1, y_1)\begin{pmatrix} x_3 - x_2 \\ y_3 - y_2 \end{pmatrix} = F d_i^T H \tag{2.99}$$

$$H = \frac{1}{l}T_i^T(x_1, y_1)\begin{pmatrix} x_3 - x_2 \\ y_3 - y_2 \end{pmatrix} = F d_i^T \begin{pmatrix} e_1 \\ e_2 \\ e_3 \\ e_4 \\ e_5 \\ e_6 \end{pmatrix} \tag{2.100}$$

The derivatives of $\Pi_f$ when $d = 0$ is:

$$f_r = -\frac{\partial \Pi_f(0)}{\partial d_{ri}} = -F \frac{\partial}{\partial d_{ri}} \left( \sum_{k=1}^{6} e_k d_{ki} \right) = -F \begin{pmatrix} e_1 \\ e_2 \\ e_3 \\ e_4 \\ e_5 \\ e_6 \end{pmatrix} \rightarrow F_i \qquad (2.101)$$

$f_r$ $(r = 1, \ldots, 6)$ is added to $F_i$ in the global Equation 2.4.

The potential energy of the frictional force $F_f$ at $P_0$ on block $j$ is:

$$\Pi_f = -\frac{F}{l}(u_0 \quad v_0)\begin{pmatrix} x_3 - x_2 \\ y_3 - y_2 \end{pmatrix} = -\frac{F}{l} d_j^T T_j^T (x_0, y_0)\begin{pmatrix} x_3 - x_2 \\ y_3 - y_2 \end{pmatrix} = -F d_j^T G \qquad (2.102)$$

$$G = \frac{1}{l} T_j^T (x_0, y_0)\begin{pmatrix} x_3 - x_2 \\ y_3 - y_2 \end{pmatrix} = F d_j^T \begin{pmatrix} g_1 \\ g_2 \\ g_3 \\ g_4 \\ g_5 \\ g_6 \end{pmatrix} \qquad (2.103)$$

The derivatives of $\Pi_f$ when $d = 0$ is:

$$f_r = -\frac{\partial \Pi_f(0)}{\partial d_{rj}} = -F \frac{\partial}{\partial d_{rj}} \left( \sum_{k=1}^{6} e_k d_{kj} \right) = -F \begin{pmatrix} e_1 \\ e_2 \\ e_3 \\ e_4 \\ e_5 \\ e_6 \end{pmatrix} \rightarrow F_j \qquad (2.104)$$

$f_r$ $(r = 1, \ldots, 6)$ is added to $F_j$ in the global Equation 2.4.

## 2.4 TIME INTEGRATION SCHEME AND GOVERNING EQUATIONS FOR BLOCKY SYSTEMS

Time integration in DDA follows Newmark–$\beta$ method which can be written as follows:

$$d_{n+1} = d_n + \Delta t \dot{d}_n + \frac{\Delta t^2}{2}[(1 - 2\beta)\ddot{d}_n + 2\beta \ddot{d}_{n+1}] \qquad (2.105)$$

$$\dot{d}_{n+1} = \dot{d}_n + \Delta t[(1 - \gamma)\ddot{d}_n + \gamma \ddot{d}_{n+1}] \qquad (2.106)$$

$\Delta t$ is the time step size for an incremental dynamic formulation, $d_n$ and $d_{n+1}$ denote the approximation to the values $d(t)$ and $d(t + \Delta t)$ for a time step $\Delta t$, $\beta$ is an acceleration weighting parameter and $\gamma$ is a velocity weighting parameter.

It can be shown that the Newmark–$\beta$ method is unconditionally stable when the following condition is met (Hughes, 1983):

$$2\beta \geq \gamma \geq \frac{1}{2} \tag{2.107}$$

Newmark–$\beta$ method with two parameters $\beta = 0.5$ and $\gamma = 1.0$ is used in DDA (Doolin and Sitar, 2004); therefore Equations 2.105 and 2.106 can be simplified as follows:

$$\dot{d}_{n+1} = \dot{d}_n + \Delta t \ddot{d}_{n+1} = \frac{2}{\Delta t} d_{n+1} - \dot{d}_n \tag{2.108}$$

$$\ddot{d}_{n+1} = \frac{2}{\Delta t^2} d_{n+1} - \frac{2}{\Delta t} \dot{d}_n \tag{2.109}$$

It can be seen that this scheme satisfies the criterion in Equation 2.107 and therefore it is unconditionally stable. According to Hughes (1983), the condition to prevent bifurcation is as follows:

$$\beta \geq \left(\gamma + \frac{1}{2}\right)^2 / 4 \tag{2.110}$$

However, the Newmark–$\beta$ scheme with two parameters $\beta = 0.5$ and $\gamma = 1.0$ violates this condition, inducing bifurcation. For an undamped analysis, the critical sampling frequency at which bifurcation occurs is:

$$\Omega_{bif} = \left[\frac{1}{4}\left(\gamma + \frac{1}{2}\right)^2 - \beta\right]^{-\frac{1}{2}} \tag{2.111}$$

For $\beta = 0.5$ and $\gamma = 1.0$, $\Omega_{bif} = 4$. On the other hand, $\Omega = \Delta t \omega$, meeting the constraint of Equation 2.111, requires that:

$$\Delta t < \frac{4}{\omega_{max}} \tag{2.112}$$

where $\Delta t$ is the time step. $\omega$ is undamped frequency of vibration of the system.

It should be mentioned that numerical damping is essential for DDA analysis, as it allows the oscillations caused by contact forces to dissipate rapidly, resulting in a stable state, which ultimately allows open-close iterations to converge rapidly. The amount of numerical damping is also proportional to the time step size.

## 2.5 SIMPLEX INTEGRATION FOR 2-D DDA

Distinguished from other numerical methods where numerical integrations are typically used, the DDA method uses Simplex for all integrations. Simplex integrations are accurate solutions on $n$-dimensional, generally shaped, domains.

A simplex is an oriented simplest domain. Simplex integration is used to calculate ordinary integrations and it is not limited for integrations on a simplex only. A complex shape can always be sub-divided into simplex, so simplex integration can be calculated in each simplex and the sum of the simplex integrations is the ordinary integration over the complex shape.

A two-dimensional block is assumed to be a polygon with vertices arranged anticlockwise such as,

$$P_1 P_2 P_3 \cdots P_n \quad P_{n+1} = P_1$$

and

$$P_i = (x_i, y_i)$$

Let $P_0 = (0,0)$. Then,

$$\int_{Area} dx dy = \sum_{k=1}^{n} \int_{P_0 P_k P_{k+1}} 1 D(x,y) = \frac{1}{2} \sum_{k=1}^{n} \begin{vmatrix} x_k & y_k \\ x_{k+1} & y_{k+1} \end{vmatrix} \tag{2.113}$$

$$\int_{Area} x dx dy = \sum_{k=1}^{n} \int_{P_0 P_k P_{k+1}} x D(x,y) = \frac{1}{6} \sum_{k=1}^{n} \begin{vmatrix} x_k & y_k \\ x_{k+1} & y_{k+1} \end{vmatrix} (x_k + x_{k+1}) \tag{2.114}$$

$$\int_{Area} y dx dy = \sum_{k=1}^{n} \int_{P_0 P_k P_{k+1}} y D(x,y) = \frac{1}{6} \sum_{k=1}^{n} \begin{vmatrix} x_k & y_k \\ x_{k+1} & y_{k+1} \end{vmatrix} (y_k + y_{k+1}) \tag{2.115}$$

$$\int_{Area} x^2 dx dy = \sum_{k=1}^{n} \int_{P_0 P_k P_{k+1}} x^2 D(x,y)$$

$$= \frac{1}{12} \sum_{k=1}^{n} \begin{vmatrix} x_k & y_k \\ x_{k+1} & y_{k+1} \end{vmatrix} (x_k^2 + x_{k+1}^2 + x_k x_{k+1}) \tag{2.116}$$

$$\int_{Area} y^2 dx dy = \sum_{k=1}^{n} \int_{P_0 P_k P_{k+1}} y^2 D(x,y)$$

$$= \frac{1}{12} \sum_{k=1}^{n} \begin{vmatrix} x_k & y_k \\ x_{k+1} & y_{k+1} \end{vmatrix} (y_k^2 + y_{k+1}^2 + y_k y_{k+1}) \tag{2.117}$$

$$\int_{Area} xydxdy = \sum_{k=1}^{n} \int_{P_0P_kP_{k+1}} xyD(x,y)$$

$$= \frac{1}{24} \sum_{k=1}^{n} \begin{vmatrix} x_k & y_k \\ x_{k+1} & y_{k+1} \end{vmatrix} (2x_ky_k + 2x_{k+1}y_{k+1} + x_{k+1}y_k + x_ky_{k+1})$$

$$(2.118)$$

For more details on simplex integration for DDA see Shi (1996).

## 2.6 SUMMARY

We reviewed the main theory of 2D DDA in this chapter, including formulation of the governing equations based on the minimum potential energy theory, contact algorithm, time-step integration and simplex integration schemes used in DDA. Since the original DDA method was introduced by Shi (1988) many developments have been presented by researchers in the field, some of which have been reviewed in Chapter 1. Many unresolved issues still remain and it is expected future research will yield advances in topics such as improvement of the computational efficiency during the open-close iteration process, determination of optimal time interval in time domain integration, damping effects, and coupling with other physical processes such as water and heat flow. The basic theory of 3D DDA, which is an extension of this chapter, will be introduced in the next chapter.

# Chapter 3

# Theory of the discontinuous deformation analysis in three dimensions

## 3.1 BLOCK DISPLACEMENT APPROXIMATION AND GLOBAL EQUILIBRIUM EQUATION

The theoretical analysis of DDA in two dimensions presented in Chapter 2 reveals the following general characteristics of DDA: (1) The principle of minimum total potential energy is used to find a solution, similar to the FEM. (2) Dynamic and static problems can be solved by applying the same formulations. (3) Different constitutive laws for the materials can be incorporated. (4) Different types of contact criterion, boundary condition, loading condition, and volumetric force, can be modelled. With regard to modelling three-dimensional problems, however, the two-dimensional approach can only provide a gross approximation of the real behavior. The shortcomings of two-dimensional representation become evident when considering for example discrete block interactions in masonry structures, expected failure modes in multiple free surface conditions such as in tunnel portals, or in blocky rock masses consisting of more than two joint sets where the interaction between the finite blocks and the free surfaces control the mechanical deformation. In order to solve such problems accurately, a robust three-dimensional approach is required. In this chapter, the basic formulations of 3D DDA is presented.

### 3.1.1 Displacement weight function

From traditional 2D to further 3D discontinuous deformation analysis (DDA), the following two basic assumptions are still valid:

- Each loading step (i.e. time step) satisfies the conditions of very small displacement and deformation.
- Block stresses and strains are constant.

The displacement vector $(u, v, w)$ in a block can then be expressed by 12 independent variables:

$$\left(u_0, \quad v_0, \quad w_0, \quad \alpha_0, \quad \beta_0, \quad \gamma_0, \quad \varepsilon_x, \quad \varepsilon_y, \quad \varepsilon_z, \quad \gamma_{yz}, \quad \gamma_{zx}, \quad \gamma_{xy}\right)$$

where $(u_0, \quad v_0, \quad w_0)$ is the rigid translation vector at block centroid of $(x_0, \quad y_0, \quad z_0)$ (see Fig. 3.1(a)); $(\alpha_0, \quad \beta_0, \quad \gamma_0)$ is the rotation angle vector around the $x$-axis, $y$-axis and $z$-axis of the block centroid (see Fig. 3.1(b)); $(\varepsilon_x, \quad \varepsilon_y, \quad \varepsilon_z, \quad r_{yz}, \quad \gamma_{zx}, \quad \gamma_{xy})$

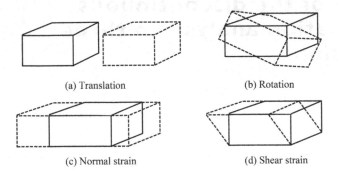

(a) Translation                    (b) Rotation

(c) Normal strain                  (d) Shear strain

*Figure 3.1* Sketch of displacement and deformation by 3D-DDA.

are normal and shear strain vectors of the block centroid (see Fig. 3.1(c) and (d)), respectively.

Assuming constant strains and constant stresses within each block, the first order approximation of displacement can be written as follows:

$$\begin{cases} u = a_0 + a_1 x + a_2 y + a_3 z \\ v = b_0 + b_1 x + b_2 y + b_3 z \\ w = c_0 + c_1 x + c_2 y + c_3 z \end{cases} \tag{3.1}$$

Substituting the coordinates of the initial position of the centroid into Equation 3.1, the displacement of the centroid can be written as:

$$\begin{cases} u_0 = a_0 + a_1 x_0 + a_2 y_0 + a_3 z_0 \\ v_0 = b_0 + b_1 x_0 + b_2 y_0 + b_3 z_0 \\ w_0 = c_0 + c_1 x_0 + c_2 y_0 + c_3 z_0 \end{cases} \tag{3.2}$$

Combining Equations 3.1 and 3.2 we get the displacement components of the block:

$$\begin{cases} u = u_0 + a_1(x - x_0) + a_2(y - y_0) + a_3(z - z_0) \\ v = v_0 + b_1(x - x_0) + b_2(y - y_0) + b_3(z - z_0) \\ w = w_0 + c_1(x - x_0) + c_2(y - y_0) + c_3(z - z_0) \end{cases}$$

The normal strains of the block are:

$$\varepsilon_x = \frac{\partial u}{\partial x} = a_1, \quad \varepsilon_y = \frac{\partial v}{\partial y} = b_2, \quad \varepsilon_z = \frac{\partial w}{\partial z} = c_3$$

The rotations of the block can be expressed as:

$$\alpha_0 = \frac{1}{2}\left(\frac{\partial w}{\partial y} - \frac{\partial v}{\partial z}\right) = \frac{1}{2}(c_2 - b_3)$$

$$\beta_0 = \frac{1}{2}\left(\frac{\partial u}{\partial z} - \frac{\partial w}{\partial x}\right) = \frac{1}{2}(a_3 - c_1)$$

$$\gamma_0 = \frac{1}{2}\left(\frac{\partial v}{\partial x} - \frac{\partial u}{\partial y}\right) = \frac{1}{2}(b_1 - a_2)$$

The shear strains are given by:

$$\gamma_{xy} = \frac{\partial u}{\partial y} + \frac{\partial v}{\partial x} = a_2 + b_1, \quad \gamma_{yz} = \frac{\partial v}{\partial z} + \frac{\partial w}{\partial y} = b_3 + c_2, \quad \gamma_{zx} = \frac{\partial u}{\partial z} + \frac{\partial w}{\partial x} = a_3 + c_1$$

The parameters can therefore be computed as follows:

$$a_1 = \varepsilon_x, \quad b_2 = \varepsilon_y, \quad c_3 = \varepsilon_z$$

$$c_2 = \frac{1}{2}\gamma_{yz} + \alpha_0, \quad b_3 = \frac{1}{2}\gamma_{yz} - \alpha_0, \quad a_3 = \frac{1}{2}\gamma_{zx} + \beta_0$$

$$c_1 = \frac{1}{2}\gamma_{zx} - \beta_0, \quad b_1 = \frac{1}{2}\gamma_{xy} + \gamma_0, \quad a_2 = \frac{1}{2}\gamma_{xy} - \gamma_0$$

Denoting: $X = x - x_0$; $Y = y - y_0$; $Z = z - z_0$, Equation 3.1 can be rewritten as:

$$
\begin{pmatrix} u \\ v \\ w \end{pmatrix} =
\begin{bmatrix}
1 & 0 & 0 & 0 & Z & -Y & X & 0 & 0 & 0 & \frac{Z}{2} & \frac{Y}{2} \\
0 & 1 & 0 & -Z & 0 & X & 0 & Y & 0 & \frac{Z}{2} & 0 & \frac{X}{2} \\
0 & 0 & 1 & Y & -X & 0 & 0 & 0 & Z & \frac{Y}{2} & \frac{X}{2} & 0
\end{bmatrix}
\begin{pmatrix}
u_0 \\ v_0 \\ w_0 \\ \alpha_0 \\ \beta_0 \\ \gamma_0 \\ \varepsilon_x \\ \varepsilon_y \\ \varepsilon_z \\ \gamma_{yz} \\ \gamma_{zx} \\ \gamma_{xy}
\end{pmatrix}
$$

$$
=
\begin{bmatrix}
t_{11} & t_{12} & t_{13} & t_{14} & t_{15} & t_{16} & t_{17} & t_{18} & t_{19} & t_{110} & t_{111} & t_{112} \\
t_{21} & t_{22} & t_{23} & t_{24} & t_{25} & t_{26} & t_{27} & t_{28} & t_{29} & t_{210} & t_{211} & t_{212} \\
t_{31} & t_{32} & t_{33} & t_{34} & t_{35} & t_{36} & t_{37} & t_{38} & t_{39} & t_{310} & t_{311} & t_{312}
\end{bmatrix}
\begin{pmatrix}
d_{1i} \\ d_{2i} \\ d_{3i} \\ d_{4i} \\ d_{5i} \\ d_{6i} \\ d_{7i} \\ d_{8i} \\ d_{9i} \\ d_{10i} \\ d_{11i} \\ d_{12i}
\end{pmatrix}
$$

$$= [T_i][D_i] = T_i d_i \tag{3.3}$$

The displacements of arbitrary points in a block can be computed using the above equation, where $d_i$ is a vector of variables representing the displacements and deformations of a block.

### 3.1.2 Global equilibrium equation

The total potential energy $\Pi$ is the summation over all the potential energy sources, namely individual stresses and forces:

$$\Pi = \Pi_{elastic} + \Pi_{initialstress} + \Pi_{pointload} + \Pi_{lineload} + \Pi_s + \Pi_{bodyforce}$$
$$+ \Pi_{inertia} + \Pi_{contact} + \Pi_{constraint} + \Pi_{fixpoint} \tag{3.4}$$

where:

- $\Pi$ is the total potential energy,
- $\Pi_{elastic}$ is the potential energy contributed by the elastic deformation of the blocks,
- $\Pi_{initialstress}$ is the potential energy contributed by initial stresses,
- $\Pi_{pointload}$ is the potential energy contributed by point loads acting on a block,
- $\Pi_{lineload}$ is the potential energy contributed by line loads acting on a block,
- $\Pi_s$ is the potential energy contributed by surface loads acting on a block,
- $\Pi_{bodyforce}$ is the potential energy contributed by body forces,
- $\Pi_{inertia}$ is the potential energy contributed by inertia forces,
- $\Pi_{contact}$ is the potential energy contributed due to contacts between blocks,
- $\Pi_{constraint}$ is the potential energy contributed by displacement constraints on a block,
- $\Pi_{fixpoint}$ is the potential energy contributed by the constrained displacement points.

In the discussion that follows, the potential energy of each force or stress and their differentiations are computed separately.

The total potential energy $\Pi$ can be written as:

$$\Pi = \frac{1}{2}(D_1^T \quad D_2^T \quad D_3^T \quad \cdots \quad D_n^T) \begin{bmatrix} K_{11} & K_{12} & K_{13} & \cdots & K_{1n} \\ K_{21} & K_{22} & K_{23} & \cdots & K_{2n} \\ K_{31} & K_{32} & K_{33} & \cdots & K_{3n} \\ \vdots & \vdots & \vdots & \ddots & \vdots \\ K_{n1} & K_{n2} & K_{n3} & \cdots & K_{nn} \end{bmatrix} \begin{pmatrix} D_1 \\ D_2 \\ D_3 \\ \vdots \\ D_n \end{pmatrix}$$

$$+ (D_1^T \quad D_2^T \quad D_3^T \quad \cdots \quad D_n^T) \begin{Bmatrix} F_1 \\ F_2 \\ F_3 \\ \vdots \\ F_n \end{Bmatrix} + C \tag{3.5}$$

where sub-matrices $K_{ii}$ depend on the material properties of block $i$ and sub-matrices $K_{ij}$ are defined by the contacts between blocks $i$ and $j$. $D_i$ and $F_i$ are $12 \times 1$ sub-matrices, and $F_i$ is the loading distributed to the twelve unknown variables of block $i$.

Based on the minimized potential energy approach we can write,

$$\frac{\partial \Pi}{\partial d_{ri}} = 0, \quad r = 1, 2, \ldots, 12 \tag{3.6}$$

where $d_{ri}$ is the variable of $i$th block.

The global equilibrium equations can be rewritten in the following form

$$
\begin{bmatrix}
K_{11} & K_{12} & K_{13} & \cdots & K_{1n} \\
K_{21} & K_{22} & K_{23} & \cdots & K_{2n} \\
K_{31} & K_{32} & K_{33} & \cdots & K_{3n} \\
\vdots & \vdots & \vdots & \ddots & \vdots \\
K_{n1} & K_{n2} & K_{n3} & \cdots & K_{nn}
\end{bmatrix}
\begin{Bmatrix}
D_1 \\ D_2 \\ D_3 \\ \vdots \\ D_n
\end{Bmatrix}
=
\begin{Bmatrix}
F_1 \\ F_2 \\ F_3 \\ \vdots \\ F_n
\end{Bmatrix}
\tag{3.7}
$$

where sub-matrices $K_{ij}$ present the $i$th row and the $j$th column stiffness which are $12 \times 12$ sub-matrices; $D_i$ is the $i$th block deformation variables $(d_{1i}, \ d_{2i}, \ d_{3i}, \ \ldots \ d_{12i})^T$; and $F_i$ is the loading on the $i$th block. Sub-matrices $K_{ij} \ (i = j)$ are determined by the block material properties, whereas submatrices $K_{ij} \ (i \neq j)$ are related to the contacts between the $i$th and the $j$th blocks.

The matrix $K_{ij} \ (i, j = 1, 2, \ldots, n)$ is derived by the differentiation:

$$
\frac{\partial^2 \Pi}{\partial d_{ri} \partial d_{sj}} \quad (r, s = 1, 2, \ldots, 12)
\tag{3.8}
$$

and $F_i \ (i = 1, 2, \ldots, n)$ is obtained from the differentiation:

$$
-\frac{\partial \Pi}{\partial d_{ri}} \bigg|_{d_{ri}=0} \quad (r = 1, 2, \ldots, 12)
\tag{3.9}
$$

## 3.2 FORMULATION OF MATRICES FOR SINGLE BLOCK

The equilibrium equations of the DDA method are derived by minimizing the total potential energy of the block system. For a single block, the total potential energy may include components from elastic strains, initial stress, point loading, line loading, body force, bolting connection, inertia force, etc., similar to those introduced in Chapter 2. To solve the derivative of deformation variables from each separated potential energy, the sub-matrices are calculated first. Then, the global equilibrium equation of the entire block system is constructed by assembling the sub-matrices.

### 3.2.1 Sub-matrices of elastic strain

The elastic strain energy $\Pi_e$ produced by the stresses of block $i$ is given by:

$$
\Pi_{elastic} = \iiint \frac{1}{2} \left( \begin{array}{l} \varepsilon_x \sigma_x + \varepsilon_y \sigma_y + \varepsilon_z \sigma_z + \gamma_{yz} \tau_{yz} \\ + \gamma_{zx} \tau_{zx} + \gamma_{xy} \tau_{xy} \end{array} \right) dx\,dy\,dz
\tag{3.10}
$$

where the integration is over the entire volume of block $i$. For each displacement step, the blocks are assumed to be linearly elastic. The relationship between stress and strain is generally expressed as follows:

$$
\sigma_i = E_i \varepsilon_i
\tag{3.11}
$$

or in a matrix form of

$$
\begin{pmatrix} \sigma_x \\ \sigma_y \\ \sigma_z \\ \tau_{yz} \\ \tau_{zx} \\ \tau_{xy} \end{pmatrix} = \frac{E}{(1-v^2)(1-2v)} \begin{bmatrix} 1-v & v & v & & & \\ v & 1-v & v & & 0 & \\ v & v_i & 1-v & & & \\ & & & 0.5-v & 0 & 0 \\ & 0 & & 0 & 0.5-v & 0 \\ & & & 0 & 0 & 0.5-v \end{bmatrix} \begin{pmatrix} \varepsilon_x \\ \varepsilon_y \\ \varepsilon_z \\ \gamma_{yz} \\ \gamma_{zx} \\ \gamma_{xy} \end{pmatrix}
$$

$$(3.12)$$

The block strain energy can then be rewritten as:

$$
\prod_{elastic} = \frac{1}{2} \iiint \begin{pmatrix} \varepsilon_x & \varepsilon_y & \varepsilon_z & \gamma_{yz} & \gamma_{zx} & \gamma_{xy} \end{pmatrix} \begin{pmatrix} \sigma_x \\ \sigma_y \\ \sigma_z \\ \tau_{yz} \\ \tau_{zx} \\ \tau_{xy} \end{pmatrix} dx\,dy\,dz
$$

$$
= \frac{1}{2} \iiint [D_i]^T [E_i][D_i] dx\,dy\,dz
$$

$$
= \frac{V_i}{2} d_i^T E_i d_i \qquad (3.13)
$$

where $V_i$ is the volume of the $i$th block.

The derivatives are calculated to minimize the strain energy $\prod_{elastic}$:

$$
k_{rs} = \frac{\partial^2 \prod_e}{\partial d_{ri} \partial d_{si}} = \frac{V_i}{2} \frac{\partial^2}{\partial d_{ri} \partial d_{si}} ([D_i]^T [E_i][D_i]) = V_i E_i, r, s = 1, \ldots, 12 \rightarrow K_{ii} \qquad (3.14)
$$

which is added to the sub-matrix $K_{ii}$ in the global Equation 3.7.

### 3.2.2  Sub-matrices of initial stress

Similar to the 2D version, 3D DDA can also consider the effect of initial stresses. Block $i$ is assumed to have initial constant stresses of $(\sigma_x^0 \quad \sigma_y^0 \quad \sigma_z^0 \quad \tau_{xy}^0 \quad \tau_{yz}^0 \quad \tau_{zx}^0)$, the potential energy due to these initial stresses is represented to be:

$$
\prod_{initial} = \iiint (\varepsilon_x \sigma_x^0 + \varepsilon_y \sigma_y^0 + \varepsilon_z \sigma_z^0 + \gamma_{yz} \tau_{yz}^0 + \gamma_{zx} \tau_{zx}^0 + \gamma_{xy} \tau_{xy}^0) dx\,dy\,dz
$$

$$
= V_i \begin{pmatrix} \varepsilon_x & \varepsilon_y & \varepsilon_y & \gamma_{yz} & \gamma_{zx} & \gamma_{xy} \end{pmatrix} \begin{pmatrix} \sigma_x^0 \\ \sigma_y^0 \\ \sigma_z^0 \\ \tau_{yz}^0 \\ \tau_{zx}^0 \\ \tau_{xy}^0 \end{pmatrix}
$$

$$
= V_i d_i^T \sigma_0 \qquad (3.15)
$$

where $\sigma_0 = [0 \quad 0 \quad 0 \quad 0 \quad 0 \quad 0 \quad \sigma_x^0 \quad \sigma_y^0 \quad \sigma_z^0 \quad \tau_{yz}^0 \quad \tau_{zx}^0 \quad \tau_{xy}^0]^T$.

The potential energy $\Pi_{initial}$ is minimized by taking the derivatives:

$$f_r = -\frac{\partial \Pi_{initial}}{\partial d_{ri}} = V_i \sigma_0 \ (r = 1, \ldots, 12) \rightarrow F_i \tag{3.16}$$

which is a $12 \times 1$ sub-matrix that is added to the sub-matrix $F_i$ in the global Equation 3.7.

### 3.2.3 Sub-matrices of point loading

Assuming the point loading force acting on a point $(x, y, z)$ of block $i$ is $(F_x, F_y, F_z)$, the potential energy of the point loading is simply expressed as:

$$\Pi_{pointload} = -(F_x u + F_y v + F_z w) = -(u \quad v \quad w) \begin{pmatrix} F_x \\ F_y \\ F_z \end{pmatrix} = -[D_i]^T [T_i(x, y, z)]^T \begin{pmatrix} F_x \\ F_y \\ F_z \end{pmatrix} \tag{3.17}$$

To minimize $\Pi_{pointload}$, the derivatives are calculated as follows:

$$f_r = -\frac{\partial \Pi_{pointload}}{\partial d_{ri}} = \frac{\partial}{\partial d_{ri}} d_i^T T_i^T \begin{pmatrix} F_x \\ F_y \\ F_z \end{pmatrix}$$

$$= T_i^T \begin{pmatrix} F_x \\ F_y \\ F_z \end{pmatrix} (r = 1, \ldots, 12) \rightarrow F_i \tag{3.18}$$

$f_r$ forms a $12 \times 1$ sub-matrix that is added to $F_i$ in the global Equation 3.7.

### 3.2.4 Sub-matrices of line loading

Assume the loading is distributed on a straight line from point $(x_1, y_1, z_1)$ to point $(x_2, y_2, z_2)$ (see Figure 3.2). The length of this line segment is:

$$l = \sqrt{(x_2 - x_1)^2 + (y_2 - y_1)^2 + (z_2 - z_1)^2}$$

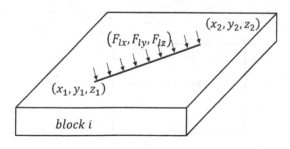

Figure 3.2 Line loading.

The line loading is:

$$\begin{cases} F_{lx} = F_{lx}(x, y, z) \\ F_{ly} = F_{ly}(x, y, z) \\ F_{lz} = F_{lz}(x, y, z) \end{cases} \tag{3.19}$$

The potential energy of the line loading $(F_{lx}, F_{ly}, F_{lz})$ is:

$$\Pi_{lineload} = -\int_l (u \quad v \quad w) \begin{pmatrix} F_{lx} \\ F_{ly} \\ F_{lz} \end{pmatrix} dl = -[D_i]^T \int_l [T_i(x, y, z)]^T \begin{pmatrix} F_{lx} \\ F_{ly} \\ F_{lz} \end{pmatrix} dl \tag{3.20}$$

The derivatives of $f_r$ are calculated to minimize the potential energy $\Pi_{lineload}$:

$$f_r = -\frac{\partial \Pi_{lineload}}{\partial d_{ri}} = \frac{\partial}{\partial d_{ri}} [D_i]^T \int_l [T_i(x, y, z)]^T \begin{pmatrix} F_{lx} \\ F_{ly} \\ F_{lz} \end{pmatrix} dl = \int_l T_i^T \begin{pmatrix} F_{lx} \\ F_{ly} \\ F_{lz} \end{pmatrix} dl \rightarrow F_i \tag{3.21}$$

$f_r$ forms a $12 \times 1$ sub-matrix and it is added to $F_i$ in the global Equation 3.7.

### 3.2.5 Sub-matrices of surface loading

Assuming a surface loading on the plane zone $\Sigma$ of the block $i$ (see Figure 3.3), the plane equation can be expressed as:

$$Ax + By + Cz + D = 0 \tag{3.22}$$

The surface loading is:

$$\begin{cases} F_{sx} = F_{sx}(x, y, z) \\ F_{sy} = F_{sy}(x, y, z) \\ F_{sz} = F_{sz}(x, y, z) \end{cases} \tag{3.23}$$

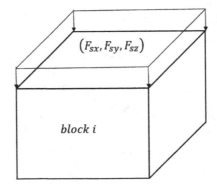

$(F_{sx}, F_{sy}, F_{sz})$

block i

Figure 3.3 Surface loading.

The potential energy of the surface loading is:

$$\Pi_s = -\iint (F_{sx}u + F_{sy}v + F_{sz}w)ds = [D_i]^T \iint [T_i]^T \begin{pmatrix} F_{sx} \\ F_{sy} \\ F_{sz} \end{pmatrix} ds \tag{3.24}$$

The derivatives of $f_s$ are calculated to minimize the potential energy $\Pi_s$:

$$f_s = -\frac{\partial \Pi_s}{\partial d_{ri}} = \iint T_i^T \begin{pmatrix} F_{sx} \\ F_{sy} \\ F_{sz} \end{pmatrix} ds \rightarrow F_i \tag{3.25}$$

$f_s$ forms a $12 \times 1$ sub-matrix and it is added to $F_i$ in the global Equation 3.7.

## 3.2.6   Sub-matrices of body force

If the body force $(f_x, f_y, f_z)$ is constant and acts uniformly on the volume of block $i$, the potential energy is:

$$\Pi_{bodyforce} = -\iiint (f_x u + f_y v + f_z w)dxdydz = -[D_i]^T \left( \iiint [T_i]^T dxdydz \right) \begin{pmatrix} f_x \\ f_y \\ f_z \end{pmatrix} \tag{3.26}$$

Since the coordinates of the centre of gravity of block $i$ are assumed to be $(x_0 \; y_0 \; z_0)$,

$$x_0 = \frac{S_x}{V_i}, \quad y_0 = \frac{S_y}{V_i}, \quad z_0 = \frac{S_z}{V_i} \tag{3.27}$$

where $S_x = \iiint xdxdydz$, $S_y = \iiint ydxdydz$, $S_z = \iiint zdxdydz$, $V_i = \iiint dxdydz$.
Then,

$$\iiint [T_i]^T dxdydz = \begin{pmatrix} V_i & 0 & 0 \\ 0 & V_i & 0 \\ 0 & 0 & V_i \\ 0 & -(S_z - z_0 V_i) & (S_y - y_0 V_i) \\ (S_z - z_0 V_i) & 0 & -(S_x - x_0 V_i) \\ -(S_y - y_0 V_i) & (S_x - x_0 V_i) & 0 \\ (S_x - x_0 V_i) & 0 & 0 \\ 0 & (S_y - y_0 V_i) & 0 \\ 0 & 0 & (S_z - z_0 V_i) \\ 0 & (S_z - z_0 V_i) & (S_y - y_0 V_i)/2 \\ (S_z - z_0 V_i)/2 & 0 & (S_x - x_0 V_i)/2 \\ (S_y - y_0 V_i)/2 & (S_x - x_0 V_i) & 0 \end{pmatrix} = \begin{pmatrix} V_i & 0 & 0 \\ 0 & V_i & 0 \\ 0 & 0 & V_i \\ 0 & 0 & 0 \\ 0 & 0 & 0 \\ 0 & 0 & 0 \\ 0 & 0 & 0 \\ 0 & 0 & 0 \\ 0 & 0 & 0 \\ 0 & 0 & 0 \\ 0 & 0 & 0 \\ 0 & 0 & 0 \end{pmatrix} \tag{3.28}$$

The derivatives of $f_r$ are calculated to minimize the potential energy $\Pi_{bodyforce}$:

$$f_r = -\frac{\partial \Pi_g}{\partial d_{ri}} = \begin{pmatrix} f_x V_i & f_y V_i & f_z V_i & 0 & 0 & 0 & 0 & 0 & 0 & 0 & 0 & 0 \end{pmatrix}^T \rightarrow F_i \quad (3.29)$$

$f_r$ forms a $12 \times 1$ sub-matrix and it is added to $F_i$ in the global Equation 3.7.

### 3.2.7 Sub-matrices of bolting connection

Consider a bolt connecting a point $(x_1, y_1, z_1)$ in block $i$ and a point $(x_2, y_2, z_2)$ in block $j$ (see Figure 3.4). Both points are not necessarily the vertices of the two blocks. The length of the bolt is:

$$l = \sqrt{(x_1 - x_2)^2 + (y_1 - y_2)^2 + (z_1 - z_2)^2}$$

and

$$dl = \frac{1}{l}[(x_1 - x_2)(dx_1 - dx_2) + (y_1 - y_2)(dy_1 - dy_2) + (z_1 - z_2)(dz_1 - dz_2)]$$

$$= \frac{1}{l}[(x_1 - x_2)(u_1 - u_2) + (y_1 - y_2)(v_1 - v_2) + (z_1 - z_2)(w_1 - w_2)]$$

$$= [D_i]^T [T_i(x, y, z)]^T \begin{pmatrix} l_x \\ l_y \\ l_z \end{pmatrix} - [D_j]^T [T_j(x, y, z)]^T \begin{pmatrix} l_x \\ l_y \\ l_z \end{pmatrix} \quad (3.30)$$

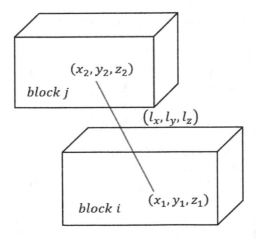

Figure 3.4 Bolt connection.

where $(l_x, l_y, l_z)$ are the direction cosines of the bolt:

$$l_x = \frac{1}{l}(x_1 - x_2)$$
$$l_y = \frac{1}{l}(y_1 - y_2) \tag{3.31}$$
$$l_z = \frac{1}{l}(z_1 - z_2)$$

Assuming the stiffness of the bolt is $s$, the bolt force is:

$$f = -s\frac{dl}{l} \tag{3.32}$$

The strain energy of the bolt is:

$$\Pi_b = -\frac{1}{2}fdl = \frac{s}{2l}dl^2 = \frac{s}{2l}\left([D_i]^T[T_i(x,y,z)]^T\begin{pmatrix}l_x\\l_y\\l_z\end{pmatrix} - [D_j]^T[T_j(x,y,z)]^T\begin{pmatrix}l_x\\l_y\\l_z\end{pmatrix}\right)^2$$
$$= \frac{s}{2l}d_i^T E_i E_i^T d_i - \frac{s}{l}d_i^T E_i G_j^T d_j + \frac{s}{2l}d_j^T G_j G_j^T d_j \tag{3.33}$$

where

$$E_i = T_i^T\begin{pmatrix}l_x\\l_y\\l_z\end{pmatrix}$$

$$G_j = T_j^T\begin{pmatrix}l_x\\l_y\\l_z\end{pmatrix}$$

The derivatives of $\Pi_b$ include a few components, they are:

$$k_{rs} = \frac{\partial^2 \Pi_b}{\partial d_{ri}\partial d_{si}} = \frac{s}{2l}\frac{\partial^2}{\partial d_{ri}\partial d_{si}}d_i^T E_i E_i^T d_i = \frac{s}{l}E_i E_i^T \rightarrow K_{ii} \tag{3.34}$$

which is added to the sub-matrix $K_{ii}$ in the global Equation 3.7.

$$k_{rs} = \frac{\partial^2 \Pi_b}{\partial d_{ri}\partial d_{sj}} = -\frac{s}{l}\frac{\partial^2}{\partial d_{ri}\partial d_{sj}}d_i^T E_i G_j^T d_j = -\frac{s}{l}E_i G_j^T \rightarrow K_{ij} \tag{3.35}$$

which is added to the sub-matrix $K_{ij}$ in the global Equation 3.7.

$$k_{rs} = \frac{\partial^2 \Pi_b}{\partial d_{rj}\partial d_{si}} = -\frac{s}{l}\frac{\partial^2}{\partial d_{rj}\partial d_{si}}d_i^T E_i G_j^T d_j = -\frac{s}{l}G_j E_i^T \rightarrow K_{ji} \tag{3.36}$$

which is added to the sub-matrix $K_{ji}$ in the global Equation 3.7.

$$k_{rs} = \frac{\partial^2 \Pi_b}{\partial d_{rj} \partial d_{sj}} = \frac{s}{2l} \frac{\partial^2}{\partial d_{rj} \partial d_{sj}} d_j^T G_j G_j^T d_j = \frac{s}{l} G_j G_j^T \rightarrow K_{jj} \qquad (3.37)$$

which is added to the sub-matrix $K_{jj}$ in the global Equation 3.7.

### 3.2.8 Sub-matrices of inertia force

Using $(u(t), v(t), w(t))$ to denote the time dependent displacement of any point $(x, y, z)$ of block $i$ and $M$ to represent the mass per unit volume, the inertia force of block $i$ is as follows:

$$\begin{pmatrix} f_x \\ f_y \\ f_z \end{pmatrix} = -M \begin{pmatrix} \frac{\partial^2 u(t)}{\partial t^2} \\ \frac{\partial^2 v(t)}{\partial t^2} \\ \frac{\partial^2 w(t)}{\partial t^2} \end{pmatrix} = -M T_i \frac{\partial^2 d_i(t)}{\partial t^2} \qquad (3.38)$$

The potential energy of the inertia force of block $i$ is written as follows:

$$\Pi_{inertia} = - \iiint (u \quad v \quad w) \begin{pmatrix} f_x \\ f_y \\ f_z \end{pmatrix} dxdydz = M \iiint d_i^T T_i^T T_i \frac{\partial^2 d(t)}{\partial t^2} dxdydz \quad (3.39)$$

Assume $d_i(0) = 0$ is the block displacement at the beginning of the time step, $\Delta$ is the time step, and $d_i(\Delta) = d_i$ is the block displacement at the end of the time step. Using the time integration we get:

$$d_i = d_i(\Delta) = d_i(0) + \Delta \frac{\partial d_i(0)}{\partial t} + \frac{\Delta^2}{2} \frac{\partial^2 d_i(0)}{\partial t^2} = \Delta \frac{\partial d_i(0)}{\partial t} + \frac{\Delta^2}{2} \frac{\partial^2 d_i(0)}{\partial t^2} \qquad (3.40)$$

Assuming the acceleration in each time step is constant:

$$\frac{\partial^2 d(t)}{\partial t^2} = \frac{\partial^2 d(0)}{\partial t^2} = \frac{2}{\Delta^2} d_i - \frac{2}{\Delta} \frac{\partial d_i(0)}{\partial t} \qquad (3.41)$$

The potential energy of the inertia force at the end of each time step is:

$$\Pi_{inertia} = M \iiint d_i^T T_i^T T_i \frac{\partial^2 d(t)}{\partial t^2} dxdydz$$

$$= M d_i^T \iiint T_i^T T_i dxdydz \left( \frac{2}{\Delta^2} d_i - \frac{2}{\Delta} \frac{\partial d_i(0)}{\partial t} \right) \qquad (3.42)$$

To reach equilibrium, $\Pi_{inertia}$ is minimized by taking derivatives with respect to the block displacement variables:

$$f_r = -\frac{\partial \Pi_{inertia}}{\partial d_{ri}} = -\frac{\partial}{\partial d_{ri}}\left(Md_i^T\iiint T_i^T T_i dxdydz\left(\frac{2}{\Delta^2}d_i - \frac{2}{\Delta}\frac{\partial d_i(0)}{\partial t}\right)\right)$$
$$r = 1, 2, \ldots, 12 \qquad (3.43)$$

$f_r$ forms a $12 \times 1$ sub-matrix, and it is added into $F_i$ in the global Equation 3.7.

As there is an unknown $d_i$ in Equation 3.43, this equation should be transformed into two as follows:

$$\frac{2M}{\Delta^2}\iiint T_i^T T_i dxdydz \rightarrow K_{ii} \qquad (3.44)$$

$$\frac{2M}{\Delta}\left(\iiint T_i^T T_i dxdydz\right)v_0 \rightarrow F_i \qquad (3.45)$$

where $v_0 = \frac{\partial d_i(0)}{\partial t}$.

## 3.2.9 Sub-matrices of displacement constraints in a direction

In some engineering problems, some blocks are fixed at specific points in specified directions. More generally, the displacement $\delta$ along a direction at a point could be a known value. The measured displacements can be applied to the blocky system by using a very stiff spring which has a pre-described displacement. Denote $(l_x, l_y, l_z)(l_x^2 + l_y^2 + l_z^2 = 1)$ as the direction along which the displacement $\delta$ exists. The spring displacement is $d = \delta - (l_x u + l_y v + l_z w)$ and the spring force is:

$$f = -pd = -p(\delta - (l_x u + l_y v + l_z w)) \qquad (3.46)$$

where $p$ is the stiffness of the spring, which is a very large value, normally from $10E$ to $1000E$, to guarantee the displacement of the spring is $10^{-1}$ to $10^{-3}$ times the total displacement. If $p$ is large enough, the computation results will be independent of the selections of $p$.

The strain energy of the spring is:

$$\Pi_{constraint} = \frac{p}{2}d^2$$

$$= \frac{p}{2}(u \quad v \quad w)\begin{pmatrix} l_x \\ l_y \\ l_z \end{pmatrix}(l_x \quad l_y \quad l_z)\begin{pmatrix} u \\ v \\ w \end{pmatrix} - p\delta(u \quad v \quad w)\begin{pmatrix} l_x \\ l_y \\ l_z \end{pmatrix} + \frac{p}{2}\delta^2 \qquad (3.47)$$

Let $L = \begin{pmatrix} l_x \\ l_y \\ l_z \end{pmatrix}$ and $c = T^T L$

Then:

$$(u \quad v \quad w) \begin{pmatrix} l_x \\ l_y \\ l_z \end{pmatrix} = d_i^T T_i^T L = d_i^T c_i$$

$$(l_x \quad l_y \quad l_z) \begin{pmatrix} u \\ v \\ w \end{pmatrix} = L_i^T T_i d_i = c_i^T d_i$$

Therefore:

$$\Pi_{constraint} = \frac{p}{2} d_i^T c_i c_i^T d_i - p\delta d_i^T c_i + \frac{p}{2}\delta^2 \tag{3.48}$$

A 12 × 12 matrix is obtained by taking the derivatives of the strain energy of the stiff spring:

$$k_{rs} = \frac{\partial^2 \Pi_{constraint}}{\partial d_{ri} \partial d_{si}} = \frac{p}{2} \frac{\partial^2}{\partial d_{ri} \partial d_{si}} d_i^T c_i c_i^T d_i = p c_i c_i^T \rightarrow K_{ii} \tag{3.49}$$

$k_{rs}$ $(r, s = 1, \ldots, 12)$ forms a 12 × 12 matrix and it is added to the sub-matrix $K_{ii}$ in the global Equation 3.7.

The spring strain energy $\Pi_{constraint}$ is minimized by taking the derivatives of $\Pi_c$ at $d = 0$:

$$f_r = -\frac{\partial \Pi_{constraint}(0)}{\partial d_{ri}} = p\delta \frac{\partial}{\partial d_{ri}} d_i^T c_i = p\delta c_i \rightarrow F_i \tag{3.50}$$

$f_r$ $(r = 1, \ldots, 12)$ is added to $F_i$ in the global Equation 3.7.

### 3.2.10  Sub-matrices of fixed point

Under certain boundary conditions, the displacements of the blocks are constrained. Assume that the specified displacements at point $P$ in block $i$ are given by $(u_{fs}, v_{fs}, w_{fs})$, while the computed displacements at the same point are $(u_f, v_f, w_f)$. To constrain the displacement of point $P$ to the ones specified, DDA applies springs with stiffness $k_{fix}$ as shown in Fig. 3.5 to move point $P''$ to the specific location $P'$. The resulting contributions to the global matrix are as follows:

$$\begin{pmatrix} f_{fx} \\ f_{fy} \\ f_{fz} \end{pmatrix} = \begin{bmatrix} -k_{fix}(u_f - u_{fs}) \\ -k_{fix}(v_f - v_{fs}) \\ -k_{fix}(w_f - w_{fs}) \end{bmatrix} \tag{3.51}$$

Therefore, the strain energy $\Pi_{\text{fixpoint}}$ caused by the constrained springs can be expressed as:

$$\Pi_{\text{fixpoint}} = \frac{k_{fix}}{2}[(u_f - u_{fs})^2 + (v_f - v_{fs})^2 + (w_f - w_{fs})^2]$$

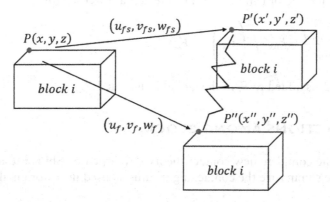

*Figure 3.5* Fixed point.

$$= \frac{k_{fix}}{2}[(u_f - u_{fs})(v_f - v_{fs})(w_f - w_{fs})] \begin{bmatrix} (u_f - u_{fs}) \\ (v_f - v_{fs}) \\ (w_f - w_{fs}) \end{bmatrix}$$

$$= \frac{k_{fix}}{2}(u_f \quad v_f \quad w_f)\begin{pmatrix} u_f \\ v_f \\ w_f \end{pmatrix} - k_{fix}(u_f \quad v_f \quad w_f)\begin{pmatrix} u_{fs} \\ v_{fs} \\ w_{fs} \end{pmatrix} + \frac{k_{fix}}{2}(u_{fs} \quad v_{fs} \quad w_{fs})\begin{pmatrix} u_{fs} \\ v_{fs} \\ w_{fs} \end{pmatrix} \tag{3.52}$$

Since:

$$\begin{pmatrix} u_f \\ v_f \\ w_f \end{pmatrix} = T_i d_i$$

As a result, Equation 3.52 can be written as:

$$\Pi_{\text{fixpoint}} = \frac{k_{fix}}{2}d_i^T T_i^T T_i d_i - k_{fix}d_i^T T_i^T \begin{pmatrix} u_{fs} \\ v_{fs} \\ w_{fs} \end{pmatrix} + \frac{k_{fix}}{2}(u_{fs} \quad v_{fs} \quad w_{fs})\begin{pmatrix} u_{fs} \\ v_{fs} \\ w_{fs} \end{pmatrix} \tag{3.53}$$

By minimizing the potential energy $\Pi_{\text{fixpoint}}$, the following matrix can then be added to the submatrix $K_{ii}$ in the global stiffness matrix:

$$k_{rs} = \frac{\partial^2 \Pi_{fixpoint}}{\partial d_{ri} \partial d_{si}} = \frac{k_{fix}}{2} T_i^T T_i \rightarrow K_{ii} \tag{3.54}$$

$k_{rs}$ ($r, s = 1, \ldots, 12$) forms a $12 \times 12$ matrix and it is added to the sub-matrix $K_{ii}$ in the global Equation 3.7.

The following vector can be added to the global force vector:

$$f_r = -\frac{\partial \Pi_{fixpoint}}{\partial d_{ri}} = k_{fix} T_i^T \begin{pmatrix} u_{fs} \\ v_{fs} \\ w_{fs} \end{pmatrix} \rightarrow F_i \qquad (3.55)$$

$f_r (r = 1, \ldots, 12)$ is added to $F_i$ in the global Equation 3.7.

## 3.3 INTERACTIONS AMONG BLOCKS

In Chapter 8 the complete new contact theory developed by Shi is presented in great detail. Here we summarize the contact algorithms as used in the original 3D-DDA.

### 3.3.1 Contact detection and types

The basic idea of 3-D DDA contact detection algorithm is based upon the concept of entrance mode. There are various contact types in the 3-D view. For example, an edge cluster connected to one vertex has many possible combinations in terms of ordering, and different contact situations may be detected if two 3-D vertices are considered. The complexity of other contact types are similar. For convenience of reference, the 3-D contacts in 3-D DDA are classified into 7 basic types, as shown in Fig. 3.6. In Shi's 3-D DDA, all of these contact types are simplified into two entrance modes: vertex-to-face mode and edge-to-edge crossing-line mode. Formulation of the contact matrices are given in Section 3.3.3.

### 3.3.2 Contact state and open-close iteration

#### Contact states

Each contact has three possible contact states:

1   When the normal component $R_n$ of a contact force is tensile:

$$R_n \leq -p d_n \leq 0 \qquad (3.56)$$

no locking and no stiff springs are employed; this contact is open.

2   When $R_n$ is compressive and the shear component $R_s$ is sufficiently high:

$$R_s \geq R_n \tan \varphi + C \qquad (3.57)$$

sliding takes place between contact pairs. A stiff spring is applied in normal direction of entrance face, and a pair of frictional forces is generated between the contact pair.

3   When the normal component $R_n$ of the contact is Compressive, and the shear component $R_s$ is lower than the maximum frictional resistance as assumed by the Coulomb law:

$$R_s \leq R_n \tan \varphi + C \qquad (3.58)$$

normal and shear springs are applied at the contact point, and the contact pair is not allowed to slide. At this point the contact is in a lock state.

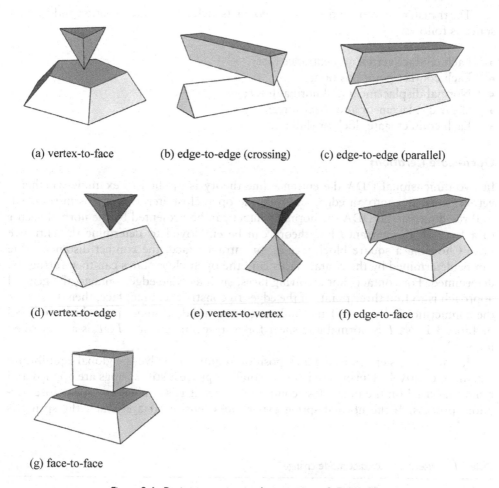

(a) vertex-to-face    (b) edge-to-edge (crossing)    (c) edge-to-edge (parallel)

(d) vertex-to-edge    (e) vertex-to-vertex    (f) edge-to-face

(g) face-to-face

Figure 3.6  Basic contact types between two 3-D blocks.

## Contact transfer

Since the calculation in discontinuous deformation analysis utilizes a time step marching scheme, the contact states at the previous time step are inherited as the initial states before the next time step begins. Therefore, all the geometrical and physical parameters in the block system must be transferred from the previous step to the next step. The transferred parameters include:

- Individual block stress;
- Individual block strain;
- Individual block velocity;
- Geometric position of each block;
- All of the closed contacts

The transferred information of close contacts includes contact position and contact state, as follows:

- Each contact vertex and entrance face;
- Each contact point position;
- Normal displacement and normal force;
- Shear displacement and shear force;
- Each contact state (lock or sliding).

### Open-close iterations

In two-dimensional DDA the entrance line theory is applied to examine whether a vertex penetrates into an edge, and then the open-close iteration law is introduced. In three-dimensional DDA the normal contact can be presented by the normal vector to a face, so the entrance line theory can be employed to determine the entrance face. Choosing a square block face as an entrance face, the contact distance value can be determined by the contact faces, and the open-close states can then further be determined. For contacts not involving faces, such as edge-edge contacts, the general approach is to find three points of the edges to construct contact face, then determine the contact normal vector. The details of contact mode changes procedure are listed in Table 3.1. $N$, $T$ is normal and shear forces respectively, and $TOL$ is a prescribed limit value.

In each time step, choosing lock position requires to solve the global equilibrium equation iteratively. During the iterative solution process stiff springs are applied and removed based on the open-close condition, hence it is so-called an 'open-close iteration' process. If the normal spring experiences tension at a contact, the spring is

*Table 3.1* Criteria for contact mode change.

| Mode change | Condition | Operation | |
|---|---|---|---|
| Open-open | $N > 0$ | No change | $N$ is tension force |
| Open-sliding | $N < 0$, $|T| > \tan \varphi |N|$ | Apply a normal spring and a pair of friction forces | $T$ is shear force |
| Open-lock | $N < 0$, $|T| < \tan \varphi |N|$ | Apply normal and shear springs | |
| Sliding-open | $N > 0$ | Remove the normal spring and the pair of friction forces | |
| Sliding-sliding | $N < 0$, $|T| \geq TOL$ | No change | $TOL$ is prescribed limit value |
| Sliding-lock | $N < 0$, $|T| < TOL$ | Remove the pair of friction forces and apply a shear spring | |
| Lock-open | $N > 0$ | Remove the normal and shear springs | |
| Lock-sliding | $N < 0$, $|T| > \tan \varphi |N|$ | Remove the shear spring and apply a pair of friction forces. | |
| Lock-lock | $N < 0$, $|T| < \tan \varphi |N|$ | No change | |

removed; if there are penetrations between contact vertices, the spring is applied at that point. Each contact pair has three modes: open, sliding and lock, and the mode changes are summarized in Table 3.1. The open-close iteration procedure continues until the following two constraints are satisfied in the entire system:

- No penetration in "open" contact state;
- No tension in contacts.

These two conditions must be satisfied in all contacts. If these two conditions are not satisfied after 6 iterations, the time step size is reduced and the open-close iterations resume from the end of the previous time step.

### 3.3.3 Formulation of contact matrices

#### Normal contact

As shown in Fig. 3.7, when vertex $P_1$ of block $i$ contacts face $P_2P_3P_4P_5P_6$ of block $j$, assume that $P_1$ penetrates into the face $P_2P_3P_4P_5P_6$ and stops at $P_0$ in block $j$. The positions and the displacements of the vertices are $(x_i, \quad y_i, \quad z_i)$ and $(u_i, \quad v_i, \quad w_i)$, $i = 0, 1, 2, 3, 4, 5, 6$, respectively, in this case. If a penetration distance by $P_1$ into $P_2P_3P_4P_5P_6$ is detected, a stiff spring with stiffness of $k_n$ is introduced into the computation to push back the vertex to the surface along the shortest distance. This potential energy contribution is computed as follows:

$$\Pi_{nc} = \frac{k_n}{2}d_n^2 \tag{3.59}$$

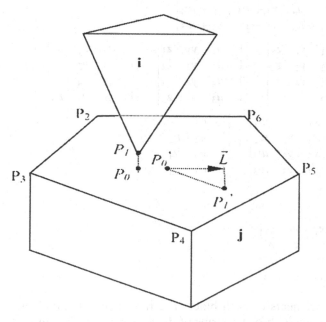

*Figure 3.7* Point-face contact (Jiang *et al.* (2004)).

where $d_n$ is the normal penetration distance from vertex $P_1$ to face $P_2P_3P_4P_5P_6$, with a positive value when $P_1$ does not penetrate into $P_2P_3P_4P_5P_6$.

On the face of $P_2P_3P_4P_5P_6$, vertices $P_2, P_3, P_4, P_5$ and $P_6$ are in a counter-clockwise sequence. Let $P'_i, i = 0-6$ be the respective displaced vertices at end of the time step. $P_1$ moves to a new updated vertex position $P'_1$. The compression of the normal contact spring can be written as:

$$d_n = \frac{\left[\overrightarrow{P'_1P'_2} \cdot \left(\overrightarrow{P'_1P'_4} \times \overrightarrow{P'_1P'_3}\right)\right]}{\left|\overrightarrow{P'_2P'_3} \times \overrightarrow{P'_2P'_4}\right|}$$

$$= \frac{1}{\left|\overrightarrow{P'_2P'_3} \times \overrightarrow{P'_2P'_4}\right|}(x_2 + u_2 - x_1 - u_1, y_2 + v_2 - y_1 - v_1, z_2 + w_2 - z_1 - w_1)$$

$$\cdot \begin{vmatrix} \overrightarrow{i} & \overrightarrow{j} & \overrightarrow{k} \\ x_4 + u_4 - x_1 - u_1 & y_4 + v_4 - y_1 - v_1 & z_4 + w_4 - z_1 - w_1 \\ x_3 + u_3 - x_1 - u_1 & y_3 + v_3 - y_1 - v_1 & z_3 + w_3 - z_1 - w_1 \end{vmatrix}$$

$$= \frac{1}{\left|\overrightarrow{P'_2P'_3} \times \overrightarrow{P'_2P'_4}\right|} \cdot \begin{vmatrix} 1 & x_1 + u_1 & y_1 + v_1 & z_1 + w_1 \\ 1 & x_2 + u_2 & y_2 + v_2 & z_2 + w_2 \\ 1 & x_4 + u_4 & y_4 + v_4 & z_4 + w_4 \\ 1 & x_3 + u_3 & y_3 + v_3 & z_3 + w_3 \end{vmatrix} = \frac{\Delta}{\left|\overrightarrow{P'_2P'_3} \times \overrightarrow{P'_2P'_4}\right|} \qquad (3.60)$$

where,

$$\Delta = \begin{vmatrix} 1 & x_1 + u_1 & y_1 + v_1 & z_1 + w_1 \\ 1 & x_2 + u_2 & y_2 + v_2 & z_2 + w_2 \\ 1 & x_4 + u_4 & y_4 + v_4 & z_4 + w_4 \\ 1 & x_3 + u_3 & y_3 + v_3 & z_3 + w_3 \end{vmatrix}$$

$$= \begin{vmatrix} 1 & x_1 & y_1 & z_1 \\ 1 & x_2 & y_2 & z_2 \\ 1 & x_4 & y_4 & z_4 \\ 1 & x_3 & y_3 & z_3 \end{vmatrix} + \begin{vmatrix} 1 & u_1 & y_1 & z_1 \\ 1 & u_2 & y_2 & z_2 \\ 1 & u_4 & y_4 & z_4 \\ 1 & u_3 & y_3 & z_3 \end{vmatrix} + \begin{vmatrix} 1 & x_1 & v_1 & v_1 \\ 1 & x_2 & v_2 & v_2 \\ 1 & x_4 & v_4 & v_4 \\ 1 & x_3 & v_3 & v_3 \end{vmatrix}$$

$$+ \begin{vmatrix} 1 & x_1 & y_1 & w_1 \\ 1 & x_2 & y_2 & w_2 \\ 1 & x_4 & y_4 & w_4 \\ 1 & x_3 & y_3 & w_3 \end{vmatrix} + \begin{vmatrix} 1 & u_1 & v_1 & w_1 \\ 1 & u_2 & v_2 & w_2 \\ 1 & u_4 & v_4 & w_4 \\ 1 & u_3 & v_3 & w_3 \end{vmatrix}$$

Let,

$$Vol = \begin{vmatrix} 1 & x_1 & y_1 & z_1 \\ 1 & x_2 & y_2 & z_2 \\ 1 & x_4 & y_4 & z_4 \\ 1 & x_3 & y_3 & z_3 \end{vmatrix}$$

If the displacements of each block in a time step are sufficiently small, the last term of $\Delta$ involving high-order terms of displacement can be ignored, and the above equation can be written as:

$$\Delta = Vol - u_1 \cdot \begin{vmatrix} 1 & y_2 & z_2 \\ 1 & y_4 & z_4 \\ 1 & y_3 & z_3 \end{vmatrix} + u_2 \cdot \begin{vmatrix} 1 & y_1 & z_1 \\ 1 & y_4 & z_4 \\ 1 & y_3 & z_3 \end{vmatrix} - u_4 \cdot \begin{vmatrix} 1 & y_1 & z_1 \\ 1 & y_2 & z_2 \\ 1 & y_3 & z_3 \end{vmatrix} + u_3 \cdot \begin{vmatrix} 1 & y_1 & z_1 \\ 1 & y_2 & z_2 \\ 1 & y_4 & z_4 \end{vmatrix}$$

$$+ v_1 \cdot \begin{vmatrix} 1 & x_2 & z_2 \\ 1 & x_4 & z_4 \\ 1 & x_3 & z_3 \end{vmatrix} - v_2 \cdot \begin{vmatrix} 1 & x_1 & z_1 \\ 1 & x_4 & z_4 \\ 1 & x_3 & z_3 \end{vmatrix} + v_4 \cdot \begin{vmatrix} 1 & x_1 & z_1 \\ 1 & x_2 & z_2 \\ 1 & x_3 & z_3 \end{vmatrix}$$

$$- v_3 \cdot \begin{vmatrix} 1 & x_1 & z_1 \\ 1 & x_2 & z_2 \\ 1 & x_4 & z_4 \end{vmatrix} - w_1 \cdot \begin{vmatrix} 1 & x_2 & y_2 \\ 1 & x_4 & y_4 \\ 1 & x_3 & y_3 \end{vmatrix} + w_2 \cdot \begin{vmatrix} 1 & x_1 & y_1 \\ 1 & x_4 & y_4 \\ 1 & x_3 & y_3 \end{vmatrix}$$

$$- w_4 \cdot \begin{vmatrix} 1 & x_1 & y_1 \\ 1 & x_2 & y_2 \\ 1 & x_3 & y_3 \end{vmatrix} + w_3 \cdot \begin{vmatrix} 1 & x_1 & y_1 \\ 1 & x_2 & y_2 \\ 1 & x_4 & y_4 \end{vmatrix}$$

In this case, $d_n$ can be expressed in the following form:

$$d_n = \frac{Vol}{A} + E_i^T d_i + G_j^T d_j \tag{3.61}$$

where:

$$E_i^T = \frac{1}{A} \left[ -\begin{vmatrix} 1 & y_2 & z_2 \\ 1 & y_4 & z_4 \\ 1 & y_3 & z_3 \end{vmatrix}, \begin{vmatrix} 1 & x_2 & z_2 \\ 1 & x_4 & z_4 \\ 1 & x_3 & z_3 \end{vmatrix}, \begin{vmatrix} 1 & x_2 & y_2 \\ 1 & x_4 & y_4 \\ 1 & x_3 & y_3 \end{vmatrix} \right] \cdot T_i^T(x_1, y_1, z_1)$$

$$G_j^T = \frac{1}{A} \left[ -\begin{vmatrix} 1 & y_1 & z_1 \\ 1 & y_4 & z_4 \\ 1 & y_3 & z_3 \end{vmatrix}, -\begin{vmatrix} 1 & x_1 & z_1 \\ 1 & x_4 & z_4 \\ 1 & x_3 & z_3 \end{vmatrix}, \begin{vmatrix} 1 & x_1 & y_1 \\ 1 & x_4 & y_4 \\ 1 & x_3 & y_3 \end{vmatrix} \right] \cdot T_j^T(x_2, y_2, z_2)$$

$$+ \frac{1}{A} \left[ -\begin{vmatrix} 1 & y_1 & z_1 \\ 1 & y_2 & z_2 \\ 1 & y_3 & z_3 \end{vmatrix}, \begin{vmatrix} 1 & x_1 & z_1 \\ 1 & x_2 & z_2 \\ 1 & x_3 & z_3 \end{vmatrix}, -\begin{vmatrix} 1 & x_1 & y_1 \\ 1 & x_2 & y_2 \\ 1 & x_3 & y_3 \end{vmatrix} \right] \cdot T_j^T(x_4, y_4, z_4)$$

$$+ \frac{1}{A} \left[ \begin{vmatrix} 1 & y_1 & z_1 \\ 1 & y_2 & z_2 \\ 1 & y_4 & z_4 \end{vmatrix}, \begin{vmatrix} 1 & x_1 & z_1 \\ 1 & x_2 & z_2 \\ 1 & x_4 & z_4 \end{vmatrix}, \begin{vmatrix} 1 & x_1 & y_1 \\ 1 & x_2 & y_2 \\ 1 & x_4 & y_4 \end{vmatrix} \right] \cdot T_j^T(x_3, y_3, z_3)$$

$$A = \left| \overrightarrow{p_2'p_3'} \times \overrightarrow{p_2'p_4'} \right| \approx \begin{vmatrix} x_2 & y_2 & z_2 \\ x_3 & y_3 & z_3 \\ x_4 & y_4 & z_4 \end{vmatrix}$$

Therefore, the strain energy of the normal contact spring forces is:

$$\Pi_{nc} = \frac{k_n}{2} d_n^2 = \frac{k_n}{2} \left( \frac{Vol}{A} + E_i^T d_i + G_j^T d_j \right)^2 \tag{3.62}$$

Four $12 \times 12$ sub-matrices and two $12 \times 1$ sub-matrices are obtained and then added to $K_{ii}$, $K_{ij}$, $K_{ji}$, $K_{jj}$, $F_i$ and $F_j$ respectively, by minimizing the strain energy $\Pi_{nc}$.

The derivative of $\Pi_{nc}$:

$$k_{rs} = \frac{\partial^2 \Pi_{nc}}{\partial d_{ri} \partial d_{si}} = k_n E_i E_i^T \rightarrow K_{ii} \tag{3.63}$$

$k_{rs}$ $(r, s = 1, \ldots, 12)$ is added to $K_{ii}$ in the global Equation 3.7.

The derivative of $\Pi_{nc}$:

$$k_{rs} = \frac{\partial^2 \Pi_{nc}}{\partial d_{ri} \partial d_{sj}} = k_n E_i G_j^T \rightarrow K_{ij} \tag{3.64}$$

$k_{rs}$ $(r, s = 1, \ldots, 12)$ is added to $K_{ij}$ in the global Equation 3.7.

The derivative of $\Pi_{nc}$:

$$k_{rs} = \frac{\partial^2 \Pi_{nc}}{\partial d_{rj} \partial d_{si}} = k_n G_j E_i^T \rightarrow K_{ji} \tag{3.65}$$

$k_{rs}$ $(r, s = 1, \ldots, 12)$ is added to $K_{ji}$ in the global Equation 3.7.

The derivative of $\Pi_{nc}$:

$$k_{rs} = \frac{\partial^2 \Pi_{nc}}{\partial d_{rj} \partial d_{sj}} = k_n G_j G_j^T \rightarrow K_{jj} \tag{3.66}$$

$k_{rs}$ $(r, s = 1, \ldots, 12)$ is added to $K_{jj}$ in the global Equation 3.7.

The derivative of $\Pi_{nc}$:

$$f_r = -\frac{\partial \Pi_{nc}(0)}{\partial d_{ri}} = -k_n \frac{Vol}{A} E_i \rightarrow F_i \tag{3.67}$$

$f_r$ $(r = 1, \ldots, 12)$ is added to $F_i$ in the global Equation 3.7.

The derivative of $\Pi_{nc}$:

$$f_r = -\frac{\partial \Pi_{nc}(0)}{\partial d_{rj}} = -k_n \frac{Vol}{A} G_j \rightarrow F_j \tag{3.68}$$

$f_r$ $(r = 1, \ldots, 12)$ is added to $F_j$ in the global Equation 3.7.

With respect to the edge-to-edge contact type, assume the direction of contact between block $i$ and $j$ is from block $i$ toward block $j$. As shown in Fig. 3.8 the contact faces intersect at edges $P_1 P_2$ and $P_3 P_4$, respectively in block $i$ and block $j$. The normal contact matrices are derived similarly to that of the vertex-to-face contact type.

### Shear contact

When the contact vertex and entrance face are in a "no sliding" mode the shear spring is adopted in addition to the normal spring, to obtain a "lock" mode. As shown in Fig. 3.7, assuming the vertex $P_0$ moves to a new point $P_0'$ at the end of a time step, $\vec{L}$ is

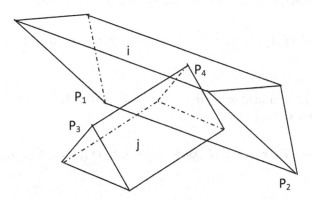

*Figure 3.8*  edge-edge contact.

the projection of the vector $\overrightarrow{P_0'P_1'}$ on the contact face $P_2P_3P_4P_5P_6$. The magnitude of $\overrightarrow{L}$ is given by:

$$| \overrightarrow{L} | = d_s = \sqrt{\left| \overrightarrow{P_0'P_1'} \right|^2 - d_n^2} \tag{3.69}$$

Assuming there is a shear spring between points $P_1$ and $P_0$ in a direction parallel to $\overrightarrow{L}$ and letting $k_s$ be the stiffness of the shear spring, the potential energy of the shear spring is given by:

$$
\begin{aligned}
\Pi_{sc} &= \frac{k_s}{2} d_s^2 = \frac{k_s}{2} \left( \left| \overrightarrow{P_0'P_1'} \right|^2 - d_n^2 \right) \\
&= \frac{k_s}{2} \begin{pmatrix} x_1 + u_1 - x_0 - u_0 \\ y_1 + v_1 - y_0 - v_0 \\ z_1 + w_1 - z_0 - w_0 \end{pmatrix}^T \begin{pmatrix} x_1 + u_1 - x_0 - u_0 \\ y_1 + v_1 - y_0 - v_0 \\ z_1 + w_1 - z_0 - w_0 \end{pmatrix} - \frac{k_s}{2} d_n^2
\end{aligned} \tag{3.70}
$$

From Equation 3.61 and Equation 3.70,

$$
\begin{aligned}
\Pi_{sc} &= \frac{k_s}{2} \left( \begin{bmatrix} x_1 - x_0 & y_1 - y_0 & z_1 - z_0 \end{bmatrix} + d_i^T T_i^T - d_j^T T_j^T \right) \left( \begin{bmatrix} x_1 - x_0 \\ y_1 - y_0 \\ z_1 - z_0 \end{bmatrix} + T_i d_i - T_j d_j \right) \\
&\quad - \frac{k_s}{2} \left( \frac{Vol}{A} + E_i^T d_i + G_j^T d_j \right)^2
\end{aligned} \tag{3.71}
$$

Four $12 \times 12$ sub-matrices and two $12 \times 1$ sub-matrices are obtained and then added to $K_{ii}$, $K_{ij}$, $K_{ji}$, $K_{jj}$, $F_i$ and $F_j$ respectively, by minimizing the strain energy $\Pi_{sc}$.

Ignoring the second-order infinitesimally small terms, the derivative of $\Pi_{sc}$:

$$k_{rs} = \frac{\partial^2 \Pi_{sc}}{\partial d_{ri} \partial d_{si}} = k_s T_i^T(x_1, y_1, z_1) T_i(x_1, y_1, z_1) - k_s E_i^T(x_1, y_1, z_1) E_i(x_1, y_1, z_1) \rightarrow K_{ii}$$

(3.72)

$k_{rs}$ $(r, s = 1, \ldots, 12)$ is added to $K_{ii}$ in the global Equation 3.7.
The derivative of $\Pi_{sc}$:

$$k_{rs} = \frac{\partial^2 \Pi_{sc}}{\partial d_{ri} \partial d_{sj}} = -k_s T_i^T(x_1, y_1, z_1) T_j(x_0, y_0, z_0) - k_s E_i^T(x_1, y_1, z_1) G_j(x_0, y_0, z_0) \rightarrow K_{ij}$$

(3.73)

$k_{rs}$ $(r, s = 1, \ldots, 12)$ is added to $K_{ij}$ in the global Equation 3.7.
The derivative of $\Pi_{sc}$:

$$k_{rs} = \frac{\partial^2 \Pi_{sc}}{\partial d_{rj} \partial d_{si}} = -k_s T_j^T(x_0, y_0, z_0) T_i(x_1, y_1, z_1) - k_s G_j^T(x_0, y_0, z_0) E_i(x_1, y_1, z_1) \rightarrow K_{ji}$$

(3.74)

$k_{rs}$ $(r, s = 1, \ldots, 12)$ is added to $K_{ji}$ in the global Equation 3.7.
The derivative of $\Pi_{sc}$:

$$k_{rs} = \frac{\partial^2 \Pi_{sc}}{\partial d_{rj} \partial d_{sj}} = k_s T_j^T(x_0, y_0, z_0) T_j(x_0, y_0, z_0) - k_s G_j^T(x_0, y_0, z_0) G_j(x_0, y_0, z_0) \rightarrow K_{jj}$$

(3.75)

$k_{rs}$ $(r, s = 1, \ldots, 12)$ is added to $K_{jj}$ in the global Equation 3.7.
The derivative of potential energy $\Pi_{sc}$ for the force at $P_0$ on block $i$ is:

$$f_r = -\frac{\partial \Pi_{sc}(0)}{\partial d_{ri}} = -k_s T_i^T(x_1, y_1, z_1) \begin{pmatrix} x_1 - x_0 \\ y_1 - y_0 \\ z_1 - z_0 \end{pmatrix} + \frac{k_s Vol}{A} E_i^T(x_1, y_1, z_1) \rightarrow F_i \quad (3.76)$$

$f_r$ $(r = 1, \ldots, 12)$ is added to $F_i$ in the global Equation 3.7.
The derivative of $\Pi_{sc}$ for the force at $P_0$ on block $j$ is:

$$f_r = -\frac{\partial \Pi_{sc}(0)}{\partial d_{rj}} = k_s T_j^T(x_0, y_0, z_0) \begin{pmatrix} x_1 - x_0 \\ y_1 - y_0 \\ z_1 - z_0 \end{pmatrix} + \frac{k_s Vol}{A} G_j^T(x_0, y_0, z_0) \rightarrow F_j \quad (3.77)$$

$f_r$ $(r = 1, \ldots, 12)$ is added to $F_j$ in the global Equation 3.7.

### Frictional force at sliding state

Frictional force exists when a contact is at the sliding state. It is responsible for the majority of energy dissipation through the mechanical deformation of the block system.

For the sliding mode, in addition to the normal spring, a pair of frictional forces instead of a shear spring needs be added. Based on the Coulomb's friction law, the frictional force is:

$$F_f = k_n |d_n| \tan(\varphi) + c \tag{3.78}$$

where $k_n$ is normal spring stiffness, $d_n$ is normal penetration distance, $\varphi$ is the friction angle, and $c$ is cohesion. The direction of frictional force $L$ is determined as follows:

$$\left| \overrightarrow{P_0' P_1'} \right|^2 = (x_1 + u_1 - x_0 - u_0)i + (y_1 + v_1 - y_0 - v_0)j + (z_1 + w_1 - z_0 - w_0)k$$
$$= ai + bj + ck \tag{3.79}$$

Assuming $N$ is the unit normal vector of face $P_2 P_3 P_4 P_5 P_6$, the frictional force $F_f$ acts along the direction:

$$\delta_\tau = \delta - \delta_n = \overrightarrow{P_0' P_1'} - d_n N = (1 - N^2) \overrightarrow{P_0' P_1'} \tag{3.80}$$

$\delta_\tau$ is the direction of the frictional force. Substituting Equation 3.79 into Equation 3.80, we get:

$$\delta_\tau = \left( 1 - \begin{bmatrix} n_1 n_1 & n_1 n_2 & n_1 n_3 \\ n_2 n_1 & n_2 n_2 & n_2 n_3 \\ n_3 n_1 & n_3 n_2 & n_3 n_3 \end{bmatrix} \right) \begin{pmatrix} a \\ b \\ c \end{pmatrix} = r_1 i + r_2 j + r_3 k \tag{3.81}$$

Let $R = \sqrt{r_1^2 + r_2^2 + r_3^2}$, the potential energy of the frictional force $F_f$ at $P_1$ at block $i$ is:

$$\Pi_f = \frac{F_f}{R} (u_1 \quad v_1 \quad w_1) \begin{pmatrix} r_1 \\ r_2 \\ r_3 \end{pmatrix} = \frac{F_f}{R} d_i^T T_i^T (x_1, y_1, z_1) \begin{pmatrix} r_1 \\ r_2 \\ r_3 \end{pmatrix} = F_f d_i^T H \tag{3.82}$$

where:

$$H = \frac{1}{R} T_i^T (x_1, y_1, z_1) \begin{pmatrix} r_1 \\ r_2 \\ r_3 \end{pmatrix}$$

The derivative of $\Pi_f$ when $d = 0$ is:

$$f_r = -\frac{\partial \Pi_f(0)}{\partial d_{ri}} = -F_f H \rightarrow F_i \tag{3.83}$$

$f_r (r = 1, \ldots, 12)$ is added to $F_i$ in the global Equation 3.7.

The potential energy of the frictional force $F_f$ at $P_0$ on block $j$ is:

$$\Pi_f = -\frac{F_f}{R} (u_0 \quad v_0 \quad w_0) \begin{pmatrix} r_1 \\ r_2 \\ r_3 \end{pmatrix} = -\frac{F_f}{R} d_j^T T_j^T (x_0, y_0, z_0) \begin{pmatrix} r_1 \\ r_2 \\ r_3 \end{pmatrix} = -F_f d_j^T G \tag{3.84}$$

where

$$G = \frac{1}{R} T_j^T(x_0, y_0, z_0) \begin{pmatrix} r_1 \\ r_2 \\ r_3 \end{pmatrix}$$

The derivative of $\Pi_f$ when $d = 0$ is:

$$f_r = -\frac{\partial \Pi_f(0)}{\partial d_{rj}} = F_f G \to F_j \tag{3.85}$$

$f_r (r = 1, \ldots, 12)$ is added to $F_j$ in the global Equation 3.7.

## 3.4 SIMPLEX INTEGRATION FOR 3D DDA

One of the advantage of DDA is to use the closed-form simplex integration method over the blocks. This section will focus on the derivation of analytical formula of the 2D polygon and 3D polyhedron. It is a straightforward extension from the 2D version of DDA.

Assume the vertex list of the $i$-th plane loop is arranged anticlockwise as defined in Shi (1996), as shown in Fig. 3.9:

$$P_1^{[i]} P_2^{[i]} P_3^{[i]} \ldots P_{n(i)}^{[i]} \quad P_{n(i)+1}^{[i]} = P_1^{[i]}$$

and

$$P_j^{[i]} = (x_j^{[i]}, y_j^{[i]}, z_j^{[i]}) \quad j = 1, 2, \ldots, n(i)$$

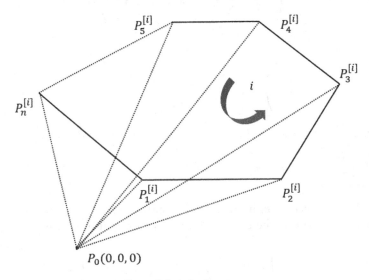

$$P_0(0, 0, 0)$$

*Figure 3.9  i-th plane loop.*

Then, total $s$ polygon loops are assumed to compose the current block and set $P_0 = (0, 0, 0)$. Computed by simplex integrations, integrals for $1, x, y, z, x^2, y^2, z^2, xy, yz, xz$ are represented by the coordinates of the boundary vertices only.

$$S^1 = \iiint_V 1 \, dx \, dy \, dz = \sum_{i=1}^{s} \sum_{k=1}^{n(i)} \int_{P_0 P_k^{[i]} P_{k+1}^{[i]}} 1 D(x, y, z)$$

$$= \frac{1}{6} \sum_{i=1}^{s} \sum_{k=1}^{n(i)} \begin{vmatrix} x_1^{[i]} & y_1^{[i]} & z_1^{[i]} \\ x_k^{[i]} & y_k^{[i]} & z_k^{[i]} \\ x_{k+1}^{[i]} & x_{k+1}^{[i]} & z_{k+1}^{[i]} \end{vmatrix} \tag{3.86}$$

$$S^x = \iiint_V x \, dx \, dy \, dz = \sum_{i=1}^{s} \sum_{k=1}^{n(i)} \int_{P_0 P_k^{[i]} P_{k+1}^{[i]}} x D(x, y, z)$$

$$= \frac{1}{24} \sum_{i=1}^{s} \sum_{k=1}^{n(i)} \begin{vmatrix} x_1^{[i]} & y_1^{[i]} & z_1^{[i]} \\ x_k^{[i]} & y_k^{[i]} & z_k^{[i]} \\ x_{k+1}^{[i]} & x_{k+1}^{[i]} & z_{k+1}^{[i]} \end{vmatrix} (x_1 + x_k + x_{k+1}) \tag{3.87}$$

$$S^y = \iiint_V y \, dx \, dy \, dz = \sum_{i=1}^{s} \sum_{k=1}^{n(i)} \int_{P_0 P_k^{[i]} P_{k+1}^{[i]}} y D(x, y, z)$$

$$= \frac{1}{24} \sum_{i=1}^{s} \sum_{k=1}^{n(i)} \begin{vmatrix} x_1^{[i]} & y_1^{[i]} & z_1^{[i]} \\ x_k^{[i]} & y_k^{[i]} & z_k^{[i]} \\ x_{k+1}^{[i]} & x_{k+1}^{[i]} & z_{k+1}^{[i]} \end{vmatrix} (y_1 + y_k + y_{k+1}) \tag{3.88}$$

$$S^z = \iiint_V z \, dx \, dy \, dz = \sum_{i=1}^{s} \sum_{k=1}^{n(i)} \int_{P_0 P_k^{[i]} P_{k+1}^{[i]}} z D(x, y, z)$$

$$= \frac{1}{24} \sum_{i=1}^{s} \sum_{k=1}^{n(i)} \begin{vmatrix} x_1^{[i]} & y_1^{[i]} & z_1^{[i]} \\ x_k^{[i]} & y_k^{[i]} & z_k^{[i]} \\ x_{k+1}^{[i]} & x_{k+1}^{[i]} & z_{k+1}^{[i]} \end{vmatrix} (z_1 + z_k + z_{k+1}) \tag{3.89}$$

$$S^{x^2} = \iiint_V x^2 dx dy dz = \sum_{i=1}^{s} \sum_{k=1}^{n(i)} \int_{P_0 P_k^{[i]} P_{k+1}^{[i]}} x^2 D(x,y,z)$$

$$= \frac{1}{60} \sum_{i=1}^{s} \sum_{k=1}^{n(i)} \begin{vmatrix} x_1^{[i]} & y_1^{[i]} & z_1^{[i]} \\ x_k^{[i]} & y_k^{[i]} & z_k^{[i]} \\ x_{k+1}^{[i]} & x_{k+1}^{[i]} & z_{k+1}^{[i]} \end{vmatrix}$$

$$\times (x_1 x_1 + x_1 x_k + x_1 x_{k+1} + x_k x_k + x_k x_{k+1} + x_{k+1} x_{k+1})$$

$$(3.90)$$

$$S^{y^2} = \iiint_V y^2 dx dy dz = \sum_{i=1}^{s} \sum_{k=1}^{n(i)} \int_{P_0 P_k^{[i]} P_{k+1}^{[i]}} y^2 D(x,y,z)$$

$$= \frac{1}{60} \sum_{i=1}^{s} \sum_{k=1}^{n(i)} \begin{vmatrix} x_1^{[i]} & y_1^{[i]} & z_1^{[i]} \\ x_k^{[i]} & y_k^{[i]} & z_k^{[i]} \\ x_{k+1}^{[i]} & x_{k+1}^{[i]} & z_{k+1}^{[i]} \end{vmatrix}$$

$$\times (y_1 y_1 + y_1 y_k + y_1 y_{k+1} + y_k y_k + y_k y_{k+1} + y_{k+1} y_{k+1})$$

$$(3.91)$$

$$S^{z^2} = \iiint_V z^2 dx dy dz = \sum_{i=1}^{s} \sum_{k=1}^{n(i)} \int_{P_0 P_k^{[i]} P_{k+1}^{[i]}} z^2 D(x,y,z)$$

$$= \frac{1}{60} \sum_{i=1}^{s} \sum_{k=1}^{n(i)} \begin{vmatrix} x_1^{[i]} & y_1^{[i]} & z_1^{[i]} \\ x_k^{[i]} & y_k^{[i]} & z_k^{[i]} \\ x_{k+1}^{[i]} & x_{k+1}^{[i]} & z_{k+1}^{[i]} \end{vmatrix}$$

$$\times (z_1 z_1 + z_1 z_k + z_1 z_{k+1} + z_k z_k + z_k z_{k+1} + z_{k+1} z_{k+1})$$

$$(3.92)$$

$$S^{xy} = \iiint_V xy dx dy dz = \sum_{i=1}^{s} \sum_{k=1}^{n(i)} \int_{P_0 P_k^{[i]} P_{k+1}^{[i]}} xy D(x,y,z)$$

$$= \frac{1}{120} \sum_{i=1}^{s} \sum_{k=1}^{n(i)} \begin{vmatrix} x_1^{[i]} & y_1^{[i]} & z_1^{[i]} \\ x_k^{[i]} & y_k^{[i]} & z_k^{[i]} \\ x_{k+1}^{[i]} & x_{k+1}^{[i]} & z_{k+1}^{[i]} \end{vmatrix}$$

$$\times (2x_1 y_1 + x_1 y_k + x_1 y_{k+1} + x_k y_1 + 2x_k y_k + x_k y_{k+1} + x_{k+1} y_1 + x_{k+1} y_k + 2x_{k+1} y_{k+1})$$

$$(3.93)$$

$$S^{xz} = \iiint_V xzdxdydz = \sum_{i=1}^{s}\sum_{k=1}^{n(i)} \int_{P_0 P_k^{[i]} P_{k+1}^{[i]}} xzD(x,y,z)$$

$$= \frac{1}{120} \sum_{i=1}^{s}\sum_{k=1}^{n(i)} \begin{vmatrix} x_1^{[i]} & y_1^{[i]} & z_1^{[i]} \\ x_k^{[i]} & y_k^{[i]} & z_k^{[i]} \\ x_{k+1}^{[i]} & x_{k+1}^{[i]} & z_{k+1}^{[i]} \end{vmatrix}$$

$$\times (2x_1z_1 + x_1z_k + x_1z_{k+1} + x_kz_1 + 2x_kz_k + x_kz_{k+1} + x_{k+1}z_1 + x_{k+1}z_k + 2x_{k+1}z_{k+1})$$

$$(3.94)$$

$$S^{yz} = \iiint_V yzdxdydz = \sum_{i=1}^{s}\sum_{k=1}^{n(i)} \int_{P_0 P_k^{[i]} P_{k+1}^{[i]}} yzD(x,y,z)$$

$$= \frac{1}{120} \sum_{i=1}^{s}\sum_{k=1}^{n(i)} \begin{vmatrix} x_1^{[i]} & y_1^{[i]} & z_1^{[i]} \\ x_k^{[i]} & y_k^{[i]} & z_k^{[i]} \\ x_{k+1}^{[i]} & x_{k+1}^{[i]} & z_{k+1}^{[i]} \end{vmatrix}$$

$$\times (2y_1z_1 + y_1z_k + y_1z_{k+1} + y_kz_1 + 2y_kz_k + y_kz_{k+1} + y_{k+1}z_1 + y_{k+1}z_k + 2y_{k+1}z_{k+1})$$

$$(3.95)$$

## 3.5  SUMMARY

Three-dimensional DDA is still under development mainly due to the complex nature of the contact detection algorithm, contact status conversion, and convergence efficiency of the open-close iteration process in three dimensions. The basic theory of two-dimensional DDA was introduced in Chapter 2, and is extended here to three dimensions. It should therefore be easier to follow the concepts presented in this chapter after studying Chapter 2. In addition to the complex contact algorithm in 3D DDA, challenges also exist in conducting failure and multi-physics analysis with 3D DDA; some recent advances are reviewed in Chapter 1. With the new contact theory recently proposed by Shi (see Chapter 8), we expect that further development of 3D DDA will be accelerated and numerical simulations of engineering problems involving highly fractured rock masses may become possible in the future.

# Chapter 4

# Geological input parameters for realistic DDA modeling

## 4.1 INTRODUCTION

We have seen in the previous chapters that DDA is based on rigorous mathematical principles, but it is important to realize that its realistic application requires good understanding of the geology of the rock mass that is being modeled numerically. Engineers that are working on the mathematical and/or computational development of DDA are not always familiar with the basic geological principles that control rock mass structure. However, such an understanding is essential for realistic representation of the rock mass structure in the field by a DDA block system that will produce meaningful results after forward numerical modeling. The most important thing to realize at the block construction stage is what should be included, and what could be safely omitted, when attempting to represent a complex geological structure with sets of lines that intersect one another to form a discrete element system. This would be the main purpose of this chapter. We will begin by addressing the issue of scale and the geometrical parameters that form the DDA block system (number of joint sets, joint orientation, spacing, and length) and will continue to discuss realistic ranges for input mechanical parameters of both intact elements (Young' modulus and Poisson's ratio) as well as discontinuities (cohesion, tensile strength, and friction angle). Optimizing the numerical control parameters, with particular emphasis on time step size and contact spring stiffness, will be discussed in Chapter 5 in the context of benchmarking of the DDA method.

## 4.2 REALISTIC REPRESENTATION OF ROCK MASS STRUCTURE

As briefly discussed in Chapter 1, rock masses are never continuous, at any scale of reference. Choices must be made, therefore, as to which discontinuities to represent in the modeled domain and which to omit, because with a finite computational capability, regardless of the computational platform, a complete numerical simulation incorporating all discontinuities at all scales is simply not feasible. It is also not necessary because the solution to most rock engineering problems in discontinuous rock masses can be obtained by correctly and selectively representing in the modeled system the relevant discontinuities that control the rock mass deformation.

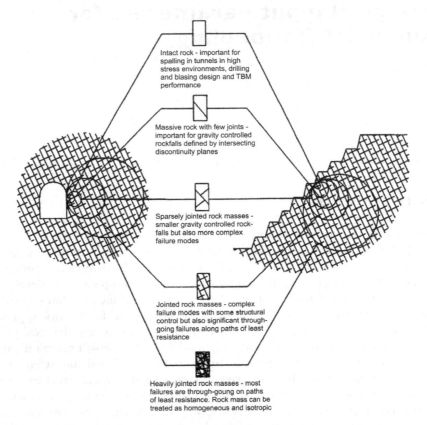

Intact rock - important for spalling in tunnels in high stress environments, drilling and blasing design and TBM performance

Massive rock with few joints - important for gravity controlled rockfalls defined by intersecting discontinuity planes

Sparsely jointed rock masses - smaller gravity controlled rock-falls but also more complex failure modes

Jointed rock masses - complex failure modes with some structural control but also significant through-going failures along paths of least resistance

Heavily jointed rock masses - most failures are through-goung on paths of least resistance. Rock mass can be treated as homogeneous and isotropic

*Figure 4.1*  Scales of interest in underground (left) and slope (right) engineering (Hoek, 2007).

The issue of relative scale is clearly of the essence here. Consider the famous chart drawn by Professor Evert Hoek (Hoek, 2007) that delineates the scales of reference in underground or rock slope engineering problems (Figure 4.1).

If the factor of safety against failure in compression in the sidewalls of a deep underground excavation is of interest, then the microstructure of the rock is important, because as has been shown by many (e.g. Fredrich *et al.*, 1990; Hatzor and Palchik, 1997; Sammis and Ashby, 1986), microstructure, and particularly the size of initial flaws such as grain boundaries and pores, controls brittle strength of crystalline rocks (see Figure 4.2). Moreover, if a fracture evolution study is intended, for example in order to obtain the extent of the excavation damage zone (EDZ) as controlled by micromechanical processes (e.g. Cai and Kaiser, 2005), then the micro scale cannot be ignored.

Indeed, when the level of *in situ* stress approaches the strength of intact rock around the excavation it would be prudent to employ continuum-based methods so that the influence of microstructure on fracture evolution can be modeled accurately. In most instances, however, the influence of microstructure on intact rock strength will be studied experimentally in the lab to obtain the ultimate strength of the rock

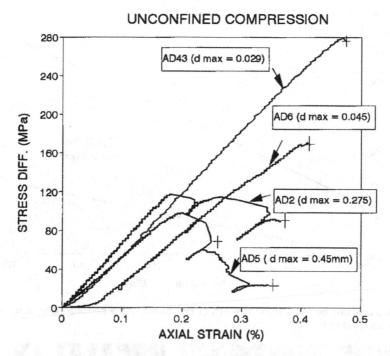

Figure 4.2 Influence of maximum grain size on unconfined compressive strength of dolomites.

and the results will be used in the numerical model as known magnitudes, to enable computation of factor of safety around the excavation.

Fracture mechanics approaches are appropriate when the rock mass is continuous at the relevant field scale, namely no discontinuities transect the rock mass to a distance of several opening diameters in the case of an underground opening or to a distance into the excavation that roughly equals the slope height, as delineated by the outermost circles in Figure 4.1. If this indeed is the case, the stability of the project may be studied using continuum approaches allowing complicated fracture propagation and coalescence processes in which the microscale is important. If however the relevant field scale is transected by discontinuities, it would be safe to assume that most of the deformation, either shear sliding, opening, or rotation, will take place along preexisting joints because of the much lower shear, cohesive, or tensile strength they possess in comparison to the intact rock material (see Figure 4.3). In such cases, it would be more realistic to adopt a discontinuous approach, and the microscale flaws may be safely ignored.

It is well established that the zone of influence of underground excavations extends to several opening diameters, as also portrayed in the left panel of Figure 4.1. This is the zone where stress relaxation takes place followed by loosening and where the excavation damage zone develops (e.g. Li *et al.*, 2011). With regard to rock loads that might develop on passive support elements such as steel arches or concrete segments, Terzaghi (1946) has already predicted, based on many tunneling case studies in the Alps, that the height of loosening would be roughly one half the excavation span in

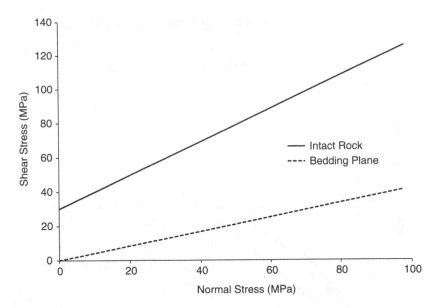

*Figure 4.3* Comparison between shear strength of intact rock and a bedding plane in the dolomitic rock mass of Masada.

*Figure 4.4* Left – Rockmass consisting of columnar jointed basalt (photo courtesy of Dr. Li Shao-Jun, Institute of Soil and Rock Mechanics, Chinese Academy of Science); Right – Detail of a DDA model representing a tunnel in columnar jointed basalt rock mass. The opening span in the model is 19.7 m, the mean spacing between the lava basal planes is 5 m, and between the columnar joints is 0.3 m.

blocky rock masses. This prediction by the famous geotechnical engineer proves to be valid and is confirmed by results of numerical analyses (see Hatzor and Benary, 1998). To demonstrate this, consider a blocky rock mass model created with DDA for a columnar jointed basalt rock mass based on the geometrical properties of the rock mass encountered during construction of the Baihetan hydropower station in south west China (Jiang *et al.*, 2014) as shown in Figure 4.4. The extent of the loosening

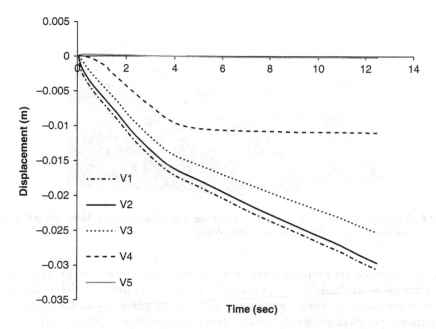

*Figure 4.5* Vertical displacements in the roof of the tunnel shown in Figure 4.4 as computed with DDA. Point V1 located in immediate roof, V2 @ 2.5 m, V3 @ 9 m, V4 @ 25 m, V5 @ 50 m above immediate roof.

zone can be inferred from the vertical displacement output obtained from five measurement points that are positioned in the roof (Figure 4.5). All measurement points up to a vertical distance of 10 meters from the immediate roof output ongoing vertical displacement, namely the loosening zone in this rock mass, according to DDA predictions, extends to at least half the excavation span which here is 20 meters, as indeed predicted by Terzaghi for blocky rock masses, on the basis of many field observations.

To be able to generate such a realistic block system that will yield meaningful numerical results attention must be paid to the correct characterization of the rock mass structure, with particular emphasis on the number of joint sets, the mean orientation, spacing, and length of each joint set in the rock mass.

## 4.2.1    Number of joint sets

The best way to discern if the rock mass consists of several principal joint sets of common orientation is to perform a joint survey, using either field exposures or retrieved cores from boreholes. The term "joints" is usually used in this text as a generic term for all rock discontinuities, regardless of type and origin. Sometimes however a distinction will be made between joints in the sense of tensile fractures, and shears, bedding planes, or foliations.

Joints are represented by either the dip vector or the normal vector to the plane. An illustration of joint plane with the dip and strike vectors is shown in Figure 4.6. The dip vector is the line on the plane with the steepest inclination whereas the strike vector is the line on the plane with zero inclination. The dip and strike vectors both

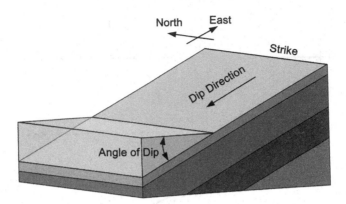

*Figure 4.6* Schematic illustration of a plane with the dip and strike vectors. Here the strike trends North–South and the dip points to the West.

lie on the plane and are perpendicular to one another. The normal to the plane (sometimes referred to as "pole") is perpendicular to both the dip and the strike vectors and can point either upwards or downwards from the plane. By convention, the dip vector always points downwards (geologists always tend to look down to the ground). When discussing poles however it is necessary to specify if upward or downward poles are considered. To provide complete information on the dip that fully characterizes the plane both the amount of inclination of the dip vector with respect to a horizontal plane and the direction of the dip vector (the azimuth as measured on a horizontal plane) must be specified.

The dip and dip direction of each detected joint should be recorded and the poles to all joints should be plotted using stereographic projection. This way the poles of joints belonging to the same set will plot in close proximity to one another, thus making it easy for the geological engineer or the engineering geologist to quickly identify the principal joint sets that comprise the rock mass structure, the degree of dispersion in each set, and the mean orientation of each set. Either 'lower' or 'upper' hemisphere projections can be used for stereographic projection of poles, although the upper hemisphere (UH) projection has the advantage that the poles orientation as plotted on the reference circle has the same attitude as the dip direction. The principles of stereographic projection as used in rock engineering are reviewed by several authors (Goodman, 1976, 1989; Priest, 1993) and readers that are not familiar with this useful technique are strongly advised to consult these excellent sources.

Examples of well-organized rock mass structures as inferred from joint surveys performed in the field on rock exposures are shown in Figure 4.7 and Figure 4.8.

In Figure 4.7 only the joints (tensile fractures) are plotted, excluding bedding planes. Quick inspection of Figure 4.7 clearly reveals that the rock mass is transected by two principal joint sets labeled in the Figure as $J_2$ and $J_3$. The two joint sets are sub-vertical with orthogonal strike directions: $J_2$ strikes ESE–WNW and $J_3$ strikes NNE–SSW. In Figure 4.8 all discontinuities (including joints and bedding planes) measured in the exposed rock slopes of King Herod's palace at the northern face of Masada mountain are plotted using upper hemisphere projection of poles. The principal joint

*Figure 4.7* Left – Joint survey in Ramleh open pit mine, Right – Principal joint sets in the rock mass as inferred from lower hemisphere projection of poles (joint survey performed by Dagan Bakun-Mazor).

*Figure 4.8* The rock mass structure at the north face of Masada mountain (left), as represented by upper hemisphere projection of poles (right). Joint survey performed by Michael Tsesarsky.

sets are again sub vertical striking NNE–SSW (J$_2$) and ESE–WNW (J$_3$). The bedding planes dip shallowly to the north.

## 4.2.2   Types of joint sets

Rock masses exhibit several characteristic sets of discontinuities, the intersections of which form what is commonly referred to as the "rock mass structure". The reason for

*Figure 4.9* Horizontal bedding planes at the old city walls foundations, Jerusalem.

the co-existence of several sets of discontinuities in any given rock mass is that over its geological history the rock has been subjected to different *in situ* stress regimes that have left their imprint on the rock mass in the form of well-defined sets of discontinuities. Rock discontinuities typically cluster into sets of common orientation that can be grouped into several common genetic categories including: bedding planes, joints (or tensile fractures), shears (or faults), folds, and foliation planes. Below we will briefly describe each of these sets.

### Bedding planes

Sedimentary rocks that form by slow deposition from water bodies will typically exhibit bedding planes of "infinite" extent with respect to the dimensions of the engineering problem at hand (see Figure 4.9). Bedding planes may be filled with infilling or alteration material (see Figure 4.10) and often exhibit irregular surface geometry. The roughness of bedding planes, and planes of discontinuities in general, adds to their shear resistance whereas the infilling material typically decreases their available shear strength. Because of their large persistence (lateral extent) bedding planes almost always participate in the formation of rock blocks in a sedimentary rock mass. Bedding plane orientation is derived from the regional structure and typically dictates the optimum orientation of the designed engineered structure. Initially bedding planes will form parallel to the direction of sedimentation and will therefore be horizontal. Crustal deformation processes however, such as folding and faulting, may lead over time to inclined bedding planes. An extreme example is the rock salt diapir on the western margins of the Dead Sea near Sedom (see Figure 4.11), where the initially horizontal bedding planes are now vertical to sub-vertical as a result of salt intrusion into the surrounding country rock.

*Figure 4.10*  Filled bedding plane at Arad open pit mine.

*Figure 4.11*  Vertically dipping bedding planes in the Sedom rock salt wall, Dead Sea.

## Joints

The term joints, when not relating to discontinuities in general, typically refers to tensile fractures formed in the rock mass parallel to the direction of the maximum

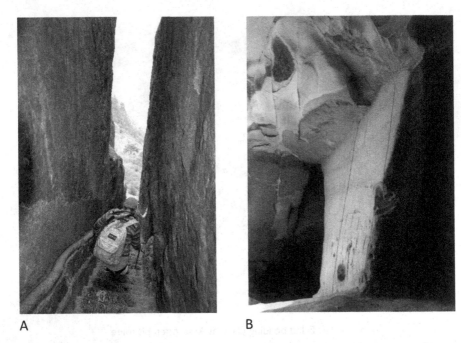

A                                                    B

*Figure 4.12* A) two parallel sub-vertical joints in Yellow Mountain, China; B) A horizontal excavation
induced tensile fracture with characteristic plumose structures. The tensile fracture was
induced by block sliding along a steeply inclined preexisting joint at the bell-shaped caverns
of Bet Guvrin.

principal compressive stress, or in direction perpendicular to the least compressive
principal stress. Tensile fractures are typically smooth (Figure 4.12A), but sometimes
exhibit distinct fracture markings called "plumose structures" (Figure 4.12B), the study
of which is the scope of a field of science known as "Tectono-fractography" (Bahat,
1991). Since tensile fractures typically do not exhibit shear striations it is assumed that
no shear displacement has taken place along these discontinuities in the past. Tensile
fractures can have a significant persistence in the rock mass, from several meters to
tens of meters.

While joint surfaces are often clean, in the presence of groundwater their aperture
may be filled over geologic times by recrystallized minerals, a process that will bind
the joint walls together resulting in a significant joint tensile strength, which otherwise
can be safely ignored for engineering purposes.

### Shears and faults

Shears and faults are, as their name implies, discontinuities that represent failure of
brittle rock by shear along well-defined planes. Shears and faults typically exhibit
slip striations known as "slickensides". Large shear displacements across this type
of discontinuities may lead to the formation and accumulation of thick gouge along
the surface, a process that typically results in marked decrease of shear resistance.

With increasing gouge thickness the shear resistance of shears and faults may be reduced to a point where the available shear strength equals the shear strength of the infilling material itself, when the infilling thickness approaches or supersedes the asperity height (Goodman, 1976). The slip striations often observed on shear planes are assumed to delineate the direction of motion which is typically parallel to the direction of the maximum shear stress active under the governing paleo tectonic stress field. Indeed, slip striations along well defined shear planes, have been used to recover the paleo *in-situ* tectonic stress tensor, by inversion (Reches, 1987). The relationship between the statistical characteristics of the roughness profile and slip history is discussed by Sagy and Brodsky (2009). Examples of slip markings on the planes of a strike slip and normal faults are shown in Figure 4.13.

### Folds

Folds are regional structural features that typically can be traced on geological maps (see Figure 4.14). Folds often represent ductile deformation of originally deep buried rocks under high temperature and pressure conditions. At the project scale all discontinuities belonging to a particular fold may have the same orientation, however in the regional scale the folds may exhibit changing orientation according to the particular structural configuration (see Figure 4.15).

### Foliations and exfoliations

Foliation planes are commonly observed in metamorphic rocks that have undergone complex and extensive history of tectonic activity. They typically exhibit weak shear resistance due to the abundance of alterations and infilling materials within their aperture. Two examples of foliation planes in mica schist (a metamorphic sediment) as exposed in Eilat area of Southern Israel are shown in Figure 4.16.

Exfoliation planes are more common in magmatic rocks and are believed to form in response to uplift and cooling of the magma. As such, they typically exhibit non-planar surfaces that are oriented parallel to the boundaries of the uplifted structure. Exfoliation planes are often clean and relatively smooth and exhibit large persistence. Examples of exfoliation planes in gneiss rocks are shown in Figure 4.17 and Figure 4.18.

## 4.2.3   Joint set orientation

Joint orientation can be expressed in Cartesian space, therefore solid geometry or stereographic projection techniques can be used to describe it quantitatively. Since any plane in Cartesian space is completely defined geometrically by its normal, we will use here the joint normal rather than the actual plane as representative of the joint plane orientation.

Each joint normal can be considered a unit vector; the orientation of the resultant of all the unit vectors in a given joint set represents the preferred, or mean, joint set orientation. The summation of the unit vectors can be accomplished by accumulating the direction cosines. Following the notation used by Goodman (1989) let $x$ be directed horizontally North, $y$ horizontally West, and $z$ Up (Figure 4.19).

A

B

C

*Figure 4.13* Faults and shears exhibiting distinct slip striations: A) Rough striated strike slip fault surface in carbonate rocks, Southern Alps, Italy (Photo by Nicholas van der Elst); B) Smooth normal fault surface in Andesite rocks, Flowers Pit Fault, Oregon; C) Striated normal fault surface in Silicate rocks, Dixie valley, Nevada. (photos and captions courtesy of Dr. Amir Sagy, The Geological Survey of Israel).

*Figure 4.14* The External Zone of the French Alps. View from Huez to the southwest across the Romanche glacially-carved valley and Le Bourg d'Osians. Folded and faulted Jurassic and Triassic sediments overlying Hercynian basement. Deformation in the area is dominated by polyphase shortening during the Alpine orogeny (photo and caption courtesy of Dr. Itai Haviv, Ben-Gurion University of the Negev).

*Figure 4.15* The Front Ranges of the southern Canadian Rocky Mountains near Banff. View from Sulphur Mountain towards the southeast. Mount Rundle stretches from left to right at the horizon. Resistant Paleozoic carbonates are exposed along the ridge crests. Sulphur Mountain Thrust stretches roughly parallel to the valley. Deformation in the area is characterized by southwest-dipping thrust sheets and is part of the Laramide orogeny (photo and caption courtesy of Dr. Itai Haviv, Ben-Gurion University of the Negev).

*Figure 4.16*  Precambrian mica schist composed of the minerals biotite, quartz, plagioclase and garnet as exposed in Eilat area, southern Israel. Top: shallow dipping foliations parallel to the layering of original bedding planes. Bottom: sub-vertical foliations parallel to layering of original bedding planes (photos and captions courtesy of Prof. Dov Avigad, Hebrew University of Jerusalem).

*Figure 4.17* Example of exfoliation planes in Gneiss, Tenaia Lake, Yosemite national park, California.

*Figure 4.18* Example of exfoliation planes in Gneiss, Sigiriya (Lion Rock) monument, Sri Lanka.

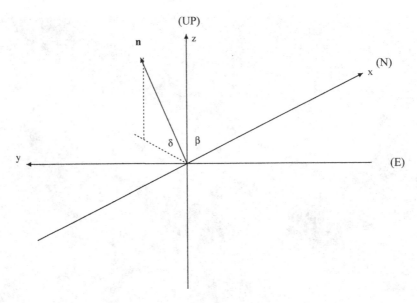

*Figure 4.19* Sign convention for upward trending joint normal.

If the upward joint normal rises at angle $\delta$ above the horizontal in direction $\beta$ measured counterclockwise from the north, then the direction cosines for the joint normal are:

$$l = \cos\delta\cos\beta$$
$$m = \cos\delta\sin\beta$$
$$n = \sin\delta \tag{4.1}$$

For a cluster of normals in a given joint set the preferred, or mean, joint set orientation will trend parallel to the resultant vector of the cluster $R$, with direction cosines $(l_R, m_R, n_R)$:

$$l_R = \frac{\sum l_i}{\left|\overline{R}\right|}$$

$$m_R = \frac{\sum m_i}{\left|\overline{R}\right|}$$

$$n_R = \frac{\sum n_i}{\left|\overline{R}\right|} \tag{4.2}$$

and the magnitude of the resultant is given by:

$$\left|\overline{R}\right| = \sqrt{\left(\sum l_i\right)^2 + \left(\sum m_i\right)^2 + \left(\sum n_i\right)^2} \tag{4.3}$$

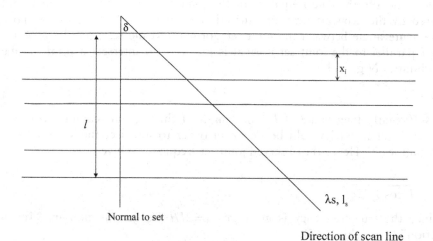

*Figure 4.20* A joint set intersected by a sampling line of general orientation (after Priest, 1985).

The angle of rise $\delta$ and the direction of the rise $\beta$ of the resultant normal are obtained as in Equation 4.1 but with some rules for the correct sign:

$$\delta_R = \sin^{-1}(n_R); \quad 0 \leq \delta_R \leq 90°$$

$$\beta_R = +\cos^{-1}\left(\frac{l_R}{\cos \delta_R}\right) \quad \text{if } m_R \geq 0$$

$$\beta_R = -\cos^{-1}\left(\frac{l_R}{\cos \delta_R}\right) \quad \text{if } m_R < 0 \tag{4.4}$$

With these simple equations, the engineer or engineering geologist can readily determine the mean joint set orientation in each set based on data from a field survey.

## 4.2.4   Joint set spacing and bias correction

The discussion below and the notation used are based on Priest (1985, 1993) in his useful discussion of joint statistics. Discontinuities, referred to here as joints for simplicity, are typically sampled in the field using scan line or borehole surveys. The orientation of the sampling device is often dictated by the conditions in the field, and is independent of the joint set attitudes. Therefore in some cases a scanning direction used in the joint survey may be completely blind with respect to a specific joint plane attitude. Consider Figure 4.20 where a linear scanline direction ($\lambda s$) in a set of parallel joints with variable spacing is shown. The scan line will be completely "blind" to the spacing of the set if it was oriented parallel to the strike of the joints, or rather, if the angle between the scan line and the normal to the joints was $\delta = 90°$. However the "true" spacing between the joints would be retrieved with no bias if the scan line was oriented parallel to the joint normal ($\delta = 0°$).

Let us denote the true frequency of the joint set $\lambda$ and the apparent frequency measured by the arbitrary scan line direction $\lambda_s$, which is less than or equal to $\lambda$. Let the acute angle between the joint set normal and the scan line be $\delta$. A line of length $l$ parallel to the joint set normal is expected to intersect a total number of discontinuities $N$ given by:

$$N = \lambda l \qquad (4.5)$$

for a sufficiently long value of $l$. The length of the general scanline with angle $\delta$ to the joint set normal would be $l/\cos\delta$ in order to intersect the same number of discontinuities $N$. Hence the observed joint set frequency is given by:

$$\lambda_s = \frac{N}{l/\cos\delta} \qquad (4.6)$$

and since the true frequency is given by $\lambda = N/l$ we get the famous "Terzaghi's correction" (Terzaghi, 1965):

$$\lambda_s = \lambda \cos\delta \qquad (4.7)$$

Ruth Terzaghi proposed this bias correction originally in 1965 in the context of drill hole sampling. The meaning of this correction is that the number of discontinuities from a given set, intersected by a sampling line that makes an acute angle $\delta$ to the set normal, reduces with increasing values of $\delta$ and approaches zero when $\delta$ approaches 90°. We now understand why orientation data from linear sampling lines may be severely biased.

Following the discussion provided by Priest (1985) we realize that in the general case there are $m$ joint sets, each of which is assumed to contain parallel joint planes, with joint set frequency $\lambda_i$ and acute angle between $n_i$ and the scanline equals $\delta_i$ where $i = 1, 2, 3, \ldots, m$. Thus using Equation 4.7 above, the frequency of the $i$th set measured along the sampling line is:

$$\lambda_{s,i} = \lambda_i \cos\delta_i \qquad (4.8)$$

and the total frequency along the sampling line $\lambda_s$ is given by the sum of the frequency components as follows:

$$\lambda_s = \sum_{i=1}^{m} \lambda_{s,i} \qquad (4.9)$$

The total sample size $N_s$, obtained from a sampling length $l_s$, is given by:

$$N_s = \lambda_s l_s \qquad (4.10)$$

The number $N_{s,i}$, of discontinuities from the $i$th set in the sample is given by:

$$N_{s,i} = \lambda_{s,i} l_s \qquad (4.11)$$

If we examine Equation 4.11 we see that the number of discontinuities in the $i$th set in the sample depends in part upon the value of $\lambda_i$ and the angle $\delta_i$ (from Equation 4.8).

While it is reasonable that the sample size would depend upon the true frequency of the set $\lambda_i$, it is unreasonable that it should depend upon the arbitrary angle $\delta_i$. Terzaghi (1965) suggested that this dependence could lead to errors in interpreting the results of discontinuity surveys. In the theoretical case where all planes in a joint set are perfectly parallel this dependence can be removed by dividing $N_{si}$ by $\cos \delta_i$ to give a weighted sample size $N_i$:

$$N_i = \frac{N_{s,i}}{\cos \delta_i} = \lambda_i l_s \tag{4.12}$$

Since each joint set consists of individual members which are not perfectly parallel we can apply this weighting factor to each individual joint by using individual weighting factor $1/\cos \delta_i$ for each mapped joint, where $\delta_i$ is the acute angle between the individual joint normal and the arbitrary scan line direction. We can thus define a "weighting factor" $\omega$ for each member in a set as:

$$\omega = \frac{1}{\cos \delta} \tag{4.13}$$

As we know, the angle $\theta$ between two lines is given by:

$$\cos \theta = \cos(\alpha_1 - \alpha_2) \cos \beta_1 \cos \beta_2 + \sin \beta_1 \sin \beta_2 \tag{4.14}$$

where $\alpha$ and $\beta$ are the dip and dip direction of each line, so we can write the weighting factor $\omega$ as:

$$\omega = \frac{1}{|\cos(\alpha_n - \alpha_s) \cos \beta_n \cos \beta_s + \sin \beta_n \sin \beta_s|} \tag{4.15}$$

where $\alpha_s$ and $\beta_s$ are the trend and plunge of the scanline and $\alpha_n$ and $\beta_n$ are the trend and plunge of the normal to the discontinuity plane. Alternatively, the angle $\delta$ can be measured directly along a great circle common to the joint normal and the scanline vector using the stereographic projection. A simple procedure to contour the joint densities on the stereographic projection is proposed by Priest (1985).

## 4.2.5  Joint set dispersion

The scatter of normals about the mean orientation may be estimated by comparing the length of the resultant $R$ with the number of individual joints in the set $N$. If the joints were all parallel the resultant would equal $N$, whereas if the joints were widely varied in orientation the resultant would be considerably less than $N$. This dispersion of the normal about the mean joint set orientation can be expressed mathematically by the Fisher constant of the distribution of poles on a sphere (Fisher, 1953):

$$K_F = \frac{N}{N - |\overline{R}|} \tag{4.16}$$

$K_F$ becomes very large as the dispersion of joint orientations becomes small. Fisher (1953) showed mathematically that for a normal hemispherical distribution, the density distribution of which has a bell shaped geometry, the probability that a normal will make an angle $\psi$ or less with the mean orientation is given by:

$$\cos\varphi = 1 + \frac{1}{K_F}\ln(1-P) \tag{4.17}$$

thus one can express the spread of values about the mean, the dispersion, corresponding to any degree of certainty. Fisher has also shown that the standard deviation of the hemispherical normal distribution is given by:

$$\overline{\varphi} = \frac{1}{\sqrt{K_F}} \tag{4.18}$$

The length of the resultant $R$ for the un-weighted case is given by Equation 4.3, and the un-weighted number of discontinuities in a set is $N$. The Fisher constant can also be found for a weighted population by application of the weighting factor $\omega$. As shown by Priest (1985), The weighted number of the population is given by:

$$N_w = \sum_{j=1}^{N} \omega_j \tag{4.19}$$

where each of the weighting factors is greater than or equal to 1.0. Consequently $N_w$ will usually be greater than $N$, with typical values of $N_w/N$ between 2 and 5. This artificial increase of sample size is of little concern for contouring as the concentrations are expressed in percent. However, the value of $N$ is of critical importance when estimating the precision of the data. Priest (1985) suggested that this obstacle can be overcome by normalizing (again ...) each of the weighting factors $\omega_j$ so the total normalized weighted sample size is equal to $N$:

$$\omega_j' = \omega_j \frac{N}{N_w} \tag{4.20}$$

so that

$$\sum_{j=1}^{N} \omega_j' = N \tag{4.21}$$

The ratio $N/N_\omega$ is a constant and does not change therefore the relative weighting values. The weighted resultant is given by:

$$R_w = \sqrt{R_x^2 + R_y^2 + R_z^2} \tag{4.22}$$

where

$$R_x = \sum_{j=1}^{N} n_{j,x}$$

$$R_y = \sum_{j=1}^{N} n_{j,y} \tag{4.23}$$

$$R_z = \sum_{j=1}^{N} n_{j,z}$$

and

$$n_{j,x} = \omega' \cos \delta \cos \beta$$
$$n_{j,y} = \omega'_j \cos \delta \sin \beta \qquad (4.24)$$
$$n_{j,z} = \omega'_j \sin \delta$$

for upper hemisphere projection. Priest (1985) shows the same method but for lower hemisphere projection with a different sign convention for the direction cosines.

The original solution by Fisher did not consider the use of weighting factors to correct for sampling bias and hence implicitly assumed that all weighting values $\omega_j$ were equal to unity. In this case, therefore, each normal vector is of unit magnitude and the total weighted sample size $N_w$ is equal to $N$ and normalization is not required for the population size. However, normalization for the resultant size must be performed so that it is never greater than the population size. The weighted Fisher constant is thus given by (Priest, 1985):

$$K_{F,w} = \frac{N}{N - R_w} \qquad (4.25)$$

It should be noted that the error introduced by normalizing the weighting factors (Equation 4.20) is much less than the error which may be introduced by ignoring the bias correction all together. Still it would be a good engineering practice to first plot un-weighted poles, examine the results, and proceed with the bias correction when necessary.

## 4.2.6  Joint length

The joint length distribution has been measured by many authors and results were summarized by a series of classic papers by Hudson and Priest (Hudson and Priest, 1979, 1983; Priest and Hudson, 1976, 1981). Hudson and Priest in their papers clearly demonstrate that both joint spacing as well as joint trace length tend to follow a negative exponential distribution in many types of geological rock masses. This observation has been confirmed by many later studies. One typical example confirming the characteristic negative exponential distribution for joint set spacing and length is shown in Figure 4.21 for the same rock mass as shown in Figure 4.8.

The most common method to obtain joint length and spacing is the liner trace line method, as discussed by Hudson and Priest. Recently new methods have been proposed to obtain the same information using circular windows on the rock mass as proposed by Mauldon and coworkers (Mauldon, 1998; Mauldon et al., 2001; Mauldon and Mauldon, 1997; Rohrbaugh et al., 2002) and independently by Einstein and coworkers (Zhang and Einstein, 1998; Zhang et al., 2002).

The engineer or engineering geologist can use any available method to obtain the necessary information from the field scans or boreholes in order to be able to represent each joint set with the characteristic spacing and length. Good and representative parameters are essential for creating a meaningful and realistic DDA block system.

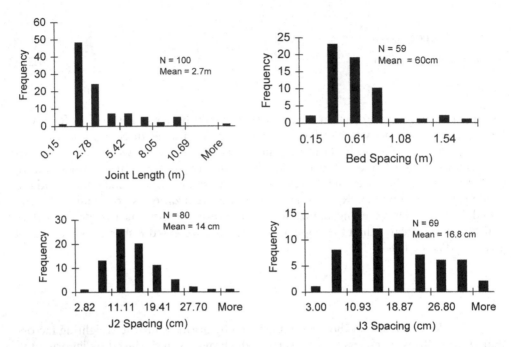

*Figure 4.21* An example of joint set spacing and length distribution from Masada (joint survey performed by Michael Tsesarsky).

### 4.2.7 Rock bridges and realistic mesh generation

Joints in the field are never indefinitely persistent and intact rock bridges may interrupt joint continuity. These rock bridges may strongly affect the strength of the rock mass and the stability of rock slopes as discussed by many authors (e.g. Einstein *et al.*, 1983; Kemeny, 2005) and various methods to incorporate them in numerical analyses of rock mass deformation have been proposed (e.g. Stead *et al.*, 2006). In DDA the user can input mean values for joint spacing, length, and bridge, and can let these vary between some upper and lower statistical bounds by means of the "degree of randomness" parameter which is coded in the line generation program. This capability enables modeling complicated geometries more realistically. By controlling the simulated joint spacing, length and bridge, it is even possible to generate a DDA block system that will simulate a structure controlled by mechanical layering (Bai and Gross, 1999; Gross, 1993; Ruf *et al.*, 1998), as typically observed in sedimentary rocks (e.g. Narr and Suppe, 1991). A DDA mesh simulating a rock mass structure governed by mechanical layering, generated by controlling the input joint length, spacing, and bridge, is shown in Figure 4.22.

Introducing rock bridges into the DDA block system does require some understanding of the geology and structure of the rock mass at the site, but it is a worthwhile effort as the result is much more realistic. Consider for example the upper terrace at the north face of Masada, one of three terraces on which king Herod erected his palace more than 2000 years ago (Figure 4.8A). The principle joint set orientations are

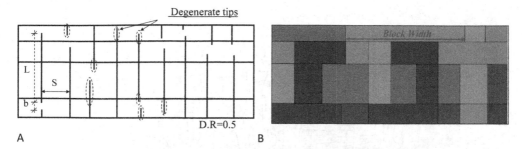

A                                    B

*Figure 4.22* Modeling mechanical layering with DDA by controlling input joint length, spacing, and bridge. A) a trace map with definition of terms, B) the generated DDA block system after block cutting and dead end trimming (after Bakun-Mazor *et al.*, 2009).

*Figure 4.23* DDA block system for the upper terrace of Herod's palace, Masada, assuming joint spacing and length distribution as measured in the field (Figure 4.21).

plotted in Figure 4.8B. Inspection of Figure 4.8A clearly reveals that the joints cannot be assumed infinitely persistent, therefore simulating this rock mass with infinite joint trace length with respect to the modeled domain assuming zero bridge length would be totally unrealistic and overly conservative. The results of the joint survey performed at the site (Figure 4.21) can be utilized to generate a jointing pattern that will reflect the finite joint persistence, even at the outcrop scale, as demonstrated in Figure 4.23. However, when attempting to run forward analysis the jointed domain disintegrates when subjected to the slightest input earthquake vibration, but we know that this slope has survived several strong earthquake events in its 2000 year history. Although the slopes geometry has been modified over the years by local block sliding and toppling

A

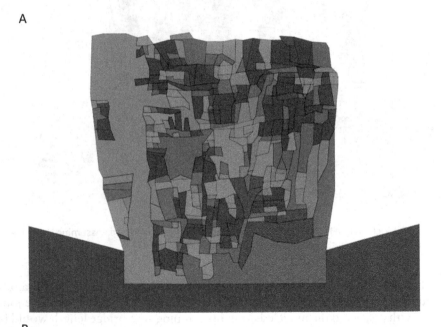

B

*Figure 4.24* A) Joint map of the upper terrace of Herod's palace in Masada as obtained from good resolution air photos of the outcrop, B) DDA mesh cut from the map shown in (A) after digitization.

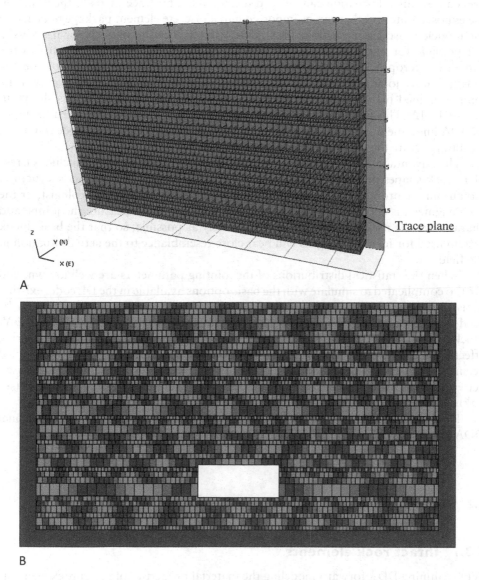

*Figure 4.25* A) DFN model incorporating mechanical layering effects used for preprocessing, B) resulting DDA mesh (after Bakun-Mazor *et al.*, 2009).

from the left and right sides (E-W) respectively, the bulk of the slope remained intact
over the years, suggesting that such a representation of the rock mass, while more
realistic than when using infinite joint traces, is still overly conservative.

When representative surface exposures are available, joint trace mapping using
advanced photogrammetric technics (e.g. Ferrero *et al.*, 2009) may be used to obtain a
digital representation of the rock mass structure. Then using adequate data processing

procedures this information can be used to obtain digital trace maps of the joints on the exposed outcrops that can be imported into a discrete element package to generate the modeled discontinuous domain, semi-automatically. Alternatively, or if access is not possible for photogrammetric measurements, high resolution air photos of the exposed outcrop may be digitized, either manually or automatically, to obtain the most accurate joint trace map of the field exposure. A map of the outcrop with its joints obtained by manually tracing the joints detected from air photos is shown in Figure 4.24A. This map was digitized manually and the line data were used for input in DDA line generation code (DL). Then the robust DDA block cutting code (DC) was used to generate the block system shown in Figure 4.24B.

The advantage of the block system shown in Figure 4.24B is that it includes irregular block shapes that are interlocked to form a more cohesive rock mass structure. No automatic procedure can produce such a result. The engineering geologist, or the rock engineer, is encouraged therefore to go out to the field, measure the joints, and characterize the rock mass structure as accurately as possible, so that the basic block system used for forward analysis will bear close resemblance to the actual situation in the field.

When the statistical distributions of the jointing parameters are well known, and are too complicated to simulate with the basic options available in the DL code, existing tools such as DFN (Dershowitz et al., 2000; Dershowitz and Einstein, 1988) may be used for preprocessing and then the resulting lines can be input directly into DDA's block cutting code DC. Such an approach has been used to simulate mechanical layering effects in the rock mass shown in Figure 4.7 (See Figure 4.25). This effort enabled accurate assessment of roof stability of natural karstic caverns discovered below an active open pit mine that needed to be preserved for future scientific investigations (see Bakun-Mazor et al., 2009).

Finally, statistical joint generation considerations pertinent to three dimensional DDA are discussed by Shi (2005) and Ma and Fu (2011).

## 4.3  MECHANICAL INPUT PARAMETERS FOR FORWARD MODELING

### 4.3.1  Intact rock elements

When running DDA forward modeling the material properties of intact rock elements are required for the solution of the equilibrium equations. These basically include the rock density, Young's modulus, and Poisson's ratio. Because we are mainly interested in block displacements and not internal block deformation, the intact rock elements are assumed isotropic and simply deformable, therefore only a single value of Young's modulus and Poisson's ratio is sufficient for computing the elastic deformation of the blocks. Introducing anisotropy in the elastic stiffness matrix would unnecessarily complicate the structure of the governing equations and is not deemed essential for most DDA applications. Material lines may be used (see manual) in order to assign different intact block parameters to blocks that are transected by different material lines. If no material lines are assigned, all blocks in the modeled domain will be assumed to possess the same elastic parameters.

The input parameters for intact rock elements should be easy to obtain from preliminary site investigations and standard laboratory tests performed on representative cores from the site. Some points to remembered when assigning intact rock parameters:

1   The introduced values should be representative of the rock in the field, therefore the engineer or engineering geologist should select samples for testing at the laboratory wisely so as not to introduce bias in the sampling and testing results.

2   Running robust simulations with DDA of a real engineering problem without first obtaining good knowledge of the actual material properties in the field is not recommended as material properties of rocks are very much site specific.

3   When running sensitivity analyses for gaining some deeper theoretical understanding of the studied problem, material properties can be obtained from publications or data bases.

4   The units for the solution are determined by the input values of the physical and mechanical parameters, namely density and Young's modulus, therefore they must be consistent. For example, if a solution in SI units is desired, then the unit weight of the elements should be in $N/m^3$ and the unit mass will be the assigned unit weight divided by gravitational acceleration in units of $m/s^2$. For consistency therefore, the Young's modulus must be in this case in units of $N/m^2$. Naturally, if unit weight is input in units of $kN/m^3$ then Young's modulus should be input in units of kPa. Any other internally consistent unit system can be used. The output values of the analysis will be in the same units.

5   As will be explored in Chapter 5, for every problem there is optimal relationship between Young's modulus and the contact spring stiffness, or penalty parameter. The issue of what is the optimal penalty value to insert is a subject of ongoing research as the accuracy of the solution very much depends on it, particularly in dynamic problems, as will be explored in Chapter 5. In the absence of any knowledge or previous experience Shi's rule of thumb (Shi, 1996) may be used: $G_0 = (E_0)(L_0)$ where $G_0$ is the stiffness of the contact spring, $E_0$ is Young's modulus of the block, and $L_0$ is the "average" block diameter. Note that by using this rule of thumb $G_0$ becomes proportional to the size of the block or to the contact force, as it should. While the value of $G_0$ may be optimized to obtain best numerical accuracy, Young's modulus should not be tweaked at all and must reflect the best representation of the physical reality in the field.

6   When there are reasons to believe that the rock mass stiffness is significantly different than that of intact rock elements it is possible to introduce rock mass modulus to the intact rock elements instead of values obtained from laboratory tests performed on intact element. Many empirical relationships between rock mass classifications and rock mass stiffness exist in the literature (e.g. Hoek and Diederichs, 2006), and these may be used for preliminary purposes. This approach could be effective also when attempting to model deformation in rock masses with filled joints, as joint infilling deformation is not incorporated in DDA other than the linear elastic deformation of the contact springs.

## 4.3.2   Shear strength of the discontinuities

DDA incorporates Coulomb friction to model sliding along the discontinuities. It therefore calls for input joint cohesion, friction angle, and tensile strength. Because most of the deformation in a discontinuous medium takes place by sliding along or opening of discontinuities (plus in plane rotations in three dimensions) the assigned parameters for the joints are of paramount importance and to a large extent control to obtained result of the computation.

The original DDA code assumes the joint parameters constant throughout the analysis. Once sliding begins the role of cohesion and tensile strength becomes less relevant as the two adjoining blocks across the interface have already detached. Therefore, joint cohesion and tensile strength are more relevant for static computations. Friction however continues to govern the dynamic deformation, both the sliding velocity and the sliding distance, to the end of the analysis. When a static analysis is of interest DDA can be used to obtain the friction angle necessary for limiting equilibrium, namely, the friction angle required to avoid sliding between blocks. If the static friction angle is known from laboratory experiments, this approach can be used to find the factor of safety against sliding.

When DDA is used to model dynamic deformation involving large and rapid displacements along discontinuities, assuming a constant friction angle throughout the analysis may lead to erroneous results that in many times may be un-conservative because of possible friction deterioration with sliding distance and/or velocity. This is particularly important when trying to model landslide runouts in response to strong earthquakes (e.g. Parker *et al.*, 2011; Tang *et al.*, 2011; Wu, 2007). In their classic paper Goodman and Seed (1966) proposed an algorithm to progressively decrease the available friction angle with ongoing cycles of shear, based on preliminary observations from soil mechanics. This has been further explored in rock mechanics context for cyclic shear along rock joints, and constitutive relationships for dynamic friction degradation in response to number and amplitude of shear cycles have been proposed (e.g. Crawford and Curran, 1981; 1982). In parallel to these efforts in the geotechnical engineering community, the seismological community studied extensively rate and state friction in the context of fault mechanics (e.g. Dieterich, 1979; Kilgore *et al.*, 1993; Marone, 1998; Ruina, 1983; Solberg and Byerlee, 1984). Indeed attempts are currently being made to implement rate dependent friction into DDA (e.g. Wang *et al.*, 2013; Wu, 2010) to enable a more realistic dynamic analysis of landslides. By doing this the advantages of the dynamic formulation of DDA may be fully realized.

# Chapter 5

# DDA verification

## 5.1 INTRODUCTION

Any numerical method, regardless of the underlying theory and robustness of the mathematical formulation, requires verification and validation once it is implemented into computer code. By *verification* we mean ensuring that there is an acceptable agreement between results obtained by running the code with the computer and by solving a closed form analytical solution of exactly the same problem. Of course if we could solve everything analytically there would not be a need for us to use numerical approaches. But we cannot, consider tunneling for example: once the immediate roof of a tunnel consists of more than three blocks, there is no analytical solution we can employ to solve the forces between the blocks and to determine the most likely mode of failure, not to mention the time dependent displacement of each and every block in the system. Similarly in a rock slope problem: when the number of interacting blocks exceeds more than two, application of analytical approaches becomes very difficult. Once the number of blocks in a slope or around a tunnel increases to realistic numbers of $10^1$ to $10^2$, depending on the intensity of the jointing pattern in the rock mass, we become completely dependent on numerical approaches. Indeed, we will never know if our numerical solution is correct when the problem we are solving does not have an analytical solution. What we can do is attempt to determine if our numerical solution is *valid*. We do this by comparing results obtained numerically to results obtained from field or laboratory experiments. Consider the previous tunnel roof example: although there is no analytical solution for say the vertical displacement of the central block in a roof consisting of three blocks, we can *measure* that displacement over time, if we can construct a similar model in the lab or if we can find such a structure in the field and monitor it properly. Then we can go back to the numerical prediction and test its validity. Such an approach is called *validation*.

Verifications and validations are extremely important for establishing the applicability of a numerical approach to real life problems. They allow us a deeper insight not only to the accuracy of the method used, but also to the sensitivity of the numerical solution to the numerical control parameters the code needs in order to run properly. Verifications are also useful as benchmark tests. They are a proving method for any numerical code, so when we develop a new code or modify an existing code, before we attempt to solve real life problems with it, we must prove that it can pass those benchmark tests satisfactorily. A comprehensive review of DDA validations summarizing

*Figure 5.1* The plane failure of Mt. Toc in the southern Italian Alps. A mass of about 270–300 million m³ of rocks slid in 1963 on a preexisting plane of weakness into the Vajont reservoir at a velocity sufficiently high to cause a wave that overtopped the Vajont dam, resulting in more than 2000 causalities and devastating the downstream town of Longarone, Italy.

work that has been done a decade after its publication was published by MacLaughlin and Doolin (2006). A set of DDA benchmark tests was published a decade later by Yagoda Biran and Hatzor (2016), and in between many more excellent work has been done in an attempt to verify and validate DDA (e.g. MacLaughlin and Hayes, 2005; Scheele and Bates, 2005; Wu *et al.*, 2007). In this chapter verifications and validation studies performed by the DDA research group at Ben-Gurion University of the Negev are discussed.

## 5.2 SINGLE PLANE SLIDING

The most common problem in rock slope engineering is that of a block resting on an inclined plane. The famous landslide of Mount Toc in the Italian Alps that overtopped the Vajont dam (see Barla and Paronuzzi, 2013) is agreed by all to have originated in a single plane sliding mechanism, the sliding surface of which is shown in Figure 5.1.

Single plane sliding is typically encountered in rock masses of sedimentary origin, where bedding planes of infinite extent transect the rock mass structure. When the bedding planes are horizontal or dipping gently we may assume there is sufficient frictional resistance to restrain any possible block motion along those weakness planes, in static conditions. If the bedding planes are smooth, clean and tight the friction angle may be assumed to be at least as high as 20 degrees, the residual friction angle of most carbonate and silicate minerals being typically higher than this value. If the surfaces

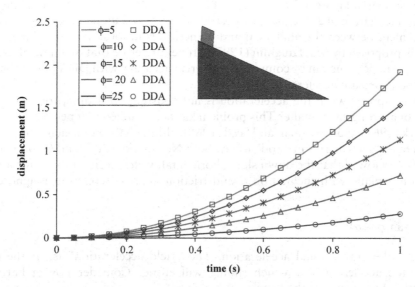

*Figure 5.2* 2D-DDA verification for dynamic sliding of a block on an inclined plane with constant friction under gravitational loading (modified after Kamai (2006) following MacLaughlin (1997)). The modeled plane inclination is 28°.

are rough, the friction angle can increase quite significantly, up to twice or three times this value, but if the bedding planes are filled with clay infilling thicker than the roughness amplitude, the friction angle can be as low as half that value, depending on the mineralogical composition of the clay and its water content. So clearly, in addition to measuring the slope inclination properly in the field, we must have a good estimate of the frictional resistance of the sliding planes, before we begin the analysis. The possible range of friction angles in rock discontinuities has been discussed by many authors (e.g. Barton, 1976; Brady and Brown, 2004; Goodman, 1989; Hoek and Bray, 1981; Jaeger *et al.*, 2007).

## 5.2.1 2D-DDA

The first to use the block on an incline problem to verify dynamic sliding under constant gravitational acceleration with 2D-DDA was MacLaughlin (1997) in her Ph.D. dissertation. This has since become a popular bench mark test for DDA because the formulation of the limit equilibrium equation as originally proposed by MacLaughlin (1997) is straight forward. The displacement $s$ at time $t$ of a block resting on a plane inclined at angle $\alpha$ with friction angle $\phi$ and subjected to constant gravitational acceleration $g$ is given by Equation 5.1 (MacLaughlin, 1997):

$$s_t = \frac{1}{2}at^2 = \frac{1}{2}g(\sin\alpha - \cos\alpha\tan\phi)t^2 \qquad (5.1)$$

A comparison between the analytical solution (Equation 5.1) and 2D-DDA as computed by Kamai (2006) in her M.Sc. thesis following the original work of MacLaughlin

(1997) is shown in Figure 5.2 for a range of friction angles, between 5 and 25 degrees, and slope inclination of 28 degrees. Inspection of Figure 5.2 immediately reveals striking agreement between the analytical and numerical solutions. This benchmark test, originally proposed by MacLaughlin (1997), proves therefore that dynamic block sliding along a single plane can be computed with great accuracy with DDA under constant friction and constant acceleration.

What happens when the acceleration is not constant with time, for example in the case of a strong earthquake? This problem has been studied independently by both Newmark (1965) and Goodman and Seed (1966), although the approach is universally referred to as "Newmark's method" today. Both Newmark (1965) and Goodman and Seed (1966) showed that the down slope, horizontal, yield acceleration $a_y$ for a block resting on a plane with inclination $\alpha$ and friction angle $\phi$ (cohesion is ignored) is given by:

$$a_y = \tan(\phi - \alpha)g \tag{5.2}$$

where $g$ is the gravitational acceleration. The "yield acceleration" is, as the name implies, the acceleration at which sliding will ensue. Consider now a harmonic acceleration function $a_t$ of the form:

$$a_t = A \sin(\omega t + \theta) \tag{5.3}$$

in which $A$ and $\omega$ are the amplitude and angular frequency of the input harmonic acceleration function. $\theta$ is the phase angle required to satisfy the initial condition $a = a_y$ at the instant sliding begins $(t = 0)$ and is found by solving Equations 5.2 and 5.3 simultaneously:

$$\theta = \frac{\sin^{-1}\left(\tan(\phi - \alpha)\, g / A\right)}{\omega} = \frac{\sin^{-1}\left(a_y/A\right)}{\omega} \tag{5.4}$$

The displacement of the block can be determined by integrating Equation 5.3 twice whenever $a_t > a_y$ using the definition of $\theta$; the first integration provides the time dependent velocity $v_t$ and the second the time dependent displacement $d_t$. A graphical illustration of this concept is shown in Figure 5.3, following the original paper by Goodman and Seed (1966).

The first who have attempted to verify DDA using Goodman and Seed's approach were Hatzor and Feintuch (2001). Note however that in the original paper by Hatzor and Feintuch the frictional resistance of the sliding interface was used to evaluate the yield acceleration $a_y$ (Equation 5.2) and therefore $\theta$ (Equation 5.4), but frictional resistance was neglected once dynamic sliding ensued, obviously resulting in less than perfect agreement between the analytical and DDA solutions. Kamai and Hatzor (2008) corrected this oversight and included the frictional resistance of the sliding interface during dynamic sliding in their DDA verification for the same problem (see Figure 5.4).

The down slope acceleration of the sliding block can be determined by subtracting the resisting forces from the driving forces:

$$a_t = \left[kg \sin(\omega t) \cos\alpha + g \sin\alpha\right] - \left[g \cos\alpha - kg \sin(wt) \sin\alpha\right] \tan\phi \tag{5.5}$$

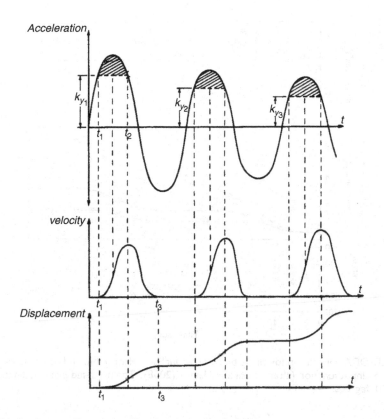

Figure 5.3 Method originally proposed by Goodman and Seed (1966) for integration of accelerograms to determine downslope displacement of a block resting on an inclined plane. Dynamic friction degradation is assumed, as implied by the decreasing value of yield acceleration between cycles. Reproduced with permission from ASCE.

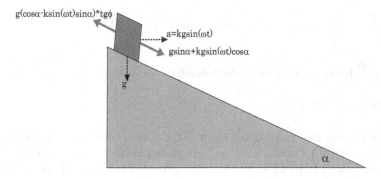

Figure 5.4 Free body diagram for the block on an incline problem subjected to harmonic horizontal acceleration loading function (Kamai, 2006).

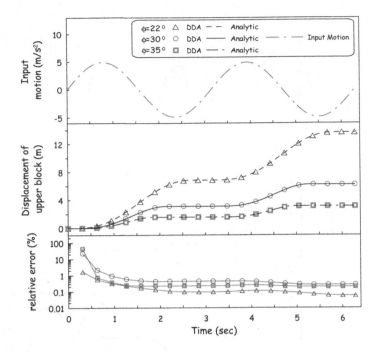

*Figure 5.5* 2D-DDA verification for single block on an inclined plane subjected to harmonic acceleration input function (after Kamai and Hatzor (2008)). The modeled plane inclination here is
20 degrees.

where $k$ calibrates the proportion between $a$ and $g$. As explained above, the displacement of the block at any time $t$ is determined by double integration on the acceleration,
with $\theta$ as reference datum, in a conditional manner:

if $a > a_y$   or   $v > 0$:

$$d_t = \int_\theta^t v = \iint_\theta a = g\left[(\sin\alpha - \cos\alpha\tan\phi)\left(t^2/2 - \theta t\right)\right]$$

$$+\frac{A}{\omega^2}[(\cos\alpha + \sin\alpha\tan\phi)\,(\omega\cos(\omega\theta)\,(t-\theta) - \sin(\omega t) + \sin(\omega\theta))] \qquad (5.6)$$

otherwise $d_t = d_{t-1}$

where the initial velocity and displacement of the sliding block are assumed to be zero.
The downslope displacements, $d(t)$, are calculated while $a_y$ is exceeded for the first
time at $\theta_1$, or when the block's velocity is positive. If neither condition is fulfilled the
block is assumed to be at rest, and sliding will commence only once $a_y$ is exceeded again
at $\theta_2$, and so on. The excellent agreement between DDA and the analytical solution
(Equation 5.6) for three different friction angles is shown in Figure 5.5. The relative

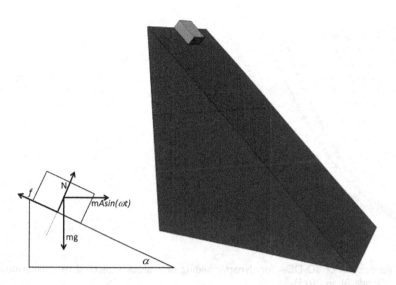

*Figure 5.6* Three dimensional DDA mesh for the block on an incline problem, the direction of loading is indicated in the inset (Yagoda-Biran, 2013).

error $(E_N)$ between the analytical and numerical solutions (Equation 5.7) proves to be less than 1% in this verification.

$$E_N = \left| \frac{d_A - d_N}{d_A} \right| \cdot 100\% \tag{5.7}$$

### 5.2.2 3D-DDA

We have seen in Chapter 3 that in 3D-DDA the number and type of contacts are significantly greater than in 2D-DDA and naturally this might affect the accuracy of the dynamic solution. To verify the accuracy of 3D-DDA when solving dynamic sliding of a single block on a single plane we use the same block on an incline problem as discussed in the previous section but this time in three dimensions, following work performed by Yagoda-Biran (2013) in her Ph.D. dissertation. The three dimensional mesh as computed by Yagoda-Biran is shown in Figure 5.6. The plane inclination is 45°, its height 10 m, and its depth 5 m. The base block is fixed in space by imposing seven fixed points, and the sliding block is a 1 m × 1 m × 0.5 m box.

The general expression for time dependent displacement under harmonic acceleration input when the initial velocity and displacement are not zero is (Yagoda-Biran, 2013):

$$d(t) = \frac{1}{2}g(\sin\alpha - \cos\alpha \tan\phi)t^2$$

$$- \frac{A}{\omega^2}\sin(\omega t)(\cos\alpha + \sin\alpha \tan\phi) + \dot{d}_0 t + d_0 \tag{5.8}$$

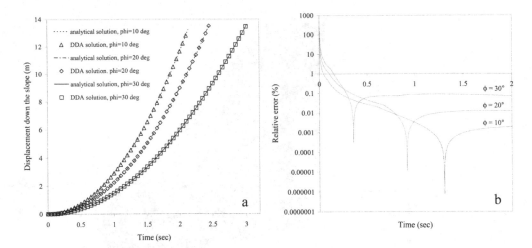

*Figure 5.7* Verification of 3D-DDA for dynamic sliding of a block subjected to gravitational loading (Yagoda-Biran, 2013).

*Table 5.1* Input numerical control parameters used in the 3D-DDA verification shown in Figure 5.7.

| Parameter | value |
|---|---|
| $dd$ – dynamic parameter | 1 |
| $g0$ – normal contact spring stiffness | $4 * 10^8$ N/m |
| $g1$ – time step interval | 0.001 s |
| $g2$ – maximum displacement ratio | 0.001 |
| Block material density | 2700 kg/m$^3$ |
| Young's modulus | 40 GPa |
| Poisson's ratio | 0.18 |

When the block is subjected to gravity loading only ($A = 0$), and begins at rest ($\dot{d}_0 = 0$, $d_0 = 0$), Equation 5.8 is reduced to Equation 5.1. The analytical vs. 3D-DDA downslope displacement history as computed by Yagoda-Biran (2013) for three values of friction angle is shown in Figure 5.7. Note that all three friction angles are smaller than the inclination of the slope and therefore sliding begins immediately the instant gravity is applied. After 0.2 s of sliding the relative error as defined by Equation 5.7 drops to below 1%, indicating the excellent agreement between the analytical and 3D-DDA solutions. The input numerical control parameters used by Yagoda-Biran (2013) are listed in Table 5.1.

While it is encouraging to learn that dynamic sliding of a single block can be studied very accurately with 3D-DDA when the sliding block is initially at rest, it would be interesting to check if the same agreement could be obtained when the sliding block has some initial velocity. This would be particularly useful for landslide studies where the sliding mass is assumed to be creeping downslope at some very low initial velocity,

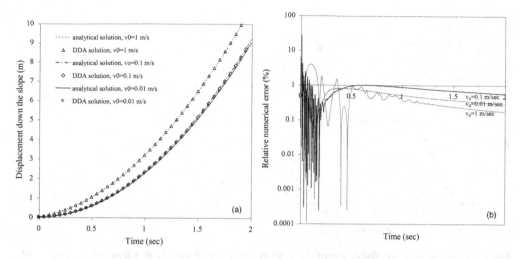

*Figure 5.8* a) Downslope displacement histories of a block on an inclined plane subjected to gravity and different initial velocities, b) The relative numerical error (Yagoda-Biran, 2013).

before some physical or mechanical barrier is removed, thus allowing the sliding mass to run out. For the analytical solution we can use again Equation 5.8 but this time with consideration of the initial velocity, namely $\dot{d}_0 \neq 0$. The agreement between the analytical and 3D-DDA solutions for initial velocities of 0.01 m/s, 0.1 m/s, and 1 m/s as computed by Yagoda-Biran (2013) is shown in Figure 5.8. The numerical control parameters remain as in Table 5.1, except for the time step interval that had to be reduced by one order of magnitude to 0.0001 s in order to obtain an agreement within 1% with the analytical solution for all initial velocities. This level of accuracy is attained after 0.5 s of sliding.

After verifying that dynamic sliding under gravitational loading can be performed extremely accurately with 3D-DDA when only a single block is considered, it would be instructive to examine the accuracy of the 3D-DDA solution for dynamic sliding under harmonic acceleration input. Yagoda-Biran (2013) subjected the block model shown in Figure 5.6 to a sinusoidal input acceleration function with amplitude of $A = 2$ m/s$^2$ and frequency of 1 Hz. A friction angle of 50° was assigned to the interface which was inclined at 45°, so that the input acceleration must have exceeded the theoretical yield acceleration before sliding began. The numerical control parameters were the same as listed in Table 5.1 with two exceptions: 1) the time step interval was further reduced by one order of magnitude to 0.0001 s as in the previous example, and 2) the normal contact spring stiffness was increased by more than one order of magnitude to $7 * 10^9$ N/m. The results of this verification are shown in Figure 5.9 where throughout the analysis the relative error never exceeds 3%.

So far in our discussion we have limited the input motion direction to be parallel to one of the axes of the coordinate system. But in a three dimensional approach we need not do that, indeed there is no reason for the dynamic input to be restricted in any way to a single direction as ground motions in all directions are recorded during

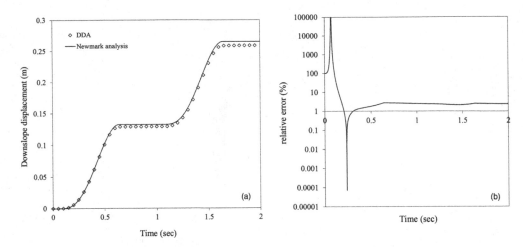

Figure 5.9 a) Downslope displacement of a block on an inclined plane subjected to sinusoidal acceleration input, b) The relative numerical error (Yagoda-Biran, 2013).

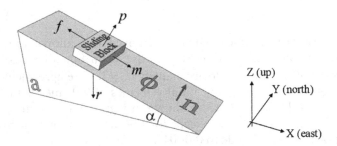

Figure 5.10 Three dimensional force representation for the block on an incline problem (Bakun-Mazor, 2011).

earthquakes. Bakun-Mazor (2011) in his Ph. D. dissertation has performed a fully three dimensional analysis of the block of incline problem using a vector approach he termed "vector analysis" and used his analytical solution to verify 3D-DDA. The model of a block on an incline used by Bakun-Mazor is shown in Figure 5.10. The dip and dip direction angles are $\alpha = 20°$ and $\beta = 90°$, respectively. A Cartesian coordinate system $(x, y, z)$ is defined where $X$ is horizontal and points to east, $Y$ is horizontal and points to north, and $Z$ is vertical and points upward. The normal vector of the inclined plane is: $\hat{n} = [n_x, n_y, n_z]$, where:

$$n_x = \sin(\alpha)\sin(\beta)$$
$$n_y = \sin(\alpha)\cos(\beta)$$
$$n_z = \cos(\alpha) \tag{5.9}$$

A block of a unit mass is assumed, therefore the force equilibrium equations presented below can be discussed in terms of accelerations. Let the resultant force vector that acts on the system at each time-step be: $\bar{r} = [r_x, r_y, r_z]$. The driving force vector that acts on the block $(\overline{m})$, namely the projection of the resultant force vector on the sliding plane at each time step, is:

$$\overline{m} = (\hat{n} \times \bar{r}) \times \hat{n} \tag{5.10}$$

The normal force vector that acts on the block at each time step is:

$$\overline{p} = (\hat{n} \cdot \bar{r})\hat{n} \tag{5.11}$$

At the beginning of a time step, if the velocity of the block is zero then the resisting force vector due to the interface friction angle $\phi$ is:

$$\bar{f} = \begin{cases} -\tan(\phi)|\overline{p}|\hat{m} & , \quad \tan(\phi)|\overline{p}| < |\overline{m}| \\ -\overline{m} & , \quad else \end{cases} \tag{5.12}$$

where $\hat{m}$ is a unit vector in direction $\overline{m}$. If at the beginning of a time step the velocity of the block is not zero, then:

$$\bar{f} = -\tan(\phi)|\overline{p}|\hat{v} \tag{5.13}$$

where $\hat{v}$ is the direction of the velocity vector.

Previous approaches to solve this problem in three dimensions (Goodman and Shi, 1985) only considered gravitational loading where the block velocity and the driving force were assumed to always have the same sign. But during earthquakes the driving force and velocity can momentarily be in opposite directions. The procedure proposed by Bakun-Mazor (2011) addresses this. The sliding force, namely the block acceleration during each time step, is $\bar{s} = [s_x\ s_y\ s_z]$ and following Goodman and Shi (1985) is calculated as the force balance between the driving and the frictional resisting forces:

$$\bar{s} = \overline{m} + \bar{f} \tag{5.14}$$

The block velocity and displacement vectors are $\overline{V} = [V_x, V_y, V_z]$ and $\overline{D} = [D_x, D_y, D_z]$, respectively. At $t = 0$, the velocity and displacement are zero. The average acceleration for time step $i$ is:

$$\overline{S}_i = \frac{1}{2}(\bar{s}_{i-1} + \bar{s}_i) \tag{5.15}$$

The velocity for time step $i$ is therefore:

$$\overline{V}_i = \overline{V}_{i-1} + \overline{S}_i \Delta t \tag{5.16}$$

It follows that the displacement for time step $i$ is:

$$\overline{D}_i = \overline{D}_{i-1} + \overline{V}_{i-1}\Delta t + \frac{1}{2}\overline{S}_i \Delta t^2 \tag{5.17}$$

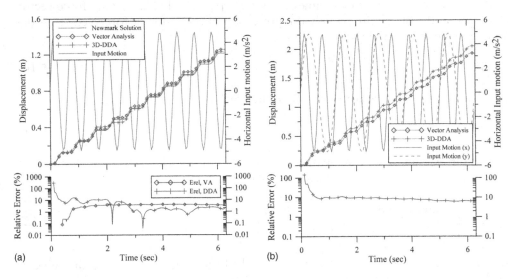

*Figure 5.11* a) Comparison between Newmark solution, Vector Analysis, and 3D-DDA for horizontal motion parallel to X axis for the model shown in Figure 5.10. The relative errors are computed against Newmark's solution, b) Comparison between VA and 3D-DDA for horizontal input motion parallel to both X and Y axes simultaneously. Here VA is used as a reference for relative error computation (after Bakun-Mazor, 2011).

Sensitivity analyses were performed by Bakun-Mazor (2011) to constrain the length of the time interval in the trapezoidal integration method without compromising accuracy. The results suggest that the accuracy is strongly affected by the length of the time interval particularly when the friction angle is greater than the slope inclination, and that accuracy is compromised when the time interval is greater than 0.001 s.

The validity of Bakun-Mazor (2011) vector approach is tested using Newmark analytical approach, considered the "exact" solution when only one direction of input acceleration is assumed. Figure 5.11a shows a comparison between Newmark, Bakun-Mazor, and 3D-DDA solutions for an inclined plane with dip and dip direction of $\alpha = 20°$ and $\beta = 90°$, respectively, and friction angle of $\phi = 30°$. A sinusoidal input motion parallel to the horizontal X axis is used, with resultant acceleration vector: $\bar{r} = [r_x \ r_y \ r_z] = [0.5 \sin(10t), 0, -1]g$. The accumulated displacements are calculated for up to 10 cycles ($t_f = 2\pi$ sec). The input horizontal acceleration is plotted as a shaded line and the values are shown on the right hand-side axis. For both Newmark and Bakun-Mazor solutions the numerical integration is calculated using a time interval of $\Delta t = 0.001$ s. The calculated displacement vector in Bakun-Mazor and 3D-DDA solutions is projected along the sliding direction to enable comparison between the three approaches. An excellent agreement is obtained between Bakun-Mazor (labeled VA in Figure 5.11) and Newmark solutions throughout the first two cycles of motion. There is a small discrepancy at the end of the second cycle, the magnitude of which will depend on the numerical input parameters and will decrease whenever the time interval decreases. The relative error of Bakun-Mazor and 3D-DDA solutions with respect to the exact Newmark solution never exceeds 3%.

*Table 5.2* Input parameters for 3D-DDA verification performed by Bakun-Mazor (2011).

|  | Block on an incline model | Tetrahedral wedge |
|---|---|---|
| **Mechanical Properties:** | | |
| Elastic Modulus, MPa | 20 | 200000 |
| Poisson's Ratio | 0.25 | 0.25 |
| Density, kg/m³ | 1000 | 1700 |
| Friction angle, Degrees | 30 | 30–36 |
| **Numerical Parameters:** | | |
| Dynamic control parameter | 1 | 1 |
| Number of time steps | 628 | 8000 |
| Time interval, Sec | 0.01 | 0.005 |
| Assumed max. disp. Ratio, m | 0.01 | 0.01 |
| Penalty stiffness, MN/m | 10 | 10000–20000 |

Once verified, Bakun-Mazor's VA approach can now be used to verify 3D-DDA for single plane sliding but under input acceleration in both the $x$ and $y$ directions as well as vertical gravitational acceleration (Figure 5.11b). The resultant input acceleration vector here is $\bar{r} = [r_x \, r_y \, r_z] = [0.5\sin(10t), 0.5\sin(5t), -1]g$, and the friction angle is again $\phi = 30°$. The two components of the input horizontal acceleration are plotted as shaded lines and the acceleration values are shown on the right-hand side axis as before. The relative error in the final position of the sliding block does not exceed 8%.

The input values used by Bakun-Mazor (2011) for the verification are listed in Table 5.2.

## 5.3   DOUBLE PLANE SLIDING

This mode of failure is commonly referred to as "Wedge Failure" and has been discussed by many authors in the past. Three dimensional limit equilibrium analysis for this failure mode using the stereographic projection is provided by Goodman (1976, 1989) for dry joints and by Londe *et al.* (1969, 1970) for water filled joints. An analytical procedure to solve the factor of safety is discussed by Hoek and Bray (1981).

Wedge failures are encountered in rock masses consisting of more than one joint set, as the intersection of two sets are necessary to produce three-dimensional wedges in the rock mass. If there are more than two joint sets in the rock mass different wedges may form, but the number of failure modes will not increase with increasing number of joint sets in the rock mass. In fact, if rotations are ignored the only possible failure modes are opening, single, and double plane sliding. Moreover, in order to create a finite block a minimum of three joint sets is required if only one free surface is considered (Goodman and Shi, 1985). As can be seen in Figure 5.12A, the "block mold" (Hatzor, 1993) that remained in the rock slope after the failure of the wedge has three joints in its boundaries, two steeply dipping and one gently dipping. The wedge slides along the line of intersection of the two steeply inclined joints and opens from the moderately dipping joint at the top of the block. The same can be seen also in Figure 5.12B, but this time the boundary planes along which sliding took place are dipping more moderately. But again, sliding only takes place on two joints in direction

*Figure 5.12* Examples of wedge failures in the field. A) a wedge in a Gypsum quarry at the Ramon crater (southern Israel), B) a wedge in the slopes of Jinping mountain, Sichuan province, China.

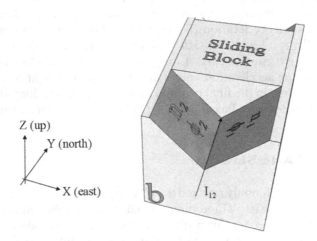

*Figure 5.13* Three dimensional representation of the wedge problem (Bakun-Mazor, 2011).

parallel to the line of intersection of the two boundary planes, and the wedge opens from the third joint at the top.

The case of wedge failure has been used by Bakun-Mazor (2011) to verify 3D-DDA. A three dimensional model of a wedge that is sliding simultaneously on two joints, plane 1 and plane 2, in direction parallel to the line of intersection ($I_{1,2}$) is shown in Figure 5.13.

The normal to plane 1 is $\hat{n}_1 = [n_{x1}, n_{y1}, n_{z1}]$ and the normal to plane 2 is $\hat{n}_2 = [n_{x2}, n_{y2}, n_{z2}]$. The line of intersection $\hat{I}_{12}$, along which sliding takes place in this failure mode is given by:

$$\hat{I}_{12} = \hat{n}_1 \times \hat{n}_2 \tag{5.18}$$

As discussed in the case of plane failure in three dimensions, the resultant force in each time step is $\bar{r} = [r_x,\ r_y,\ r_z]$, and the driving force in each time step is now:

$$\bar{m} = (\bar{r} \cdot \hat{I}_{12})\hat{I}_{12} \tag{5.19}$$

The normal force acting on plane 1 in each time step is $\bar{p} = [p_x, p_y, p_z]$, and the normal force acting on plane 2 in each time step is $\bar{q} = [q_x, q_y, q_z]$, where:

$$\bar{p} = ((\bar{r} \times \hat{n}_2) \cdot \hat{I}_{12})\hat{n}_1 \tag{5.20}$$

$$\bar{q} = ((\bar{r} \times \hat{n}_1) \cdot \hat{I}_{12})\hat{n}_2 \tag{5.21}$$

As in the case of single face sliding, the direction of the resisting force $(\bar{f})$ depends upon the direction of the velocity of the block. Therefore, as before, in each time step:

$$\bar{f} = \begin{cases} -\left(\tan(\phi_1)|\bar{p}| + \tan(\phi_2)|\bar{q}|\right)\hat{m} \ , & \bar{V} = 0 \ \text{and} \ \left(\tan(\phi_1)|\bar{p}| + \tan(\phi_2)|\bar{q}|\right) < |\bar{m}| \\ -\bar{m} & , \quad \bar{V} = 0 \ \text{and} \ \left(\tan(\phi_1)|\bar{p}| + \tan(\phi_2)|\bar{q}|\right) \geq |\bar{m}| \\ -\left(\tan(\phi_1)|\bar{p}| + \tan(\phi_2)|\bar{q}|\right)\hat{v} \ , & \bar{V} \neq 0 \end{cases} \tag{5.22}$$

where $\bar{V} = [V_x,\ V_y,\ V_z]$ is the velocity vector. The sliding force, namely the block acceleration during each time step, is $\bar{s} = [s_x\ s_y\ s_z]$ and is calculated as the force balance between the driving and the frictional resisting forces:

$$\bar{s}_i = \bar{m} + \bar{f} \tag{5.23}$$

The block velocity and displacement vectors are $\bar{V} = [V_x,\ V_y,\ V_z]$ and $\bar{D} = [D_x,\ D_y,\ D_z]$, respectively. At time $t = 0$, the velocity and displacement are zero. The average acceleration for time step $i$ is:

$$\bar{S}_i = \frac{1}{2}\left(\bar{s}_{i-1} + \bar{s}_i\right) \tag{5.24}$$

The velocity for time step $i$ is therefore:

$$\bar{V}_i = \bar{V}_{i-1} + \bar{S}_i \Delta t \tag{5.25}$$

It follows that the displacement for time step $i$ is:

$$\bar{D}_i = \bar{D}_{i-1} + \bar{V}_{i-1}\Delta t + \frac{1}{2}\bar{S}_i\Delta t^2 \tag{5.26}$$

Sensitivity analyses were performed by Bakun-Mazor (2011) to discover the maximum value of the time interval for the trapezoidal integration method without compromising accuracy, where the relative error was defined as:

$$E_{rel} = \frac{\left|D_{VectorAnalysis} - D_{3D\,DDA}\right|}{\left|D_{VectorAnalysis}\right|} \cdot 100\% \tag{5.27}$$

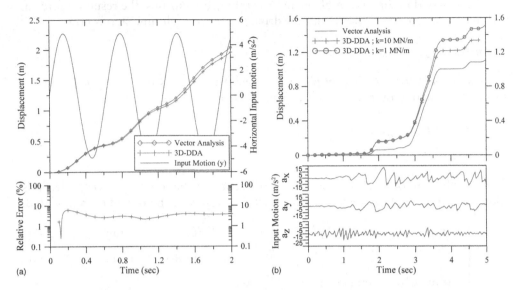

*Figure 5.14*  3D-DDA vs. analytical solution for dynamic wedge sliding. a) Wedge response to one component of horizontal sinusoidal input motion and self-weight, lower panel presents the relative error calculated according to Equation 5.27. b) Wedge response to 3D loading using data from the Imperial Valley earthquake where the three components, multiplied by a factor of 5, are shown in the lower panel (after Bakun-Mazor *et al.*, 2009).

It was found that the results are sensitive to the time interval size when the friction angle is greater than the slope inclination, and that the upper limit for time interval for acceptable accuracy was 0.001 s. The obtained accuracy with 3D-DDA for this problem can be appreciated from Figure 5.14. The input values used for the analysis are listed in Table 5.2.

## 5.4   BLOCK RESPONSE TO CYCLIC MOTION OF FRICTIONAL INTERFACE

In all verifications until now, we have introduced the force directly to the center of mass of the sliding block. During strong earthquakes however, the ground is shaking and the motion is transmitted to the overlying blocks through the frictional interface between the block and the ground or between one block to another (see Figure 5.15). Therefore, a study of block response to shaking foundation along a frictional interface is very relevant for geotechnical earthquake engineering. Verification of this failure mode was originally done with 2D-DDA by Kamai (2006) who was the first to propose a semi-analytical solution to this problem that enabled verification, which was later expanded to three dimension by Yagoda-Biran (2013), for verifying 3D-DDA. Other workers have also studied this problem following the original work of Kamai (2006) and explored various aspects of it with regards to DDA (e.g. Akao *et al.*, 2007; Sasaki *et al.*, 2007).

*Figure 5.15* Column drum response to shaking foundations at the Acropolis, Athens.

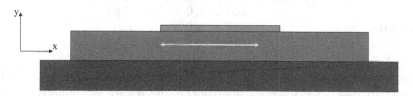

*Figure 5.16* 2D-DDA model used by Kamai (2006) to study block response to cyclic motion of frictional interface.

## 5.4.1 2D-DDA

The studied block system consists of three blocks: a fixed foundation block (Block 0), the intermediate block subjected to horizontal cyclic displacement (Block 1), and the overlying responding block (Block 2). In order to avoid rotational movements that will complicate the analytical solution the responding block is made sufficiently flat, as shown in Figure 5.16.

Block 1 is subjected to a horizontal displacement input function in the form of a cosine, starting from 0:

$$d(t) = D(1 - \cos(2\pi ft)) \tag{5.28}$$

where $D$ and $f$ are the amplitude and frequency of motion, respectively. The only force acting on Block 2 other than gravity is the frictional force, which immediately determines the acceleration of Block 2:

$$m_2 a_2 = F_{friction}$$

$$m_2 a_2 = \mu m_2 g \tag{5.29}$$

$$a_2 = \mu g$$

where $\mu$ is the friction coefficient. The direction of the driving force is determined by the direction of the relative velocity between Blocks 1 and 2 ($v_1^*$). When Block 1 moves to the right relative to Block 2 (positive $x$ direction here), the frictional force pulls Block 2 in the same direction, and determines the sign of $a_2$. When Block 2 is at rest in relation to the Block 1, the frictional force is determined by the acceleration of the bottom block ($a_1$). The threshold acceleration, under which the two blocks move in harmony, is equal to the friction coefficient multiplied by the gravitation acceleration ($\mu g$). When the acceleration of Block 1 passes this threshold value, the frictional forces act in the same direction as $a_1$. The relative velocity of Block 1 is given by:

$$v_1^* = v_1 - v_2 \tag{5.30}$$

The direction of the acceleration of Block 2 is set by the following boundary conditions:

$$
\begin{aligned}
&\text{if } v_1^* = 0 && \text{and } |a_1| < \mu g && && a_2 = a_1 \\
&&& \text{and } |a_1| > \mu g && \text{and } a_1 > 0 && a_2 = \mu g \\
&&& && \text{and } a_1 < 0 && a_2 = -\mu g \\
&\text{if } v_1^* \neq 0 && && \text{and } v_1^* > 0 && a_2 = \mu g \\
&&& && \text{and } v_1^* < 0 && a_2 = -\mu g
\end{aligned}
\tag{5.31}
$$

In the solution of Equation 5.31 Kamai (2006) employed Matlab (MATLAB, version 7) software package because the analytical solution must be computed iteratively as the relative velocity and the direction of the force are dependent upon one another. The accumulated displacement of Block 2 in response to the cyclic shaking of Block 1 as computed by Kamai (2006) using the analytical approach and 2D-DDA is shown in Figure 5.17. Three different displacement amplitudes ($D$) are modeled with a constant input frequency of 1 Hz and friction coefficient of 0.6 at the sliding interface. The obtained magnitude of accumulated displacement is directly proportional to the input amplitude, as expected. Note that the three displacement curves follow the periodic behavior of the input displacement function ($T = 1$ sec.), and that divergence between curves starts after 0.25 s where the input displacement function has an inflection point. The relative error is mostly between 1% and 2%.

## 5.4.2  3D-DDA

Yagoda-Biran (2013) expanded Kamai's approach into three dimension in order to verify 3D-DDA using the model shown in Figure 5.18. The analytical approach is similar to the one-dimensional approach originally proposed by Kamai (2006) except that here displacements, velocities and accelerations are vectors.

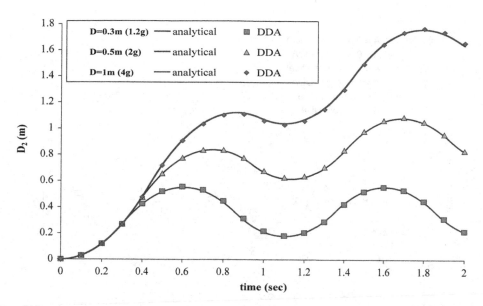

*Figure 5.17* 2D-DDA verification of block response to sliding interface problem (after Kamai, 2006).

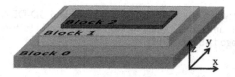

*Figure 5.18* 3D-DDA model for sliding interface used by Yagoda-Biran (2013).

Each of the two moving blocks, blocks 1 and 2, has time dependent displacements $d(t)$, velocities $\dot{d}(t)$ and accelerations $\ddot{d}(t)$. The displacement induced to block 1, $d_1$, is in the form of a cosine function similar to Equation 5.28:

$$d_1(t) = A\left(1 - \cos\left(2\pi\overline{f}t\right)\right) \tag{5.32}$$

here $A$ and $\overline{f}$ are the amplitude and frequency of motion, respectively. The acceleration of block 2 is $\left|\ddot{d}_2\right| = \mu^*g$, as before. The direction of the frictional force, and therefore of $\ddot{d}_2$, is determined by the direction of the relative velocity between the two blocks, $\dot{d}^* \equiv \dot{d}_1 - \dot{d}_2$, defined by the unit vector of the relative velocity, $\hat{\dot{d}}^*$. When $\left|\dot{d}^*\right| = 0$, the acceleration of block 2 ($\ddot{d}_2$) is determined by the acceleration of block 1 ($\ddot{d}_1$). When the acceleration of block 1 exceeds the yield acceleration $\mu^*g$, over which block 2 no longer moves in harmony with block 1, the frictional force direction is determined

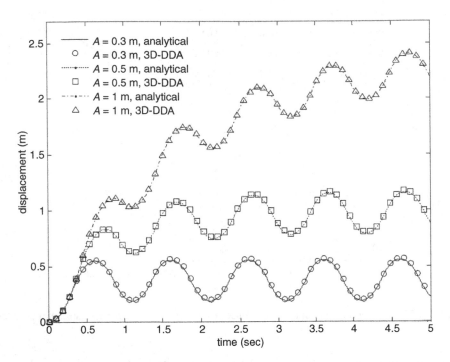

*Figure 5.19* Comparison between the semi-analytical (lines) and 3D-DDA (symbols) solutions for changing motion amplitude. Input motion frequency 1 Hz, friction coefficient of sliding interface 0.6 (Yagoda-Biran, 2013).

by the direction of $\hat{\mathbf{d}}^*$, but the magnitude of $\ddot{\mathbf{d}}_2$ is equal to $\mu^*g$. Yagoda-Biran (2013) formulated this rationale as follows:

$$
\begin{aligned}
\text{if } \left|\dot{\mathbf{d}}^*\right| = 0 \quad &\text{and } \left|\ddot{\mathbf{d}}_2\right| \le \mu^*g \quad &\text{then} \quad &\ddot{\mathbf{d}}_2 = \ddot{\mathbf{d}}_1 \\
&\text{and } \left|\ddot{\mathbf{d}}_2\right| > \mu^*g \quad &\text{then} \quad &\ddot{\mathbf{d}}_2 = (\mu^*g) \cdot \hat{\ddot{\mathbf{d}}}_1 \\
\text{if } \left|\dot{\mathbf{d}}^*\right| \neq 0 \quad & &\text{then} \quad &\ddot{\mathbf{d}}_2 = (\mu^*g) \cdot \hat{\mathbf{d}}^*
\end{aligned} \tag{5.33}
$$

The results of the validation performed by Yagoda-Biran (2013) for changing input amplitude and input friction are shown in Figure 5.19 and Figure 5.20, respectively and the numerical input parameters are listed in Table 5.3. The obtained agreement between the semi-analytical and numerical solutions is excellent, with the relative error remaining most of the time less than 1%.

The above verification is in fact the same as performed by Kamai (2006) as in both cases the input motion is constrained to one direction, namely a one dimensional solution. Yagoda-Biran (2013) added another motion component into block 1 parallel to the y-axis (see Figure 5.18) with different amplitude and frequency to enable a two-dimensional verification. The result for three different sets of x and y motions for

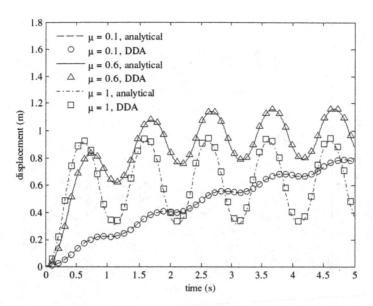

*Figure 5.20* Comparison between the semi-analytical (lines) and 3D-DDA (symbols) solutions for changing interface friction. Input motion frequency of 1 Hz and amplitude of 0.5 remained constant (Yagoda-Biran, 2013).

*Table 5.3* Physical and numerical control parameters used in the verification study of block response to cyclic motion of frictional interface with 3D-DDA.

| Parameter | Value |
|---|---|
| *dd*-dynamic parameter | 1 (fully dynamic) |
| *g0* – normal contact spring stiffness | $1*10^9$ N/m |
| *g1* – time step size | 0.0001 sec |
| *g2* – maximum displacement ratio | 0.001 |
| density | 2250 kg/m$^3$ |
| Young's modulus | 17 GPa |
| Poisson's ratio | 0.22 |

interface friction of 0.6 is shown in Figure 5.21, indicating good agreement, with the relative error remaining less than 1% throughout most of the analysis. The displacements evolution in both the $x$ and $y$ directions with time for a single set of motions ($A_x = 0.3$ m, $f_x = 2$ Hz; $A_y = 0.2$ m, $f_y = 4$ Hz) is shown in Figure 5.22.

Finally, to perform a true three dimensional verification Yagoda-Biran (2013) added a third motion component parallel to the $z$ direction. Adding time dependent displacement in the $z$ direction affects the response of block 2 as it changes the normal force between the two blocks, and therefore the frictional force between them. This in turn changes the acceleration of block 2 ($\ddot{d}_2$) and consequently yields a different

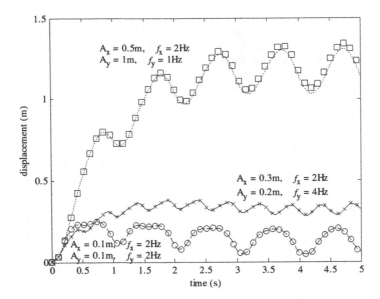

*Figure 5.21* Comparison between analytical (curves) and 3D-DDA (symbols) solutions. Each set of curves and symbols corresponds to a different set of amplitude and frequency for the input displacements, noted beside the data (Yagoda-Biran, 2013).

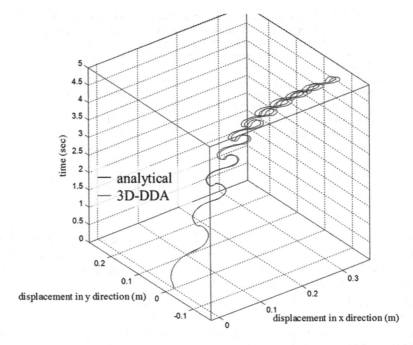

*Figure 5.22* Displacement evolution with time for x amplitude and frequency of 0.3 m and 2 Hz and y amplitude and frequency of 0.2 m and 4 Hz, respectively (Yagoda-Biran, 2013).

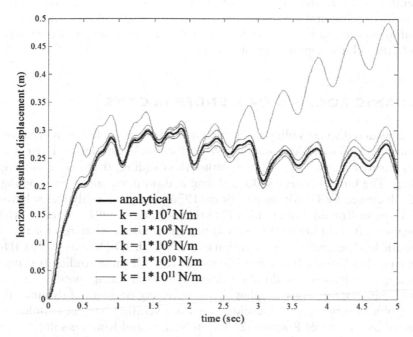

*Figure 5.23* Comparison between semi-analytical (heavy line) and 3D-DDA (light lines) solutions for three-dimensional input motion for five different values of contact spring stiffness (Yagoda-Biran, 2013).

displacement time history. Applying time-dependent displacements in the $z$ direction is actually equivalent to time-dependent changes in $g$: when block 1 has positive $z$ acceleration $(\ddot{d}_1\hat{k} > 0)$, it is added to $g$. When $\ddot{d}_1\hat{k}$ is negative, it is subtracted from $g$. The analytical solution in this case assumes no other effect of the vertical displacement of block 1 on the horizontal displacement of block 2. The induced displacement function Yagoda-Biran used for block 1 is:

$$\mathbf{d}_1(t) = 0.1(1 - \cos(2\pi 2t)) \cdot \hat{i} + 0.1(1 - \cos(2\pi 4t)) \cdot \hat{j} + 0.1(1 - \cos(2\pi t)) \cdot \hat{k}$$

$$(5.34)$$

where $\hat{i}, \hat{j}, \hat{k}$ are unit vectors parallel to the $x$, $y$, and $z$ axes, respectively. The results of the verification study with three components of induced displacements are shown in Figure 5.23 for five different values of normal contact spring stiffness ($k$).

The range of stiffness values that best fits the analytical solution is between $1 * 10^7$ and $1 * 10^9$ N/m, with stiffness of $k = 1 * 10^7$ N/m, or $0.0003\,E * L$, being the optimal selection, where $E$ is the Young's modulus of the block and $L$ is the length of the line across which the contact springs are attached. When considering 3D-DDA, it might be more relevant to compare $k$ to $E * A$, where $A$ is the area across which the contact springs are attached. In this case, $k = 1 * 10^7$ is $\sim 0.0001\,E * A$, not much different from $E * L$. For the results obtained with $k = 1 * 10^7$ N/m, the relative error stays below 3%

for the entire analysis, and the error is well below 10% for $k = 1 * 10^8$ and $1 * 10^9$ N/m as well. It is important to note here that the optimal value for normal contact spring stiffness is found in this case to be between 2 and 4 orders of magnitude less than $E * L$, the value recommended by Shi (1996).

## 5.5 DYNAMIC ROCKING OF SLENDER BLOCKS

We have considered so far sliding along frictional interfaces, a failure mechanism that is relevant mainly to rock slope stability. When the blocks are slender however, cyclic earthquake forces may induce rocking, rather than sliding, that may ultimately lead to toppling. The limits between sliding, sliding and toppling, and only toppling were originally discussed by Goodman and Bray (1976) and their analytical solution was used by Manchu Ronald Yeung in his PhD thesis (Yeung, 1991) to verify 2D-DDA. Yagoda-Biran (2013) in her PhD thesis has modified those boundaries for the case of pseudo-static loading and used her solution to verify both 2D-DDA as well as 3D-DDA methods (see also Yagoda Biran and Hatzor, 2013). All these studies are concerned with limiting equilibrium and do not address dynamic rocking motions that do not necessarily culminate in block toppling. The interesting problem of dynamic rocking of slender blocks subjected to sinusoidal input acceleration has been studied by has been studied by Makris & Roussos (2000). In Makris and Roussos solution the time dependent acceleration is input directly to the center of mass of the block, a free body of which is shown in Figure 5.24.

The centers of rotation of the freestanding column can be either 0 or 0'. Assuming there is no vertical base acceleration, the equations of motion are (Makris and Roussos, 2000):

$$I_0\ddot{\theta} + mgR\sin(-\alpha - \theta) = -m\ddot{u}_g R\cos(-\alpha - \theta), \quad \theta \leq 0 \tag{5.35}$$

$$I_0\ddot{\theta} + mgR\sin(\alpha - \theta) = -m\ddot{u}_g R\cos(\alpha - \theta), \quad \theta \geq 0 \tag{5.36}$$

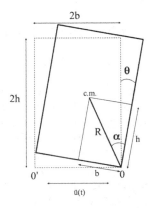

Figure 5.24 Free body diagram and sign convention for dynamic rocking of slender blocks (after Makris and Roussos, 2000). Reproduced with permission from ICE Publishing.

where $I_0$ is the mass moment of inertia, $m$ the block mass, and $\ddot{u}_g$ the ground acceleration. All the geometrical parameters are defined in Figure 5.24. Inserting the definition of $I_o$ into Equation 5.35 and Equation 5.36, introducing the parameter $p$ ($p = \sqrt{3g/4R}$), linearizing the equations due to the slender geometry of the column (small $\alpha$), using a sinusoidal input ground acceleration in the form of $\ddot{u}_g(t) = a_p \sin(\omega_p t + \psi)$ (where $a_p$, $\omega_p$ and $\psi$ are the amplitude, frequency and phase when rocking initiates, respectively) and integrating the equations yields (for complete solution see Makris and Roussos, 2000):

$$\theta(t) = A_1 \sinh(pt) + A_2 \cosh(pt) - \alpha + \frac{1}{1 + \left(\omega_p^2 / p^2\right)} \frac{a_p}{g} \sin(\omega_p t + \psi), \quad \theta \leq 0$$

(5.37)

$$\theta(t) = A_3 \sinh(pt) + A_4 \cosh(pt) + \alpha + \frac{1}{1 + \left(\omega_p^2 / p^2\right)} \frac{a_p}{g} \sin(\omega_p t + \psi), \quad \theta \geq 0$$

(5.38)

The dynamic equations for the angular velocity are (Makris and Roussos, 2000):

$$\dot{\theta}(t) = pA_1 \cosh(pt) + pA_2 \sinh(pt) + \frac{\omega_p}{1 + \left(\omega_p^2 / p^2\right)} \frac{a_p}{g} \cos(\omega_p t + \psi), \quad \theta \leq 0$$

(5.39)

$$\dot{\theta}(t) = pA_3 \cosh(pt) + pA_4 \sinh(pt) + \frac{\omega_p}{1 + \left(\omega_p^2 / p^2\right)} \frac{a_p}{g} \cos(\omega_p t + \psi), \quad \theta \geq 0$$

(5.40)

Yagoda-Biran and Hatzor (2010) used the analytical solution proposed by Makris and Roussos (2000) to verify 2D-DDA for this difficult case of bouncing angle to edge contacts. The results of this verification for input peak acceleration 1% lower and 1% higher than the peak acceleration required for toppling according to Makris and Roussos (2000) solution are shown in Figure 5.25 A and B, respectively, for column width and height of $b = 0.2$ m and $h = 0.6$ m.

In the DDA model the column rests on a fixed base and is subjected to dynamic input at its centroid. The friction angle along the interface is set to 89 degrees to avoid sliding, as the analytical solution ignores sliding. The input numerical control parameters are: energy dissipation coefficient $= 1$, maximum displacement ratio $= 0.0075$, time step size $= 0.0025$ sec, normal contact spring stiffness $= 83 * 10^6$ N/m, $E = 3$ GPa, and $\nu = 0.25$. A remarkably good agreement between the analytical and numerical solutions is indicated, as can be seen from the plotted numerical error that after initial perturbations rapidly decreases below 1%.

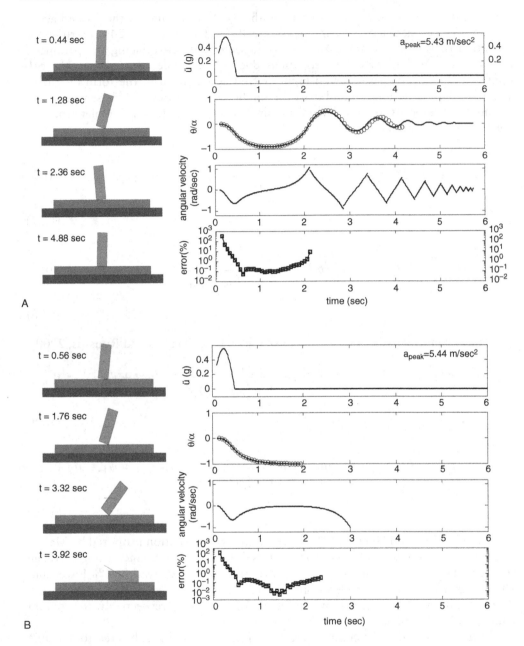

*Figure 5.25* 2D-DDA verification of dynamic rocking of slender blocks (modified after Yagoda-Biran and Hatzor, 2010). A) Peak acceleration 1% lower (A) and 1% higher (B) than peak acceleration required for toppling according to Makris and Roussos (2000) solution.

## 5.6   WAVE PROPAGATION PHENOMENA

We have shown so far that rock mechanics problems concerning stability and failure modes of discrete blocks can be addressed effectively and accurately with DDA as long as the input control parameters are correctly applied. We now turn our attention to wave propagation phenomena through a block system. This problem is important in site response analysis as well as in blasting simulations, both issues that are central to geotechnical earthquake engineering and to mining. Since the formulation of DDA is dynamic in nature, the main issue to clarify here is how restricting is DDA's simply deformable blocks assumption in such simulations? As we have seen in the introductory chapters on the theory of DDA, the blocks in the original DDA are assumed to be stiff and simply deformable, namely the computed stresses and strains are assumed to be distributed homogenously throughout the block; i.e. constant stress and strain within the block is considered in the solution. Is this assumption seriously compromising our ability to compute wave propagation accurately with DDA? We will explore this issue in this section.

### 5.6.1   P wave propagation

The ability to model properly P-wave propagation is a first prerequisite for modeling blasting effects and shock wave propagation in tunneling, rock slope, and general mining applications with DDA. When simulating a blast, pressure waves can be made to propagate radially from a point source (e.g. Zelig et al., 2015). Here, for purposes of verification only, we will constrain the pressure wave to propagate in one dimension, in direction parallel to the axis of an elastic bar divided to elements of different lengths, to allow for accurate stress analysis through the bar. An important question that comes to mind is exactly how small should those "artificial" blocks be in order to ensure solution accuracy. Furthermore, would it be correct to assume that accuracy of the numerical solution will invariably increase with decreasing element size, or perhaps there is an optimal element size, below of which the error might increase? In the finite elements method (FEM) for example it was found empirically that the optimal element side length should be smaller than approximately 1/12 the wavelength (e.g. Lysmer and Kuhlemeyer, 1969). Does this empirical rule of thumb apply also to DDA? Specifically for DDA, how does the numerical penalty value, or contact spring stiffness, affect the accuracy and finally, how sensitive is the accuracy of the solution to the time interval, or the size of the time step?

To try and obtain some answers to these key questions we use for a model an elastic bar divided into elements of varying lengths (Figure 5.26) following work performed by Bao and coworkers (Bao et al., 2014; Bao et al., 2012). The modeled bar is 100 meters long and is 1 m high (and 1 m wide). A measurement point at the midsection of the bar is used to record the solution output. The material properties used for the analysis are listed in Table 5.4. For these simulations the original DDA boundaries, rather than the modified non-reflective boundaries (Bao et al., 2012; Jiao et al., 2007), are used.

The bar is subjected to a one-cycle sinusoidal loading function as follows:

$$F(t) = 1000 \sin(200\pi t) \quad \text{(unit: kN)} \tag{5.41}$$

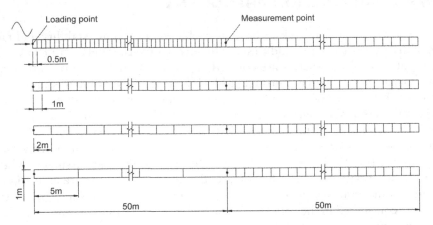

*Figure 5.26* Model of elastic bar used for P-wave propagation verification.

*Table 5.4* Input parameters for wave propagation verifications.

| Block material | Unit mass (kg/m$^3$) | 2650 |
| --- | --- | --- |
| | Young's modulus (GPa) | 50 |
| | Poisson ratio | 0.25 |
| Joint material | Friction angle | 35° |
| | Cohesion (MPa) | 24 |
| | Tensile strength (MPa) | 18 |

Since the assumed cross sectional area of the bar is $1\,\text{m} \times 1\,\text{m}$, the peak stress amplitude entering the bar from the force pulse (Equation 5.41) which is applied at the left end of the bar (Figure 5.26) is 1 MPa.

The analytical solution for wave velocity in a rod subjected to one dimensional P wave propagation is given by (Kolsky, 1964):

$$V_p = \sqrt{\frac{E}{\rho_o}} \tag{5.42}$$

where $\rho_o$ is the density and $E$ is the Young's modulus of the material. The relative error between analytical and computed stress or velocity is defined here as follows:

$$e = \frac{|A_1 - A_0|}{A_0} \times 100\% \tag{5.43}$$

where $A_1$ is the computed wave amplitude or velocity at a reference measurement point in the model, and $A_0$ is the analytical wave amplitude or velocity at a given point.

The verification is performed using four different time intervals of 0.01 ms, 0.05 ms, 0.1 ms, and 0.5 ms and four different block lengths of 0.5 m, 1 m, 2 m, and 5 m. The theoretical P-wave velocity in the bar is expected to be 4344 m/s

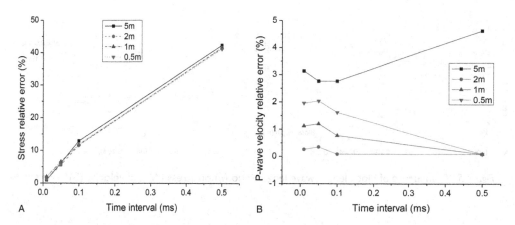

*Figure 5.27* Relative stress (A) and velocity (B) errors at center of the elastic bar as function of time interval and element size.

(Equation 5.42), and as mentioned earlier, the theoretical amplitude of the P wave is expected to be 1 MPa (Equation 5.41). The numerical P-wave velocity is calculated from the travel time of the wave between the incident point at the left end of the bar and the measurement point at the center of the bar. The numerical amplitude of the wave is computed directly by DDA as the magnitude of the horizontal normal stress component at the measurement point.

The relative stress and velocity errors are plotted in Figure 5.27 as a function of time step size and block length. Inspection of Figure 5.27A clearly reveals that the time interval has a very significant effect on the stress error, but apparently less on the velocity (Figure 5.27B). Regarding the influence of element size the reverse is true: while the stress error seems to be unaffected by block size, the velocity error clearly is. As would be intuitively expected, the velocity error indeed decreases when the element size decreases from 5 meters to 2 meters. However, when the block size further decreases to 1 meter the velocity error increases, and continues to increase when the block size is further reduced to 0.5 m (see Figure 5.27B). This result appears to be in agreement with the rule of thumb regarding the optimal ratio between the element size and wave length proposed by Lysmer and Kuhlemeyer (1969): $\eta = 1/12 = 0.08$. Considering that the period of the incident P wave used in our verification study is 0.01 s, and that the velocity of the wave is 4344 m/s, the wavelength here is 43.43 m. The optimal element size according to Lysmer and Kuhlemeyer (1969) rule of thumb should be 3.62 m and here we are getting and optimal block length of 2 m, not very different. The exact numerical errors with respect to $\eta$ are shown in Figure 5.28 where indeed it can be appreciated that near $\eta = 0.08$ both stress and velocity errors are low; the minimum error however appears to be at $\eta$ close to 0.046, at about half the recommended value by Lysmer and Kuhlemeyer (1969). The influence of time interval on the obtained waveform at the center of the bar for a block length of 1 m is shown in Figure 5.29. The complete waveform reflects both the wave stress and velocity. Clearly, the accuracy

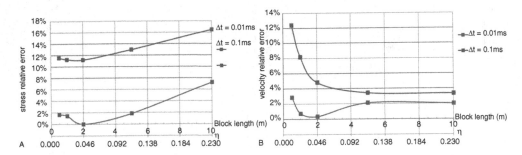

*Figure 5.28* Influence of block length/wave length ratio ($\eta$) on stress (A) and velocity (B) errors

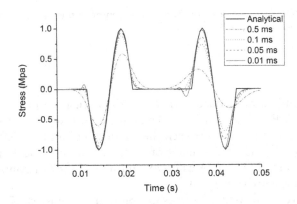

*Figure 5.29* Obtained waveform at center of bar as a function of time step size for element size of 1 m.

of the waveform increases with decreasing time interval most likely due to the strong influence of the time interval on the stress accuracy.

It has been observed by many workers that in dynamic simulations the penalty parameter, or the contact spring stiffness, has a very strong influence on the solution accuracy. Doolin and Sitar (2002) for example pointed out that increasing the stiffness of the contact springs may increase the overall accuracy of the solution in slope simulations with DDA.

Seven different contact stiffness values, between $10E$ to $640E$, are tested where $E$ is the Young's modulus of block material (here $E = 50$ MPa). The sensitivity analysis results (Figure 5.30) suggest that while the stress error is not very sensitive to the penalty parameter, the velocity error is, although not in a straightforward way. The velocity error decreases with increasing penalty parameter up to an optimal value, here $160E$, beyond which it increases again. The optimal contact spring stiffness found here is approximately three time higher than the rule of thumb suggested by Shi (1996) of $k = E_o L_o$ where $k$ is the penalty number (or contact spring stiffness), $E_o$ is the Young's modulus of the block, and $L_o$ is the block diameter.

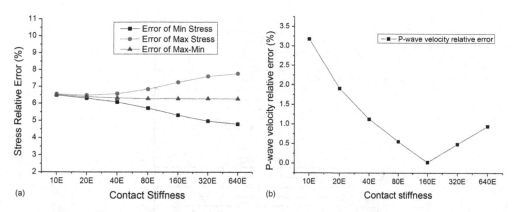

(a) Contact Stiffness
(b) Contact stiffness

*Figure 5.30* Influence of contact spring stiffness on the relative stress (a) and velocity (b) error in elastic bar with element size of 1 m. Note that while the stress error is less sensitive to contact spring stiffness, the velocity error decreases with increasing penalty value (in units of Yong's modulus E) up to an optimal value (here 160E) beyond which the error increases again.

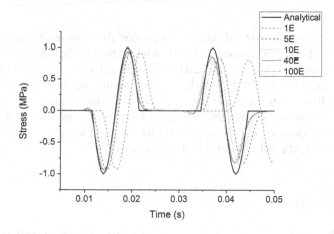

*Figure 5.31* Influence of contact spring stiffness on waveform at center of bar, element size 1 m.

The influence of the penalty parameter on the waveform is shown in Figure 5.31. Because of the great sensitivity of the velocity error to the $k$ value, simulations with low $k$ values yield artificial attenuation of the waveform resulting in lower wave velocity and amplitude. This effect is remedied well with increasing the value of the penalty parameter, but up to a point, as can be inferred from Figure 5.30.

## 5.6.2 Shear wave propagation

When studying strong ground motion effects on rock masses and built structures the vertical propagation of shear waves is of major concern. To enable DDA verification of vertical shear wave propagation we set up a simple geometrical model, as in the case of

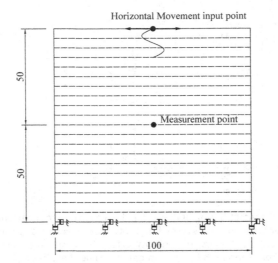

*Figure 5.32* Configuration of the 1-D S-wave propagation model, note implementation of non-reflective boundary at the bottom of the modeled domain (indicated dimensions are in meters).

P waves, but here we study vertical shear wave propagation through a stack of horizontal layers, where the layers are made very wide with respect to their height (Figure 5.32). Non reflective boundaries (Bao *et al.*, 2012) are applied at the foundation to restrain artificial reflections that might distort the measurements, taken at a point positioned at a distance of 50 m from the upper surface, where the input motion is applied.

Vertical propagation of S waves through a horizontally layered system will only induce horizontal displacements, it can therefore be considered as a one-dimensional S-wave propagation, the velocity of which is given by:

$$V_s = \sqrt{\frac{E}{2\rho_0(1+\nu)}} \qquad (5.44)$$

where $\nu$ is the Poisson's ratio. The relative error between the analytical and numerical solutions is found using Equation 5.43, as before.

The layers are 100 m wide and their height varies in the sensitivity analyses between 0.5 m, 1 m, 2 m, and 5 m. The input horizontal motion at the upper surface in the verification study is:

$$D(t) = 0.1\sin(200\pi t) \quad \text{(unit: m)} \qquad (5.45)$$

The theoretical S-wave velocity for the stacked layers model shown in Figure 5.32 is 2747 m/s (Equation 5.44) given the material properties listed in Table 5.4. The relative errors with respect to S wave velocity are shown in Figure 5.33 for varying time intervals and contact spring stiffness. The results suggest that S wave velocity errors are not sensitive to the time interval but that the deviation from the exact solution increases with decreasing block size (Figure 5.33a). The influence of contact stiffness

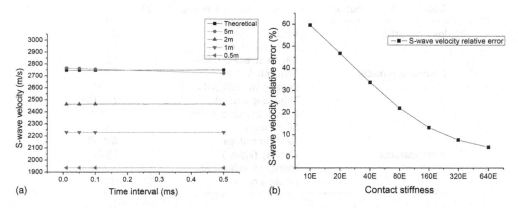

Figure 5.33 Verification of vertical S wave propagation through horizontally layered stack. a) Influence of time step size, b) influence of contact stiffness.

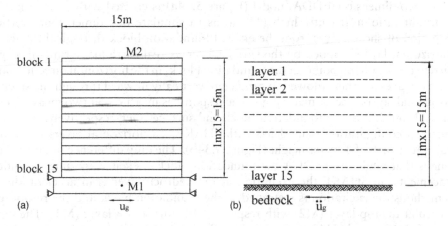

Figure 5.34 Site response analysis with DDA. a) DDA model, b) Equivalent SHAKE model.

on velocity error appears to be straightforward, with great improvement in accuracy with increasing contact stiffness (Figure 5.33b). It is difficult to relate to Shi's rule of thumb here for the optimal $k$ value as the blocks "diameter" is not well defined in the given geometry. If we consider the length of the block edge at which the contact springs are attached (100 m), then the recommended $k$ value by Shi would be $100E$, yet we see that we can continue improving accuracy by increasing the contact stiffness 6.4 times this value.

To check the possibility to perform actual site response analysis with DDA we need to compare DDA results to an alternative, well established, computational method. We chose the program SHAKE for our verification (Schnabel et al., 1972) because its algorithm has been verified by many workers and its accuracy is well established for the underlying assumptions and boundary conditions. The DDA and the equivalent SHAKE models used for this verification study are shown in Figure 5.34. The input parameters for the DDA simulations in this verification are listed in Table 5.5.

Table 5.5 DDA input parameters for verification against SHAKE.

| Joint material | Friction angle | 50° |
|---|---|---|
| | Cohesion strength (MPa) | 10 |
| | Tensile strength (MPa) | 50 |
| Control parameter | Dynamic factor | 1.0 |
| | Penalty stiffness (GN/m) | 1500 |
| | Time step size (s) | $1 \times 10^3$ |
| | Max displacement ratio | 0.0008 |
| | SOR factor | 1.5 |
| | Total time steps | 60000 |
| Block material | Density (kg/m³) | 2643 |
| | Young's modulus (GPa) | 4.788 |
| | Poisson ratio | 0.25 |

The two-dimensional DDA model (Figure 5.34a) is created with layer length to layer height ratio sufficiently high (15) so as to simulate one dimensional vertical propagation of shear waves from the excited foundation block through the stack of the horizontal layers, topped by the surface layer. A real earthquake time history is applied to the four fixed points at the foundation block, in the horizontal direction only. The shear waves are then allowed to propagate vertically upward through the stack of 15 horizontal layers, each 1 meter high. The response is measured at two measurement points M1 and M2 at the foundation block and surface layer, respectively. The same geometrical configuration is modeled with SHAKE, 15 horizontal layers of infinite lateral extent each of 1 meter height (Figure 5.34b). The only difference in the loading scheme is that while in the DDA the foundation block is excited by time dependent displacements, in SHAKE the excitation at the bedrock layer is in acceleration. In both methods the excitation is restricted to the foundation block and the response is measured at the top layer (M2) with respect to the foundation layer (M1). The input motions for DDA and SHAKE are shown in Figure 5.35.

The damping ratio which is necessary for meaningful comparisons between SHAKE and DDA is obtained here by controlling the time step size in DDA utilizing the inherent algorithmic damping (see Doolin and Sitar, 2004) as no other sort of damping is applied in the DDA version we used for verifications. A damping ratio of 2.3% thus obtained with DDA is input to the SHAKE model to enable quantitative comparison of the results.

The spectral amplifications obtained with the two different methods are plotted in Figure 5.36. With DDA the maximum amplification is 29.07 and the resonance frequency is 13.95 Hz; with SHAKE the maximum amplification is 28.93 at frequency 14.21 Hz. The agreement between the two methods is striking, suggesting that accurate site response analysis is possible with DDA, even when higher order terms are neglected due to first order approximation and the simply deformable blocks assumption. Moreover – it is clearly demonstrated here that loading the foundation block with displacement or acceleration time histories is equivalent, an issue that has focused some debate recently (e.g. Wu, 2010). Results of site response analysis for non-homogenous layers also show excellent agreement between the two methods (see Bao *et al.*, 2014).

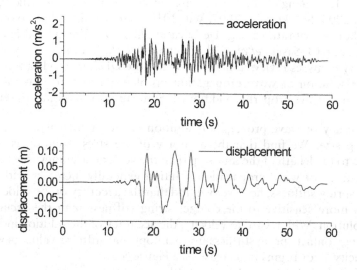

Figure 5.35 Input motion used for verification study (CHI-CHI 09/20/99).

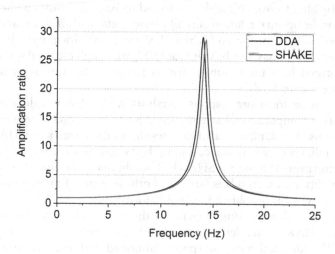

Figure 5.36 Spectral amplification obtained with DDA and SHAKE for 15 horizontal layers of homogenous material properties.

### 5.6.3 Concluding remarks regarding wave propagation accuracy

The ability to model wave propagation and to perform site response analysis with DDA is checked and verified here using simple benchmark tests. This verification paves the way for robust analysis of more complex wave propagation problems in rock mechanics practice, where the medium is fractured and discontinuous. Such applications could involve blasting (e.g. Ning *et al.*, 2007; Yang and Ning, 2005; Zhao *et al.*, 2010; Zhao *et al.*, 2011), rockbursts (e.g. He *et al.*, 2016; He *et al.*, 2012; He *et al.*, 2015;

Jiang *et al.*, 2010; Zhang *et al.*, 2013), dynamic slope stability (e.g. Miki *et al.*, 2010; Sasaki *et al.*, 2005; Sasaki *et al.*, 2007; Wu, 2010), and stability of masonry structures against earthquake vibrations (e.g. De Luca *et al.*, 2004; Kamai and Hatzor, 2008; Koyama *et al.*, 2013; Sasaki *et al.*, 2011; Stefanou *et al.*, 2006).

Sensitivity analyses reviewed here show that the DDA method can provide good accuracy for solution of wave propagation problems, provided that the numerical control parameters are properly conditioned. Some important points to remember:

- The accuracy of wave propagation solution generally increases with decreasing time step size. We find that the accuracy of the stress solution is much more sensitive to the length of the time step than the velocity accuracy (see Figure 5.27).
- The accuracy of wave propagation solution generally increases with increasing contact spring stiffness, however we find that the accuracy of the velocity solution is much more sensitive to the contact string stiffness than the accuracy of the stress solution. With regard to velocity, the accuracy of the solution increases with increasing contact spring stiffness up to an optimal stiffness value, beyond which the velocity error begins to increase (see Figure 5.30).
- Regarding the rule of thumb proposed by Shi (1996) for optimal contact spring stiffness $(k)$: $k = E_o L_o$, we find that for the dynamic wave propagation benchmark tests performed here the optimal $k$ value is within the same order of magnitude as proposed by Shi (1996), although it is found to be 3 to 6 times higher.
- We find that the optimal relationship $(\eta)$ between element size $(\Delta x)$ and wavelength $(\lambda)$: $\eta = \Delta x / \lambda = 1/12$, proposed for the FEM on an empirical basis by Lysmer and Kuhlemeyer (1969) may also be valid for DDA, although for the P wave benchmark tests performed here the optimal ratio is found to be approximately half that number (see Figure 5.28).
- The ability to perform site response analysis with DDA is validated here using an alternative computational scheme (SHAKE). To compare between DDA and SHAKE we use the damping ratio that results in the numerical DDA simulations due to the inherent algorithmic damping, by tweaking the time step size, and use that damping ratio as input for SHAKE. The obtained agreement between the two completely different methods is striking, both in terms of the obtained resonance frequency as well the amplification (see Figure 5.36).
- DDA was run with input time dependent displacements at the foundation block whereas in SHAKE time dependent accelerations were used for input (see Figure 5.35. The identical response spectra obtained with the two different loading mechanisms proves that they are equivalent, and either one can be used, as long as they are derived from the same ground motion record.

# Chapter 6

# Underground excavations

## 6.1 INTRODUCTION

Since its publication in 1993 (Shi, 1993) DDA has been applied extensively to rock engineering projects from around the world, while it was still being validated, verified and modified by the rock mechanics research community, an effort that is still on-going at the present time. Among the initial test grounds for DDA were some high profile cases, such as the Three Gorges Project (e.g. Dong et al., 1996), the Masada world heritage site (Hatzor et al., 2004), the Vajont slide (Sitar et al., 2005), the Jinping hydroelectric station (e.g. Chen and Deng, 2008), the Tanjiashan landslide triggered by the Wenchuan Earthquake (Wu et al., 2009), the Bayon temple at Angkor Thom and Pharaoh Khufu's pyramid (Ohnishi et al., 2012), to name but a few. DDA is now being applied in rock engineering projects worldwide on a routine basis, as can be appreciated by surveying the professional literature in recent years. In this chapter DDA application in the engineering of underground excavations is demonstrated using two end members representing two extreme cases of stability concern: 1) shallow underground excavations where gravitational loading controls the deformation, and 2) deep underground excavations where deformation is controlled by the level of in-situ stress. In both cases the rock mass is assumed discontinuous so that the mode of deformation is controlled by the interactions between pre-existing blocks. The chapter closes with demonstration of rockbolting design with DDA where the anisotropy of the rock mass controls the length of bolts required for safety.

## 6.2 SHALLOW UNDERGROUND EXCAVATIONS

Shallow underground excavations are at risk due to the low level of in situ stresses. In lack of high stresses, the available shear strength of discontinuities is hardly mobilized once the opening is created, and therefore loosening, sliding, and falling of blocks into the excavation space are common. Of particular concern is the height of the arching mechanism, which typically develops in the roof, below which loosening and falling of blocks is to be expected. The assumed height of the arching mechanism therefore determines the minimum rock bolt length required to ensure stability.

There is no analytical solution for the height and thickness of the stressed zone in the arching mechanism which develops in the roof of an opening excavated in an initially discontinuous rock mass. Terzaghi (1946) introduced his famous rock load on tunnel supports classification, which provides the height of the loosened zone in the roof as function of the structure of the rock mass, based on empirical observations from tunnels excavated primarily in the Austrian Alps. As much as we have found during years of research, Terzaghi's rock load classification is valid and can be used as a rule of thumb in lack of any other design tools. Although the result would typically be conservative, in some cases we found it to be accurate. Of course, if the structure of the rock mass deviates from the classes of rock structures discussed by Terzaghi, then more sophisticated design tools must be employed, namely numerical discrete element methods.

### 6.2.1 Block interactions

A semi analytical approach for determination of the height and thickness of the stressed arch in the roof of an excavation in discontinuous rock has been presented by Beer and Meek (1982) and their procedure is reviewed by Brady and Brown (2004). This procedure is applicable for a simplification of the discontinuous rock mass structure in the roof by considering a horizontally layered roof transected by vertical joints. Moreover, only the lowermost layer in the immediate roof is considered, and it is assumed to be transected by a single vertical joint in its center. The approach is applicable for continuous layers as well, the rationale being that with downward beam deflection the layer would crack in the center and a vertical joint will propagate from the immediate roof to the top of that lowermost layer. The developed tension crack will split the initial continuous beam into two blocks that will interact with one another to create the arching stresses via a hinged-beam mechanism. Joint friction and spacing are ignored in this approach, but layer thickness is not. Of course, once the rock mass structure deviates even slightly from this assumed configuration the suggested solution would not be applicable. A considerable amount of discussion of Beer and Meek approach which has been referred to as the "Voussoir beam analogue" has been conducted in the professional literature (Sofianos, 1996; Diederichs and Kaiser, 1999b, c, a; Sofianos, 1999; Bakun-Mazor et al., 2009) and indeed this approach can and should be used for example in coal mines where the excavation is performed through horizontally layered strata.

To demonstrate the limitations of the Voussoir beam methodology consider the case of Zedekiah cave, discussed in some detail by Bakun-Mazor et al. (2009). Theoretically, the layout of the underground structure would be suitable for modeling by means of the Voussoir beam analogue. The ca. 2000-year-old limestone quarry 25 meters below the old city of Jerusalem was excavated through a rock mass consisting of horizontal bedding planes that are transected by vertical joints. A representative DDA mesh is shown in Figure 6.1; note that particular attention has been given to modeling the mechanical layering structure observed in the field.

Assuming continuous roof beams, the theory of elasticity may be invoked to obtain the maximum deflection and maximum tensile stress at mid-section (see Obert and Duvall, 1967). For an excavation span of 30 m, layer thickness of 0.85 m, unit weight of $18 \, kN/m^3$, and Young's modulus of 8 GPa, the maximum tensile stress at the lowermost fiber of the immediate roof beam would be 10.5 MPa. The measured tensile strength of the rock in direction parallel to the bedding is only 2.8 MPa, less than a third, and therefore according to theory of elasticity, had the roof beam been continuous the roof

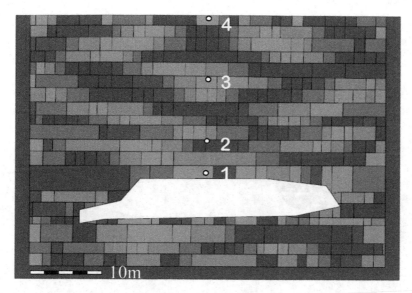

Figure 6.1 DDA model of the rock mass structure at Zedekiah cave with location of measurement points for further analysis depicted.

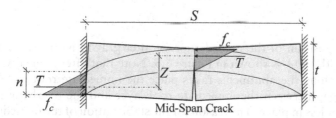

Figure 6.2 An idealization of the immediate roof after fracturing at mid-section by a tension crack.

must have been fractured in tension at mid section by a vertical crack propagating normal to the bedding direction. Note that the compressive strength of this rock in direction parallel to bedding is 16.4 MPa excluding the possibility of failure in crushing at mid-section.

If indeed that scenario had materialized and the immediate roof split at mid-section into two blocks, the resulting configuration could be analyzed by means of the Voussoir beam approach (see Figure 6.2). However, with the given geometrical and mechanical parameters the beam would have failed in a "snap through" mechanism (see Brady and Brown, 2004) when analyzed this way due to the large span (30 m) with respect to the thickness of the layer (0.85 m) and the relatively soft material stiffness.

The only way to explain the exiting stability of the model shown in Figure 6.1, and indeed of the main chamber of Zedekiah cave that has stood unsupported for over two millennia, is to allow in the analysis for the interaction between discrete blocks in the roof. To do this effectively, we must employ a numerical discrete element analysis.

*Figure 6.3* Deformation of immediate roof in Zedekiah cave as modeled with DDA. Note geometry of arching in the discontinuous roof, as delineated by the orientation of the principal stress trajectories.

The final position of the immediate roof after gravity is turned on in the model as obtained with DDA is shown in Figure 6.3. Note that after some downward shear displacement along the abutments takes place the immediate roof stabilizes, leaving a permanent gap between the immediate roof layer and the overlying layers, which remain more or less in place. The reason for the stabilization of the immediate roof after some initial shear along the abutments takes place, is the effective transfer of normal stresses through the interacting blocks, thus better mobilizing the shear strength of the joints and arresting any further downward displacement.

The dynamic deformation process can be better understood by inspection of Figure 6.4 where the vertical downward displacement and horizontal stress evolution in the four measurement points shown in Figure 6.1 are plotted in panels (a) and (b) respectively as function of real time. Consider measurement point 1 at the immediate roof marked by open diamonds in the figure. Once the opening is created at time 0 s, it begins to exhibit downward vertical displacement at a constant velocity, achieved by shearing deformation of the entire immediate roof layer along the abutments. After three seconds the vertical displacement is suddenly arrested and at the same time the horizontal stresses at that point abruptly increase from zero to 700 kPa. With the given friction coefficient for the joints in the simulation ($\mu = 0.87$) this level of horizontal stress is sufficiently high to arrest any further downward displacement under the gravitational pull on the system.

We see through this example the importance of modeling the deformation of the entire block system rather than separating the problem into few representative elements that can be handled more easily using existing analytical or semi-analytical

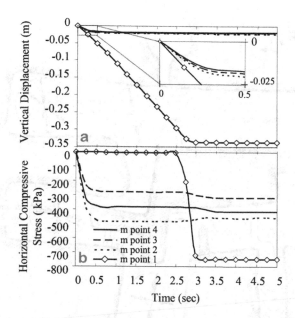

*Figure 6.4* DDA results of deflection and horizontal stress evolution in the four measurement points shown in Figure 6.1 (after Bakun-Mazor et al., 2009).

solutions. By considering the simultaneous deformation of the entire block system in the roof, a more accurate solution of the problem is obtained, thus allowing decision makers a more realistic assessment of the risk associated with the underground excavation at hand.

## 6.2.2  Joint spacing

Terzaghi (1946) has already considered the role of joint spacing in his famous rock load classification system, distinguishing between "hard and intact", "hard stratified or schistose", "massive, moderately jointed", "moderately blocky and seamy", "very block and seamy", and "completely crushed but chemically intact", before moving on to squeezing and swelling grounds, mechanical processes which are beyond the scope of this book. Similarly, all popular empirical rock mass classifications, namely the RMR (Bieniawski, 1974), Q (Barton *et al.*, 1974), and GSI (Hoek and Brown, 1997) systems, follow in Terzaghi's footsteps and assign high influence to the density of the jointing pattern in the overall rating.

While the empirical classifications realize the importance of joint spacing and address this parameter when assigning a global "ranking" to the "quality" of the rock mass, existing analytical tools often overlook it. The Voussoir beam analogue discussed in the previous section for example completely ignores joint spacing, as it considers only a single joint in the middle of the roof. But what if the immediate roof layer was transected by more than one vertical joint, as is often is the case? How would the spacing between the joints, or the relationship between the block length and the

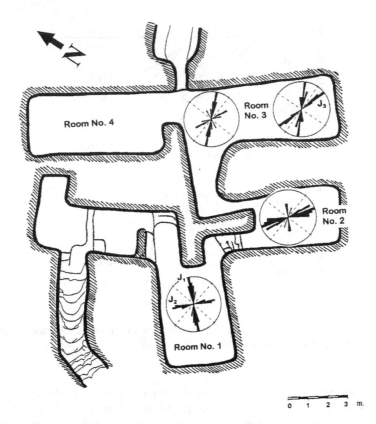

*Figure 6.5* Layout of the underground water system discovered under Tel Beer-Sheva revealing the alignment of the excavated chamber walls with the strikes of the principal joint sets shown as embedded rose diagrams (after Hatzor and Benary, 1998).

excavation span, influence the stability of the underground structure? Studying historic excavations in horizontally layered and vertically jointed rock masses can shed some light on this problem and provide constraints on predictions otherwise obtained by running countless numerical simulations on the computer.

A case in point is the underground water storage system discovered under Tel Beer-Sheva, an archeological site near the modern city of Beer-Sheva (where Ben-Gurion University of the Negev is located). The water storage system dates back to the Israelite period, ca. 3000 years before present, and is believed to have been used to store water during siege on the ancient city walls, by diverting water from a nearby stream in aqueducts that were excavated below the city walls and connected between the stream and the storage system. The system, comprised of several square chambers, was excavated in horizontally layered and vertically jointed chalk, keeping the chamber walls aligned with the strike directions of the principal joint sets that are roughly orthogonal to one another (see Figure 6.5).

The roof of the system has collapsed during excavation as attested by the application of plaster on the exposed sidewalls extending above the level of the original roof,

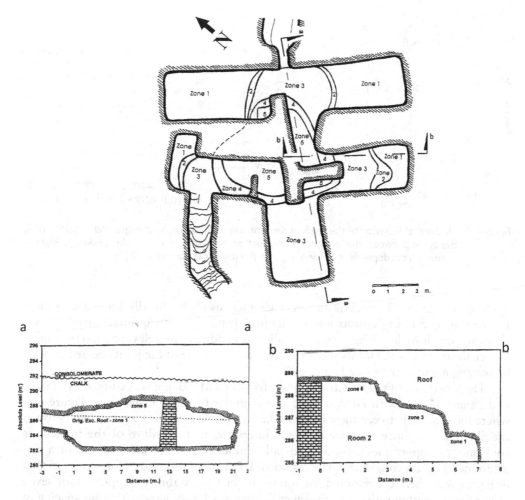

*Figure 6.6* Plan (top) and cross sections (bottom) of the collapsed roof of the underground water
system at Tel Beer Sheva (after Hatzor and Benary, 1998).

and by the massive support pillar constructed by the ancient engineers. Inspection of
the system reveals that the construction of the support pillar was sufficient to arrest
any further deformation of the roof, which indeed remained unsupported until discovered by modern archeologists several decades ago. A plan of the roof and two cross
sections (Figure 6.6) show that the roof collapsed into a more stable dome structure,
the contours of which delineate, most likely, the geometry of the arching mechanism
that developed in the roof once the original excavation was attempted.

The well documented failure of the immediate roof at Tel Beer Sheva provides us
with a good opportunity to check the predictive capability of DDA when studying the
stability of stratified and jointed rock mass structures.

The rock mass consists of horizontal bedding planes with average thickness
of 0.5 m, and two principal joint sets ($J_1$ and $J_2$ in Figure 6.5) both with mean

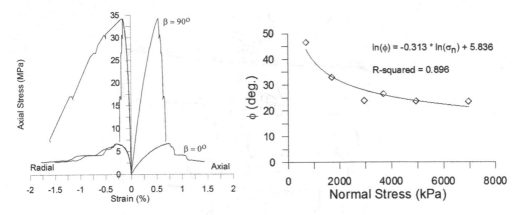

*Figure 6.7* Mechanical behavior of the chalk at the site. Left – Anisotropic strength and elasticity (β is the angle between the compression direction and the normal to the bedding), Right – normal stress dependent friction angle of the joints (Tsesarsky, 2005).

spacing of 0.25 m. The rock is an upper Cretaceous chalk, locally known as Taqiya Formation, exhibiting anisotropic mechanical behavior with ultimate strength and elasticity much higher when tested parallel to bedding, the relevant direction here. Direct shear tests of the joints show that the friction angle of the joints decreases with increasing normal stress (see Figure 6.7).

The sensitivity of the structure to joint friction and spacing was studied by Hatzor and Benary (1998) with DDA using a mesh similar to the one shown in Figure 6.8, where the spacing between the joints and the friction on the joints were varied between the simulations. Since here we are only interested in the failure of the immediate roof only the upper mesh, used for detailed modeling of the deformation of a single jointed layer, is considered. The results of DDA simulations are plotted graphically in Figure 6.9 where the required friction angle for layer stability is reported for seven joint spacing configurations. In the most slender block geometry with joint spacing of $S_j = 25$ cm, the layer consists of 28 blocks, with block dimension $S_j/t = 0.5$ and scaled block length of $S_j/S = 0.04$. In the extreme case where the layer is completely continuous, the joint spacing is $S_j = 700$ cm, the layer consists of one block only with block dimension of $S_j/t = 14$ and a scaled block length $S_j/S = 1$.

The relevant case for comparison with the Voussoir beam analogue discussed in the previous section would be for joint spacing of $S_j = 350$ cm, where the layer is split into two blocks at mid-section. As can be inferred from inspection of Figure 6.9, a friction angle of 45° would be required to keep this configuration stable. Indeed this result is in agreement with the Voussoir beam analogue which predicts that the two block beam will be safe against shear along the abutments with the given geometrical and mechanical parameters provided that the available friction angle for the joints is greater than 40° (see Hatzor and Benary, 1998). Due to the relatively high compressive strength and elasticity of this chalk in direction parallel to bedding (the direction of the thrust force in the beam), failure by crushing or by a "snap through" mechanism (see Brady and Brown, 2004) are ruled out by the Voussoir beam methodology.

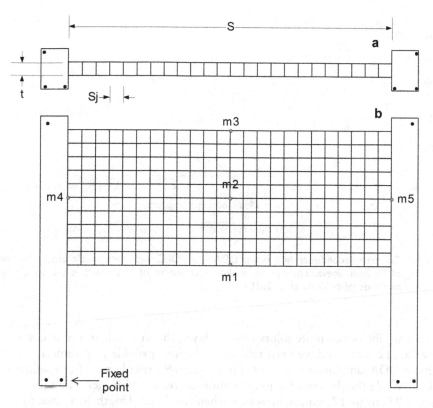

*Figure 6.8* DDA mesh similar to the one used for sensitivity analyses (after Tsesarsky and Hatzor, 2006).

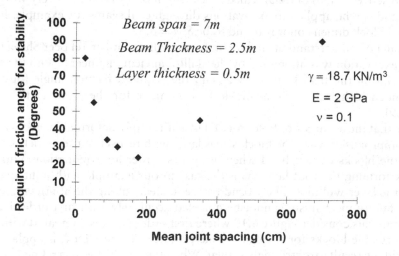

*Figure 6.9* Results of sensitivity analysis of immediate roof layer in a typical chamber at Tel Beer-Sheva (after Hatzor and Benary, 1998).

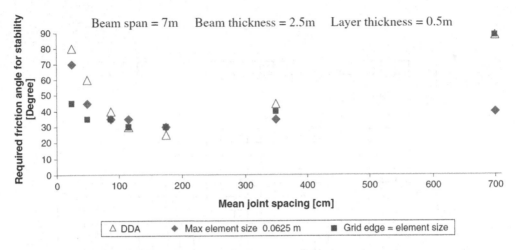

Figure 6.10 Comparison between first order DDA and UDEC with deformable blocks for the case of Tel Beer Sheva. The optimal scaled block length of $S_j/S = 0.25$ is confirmed by both methods (after Barla *et al.*, 2010).

Once we introduce more joints into the layer, the Voussoir beam analogue ceases to be valid of course, and we must rely on predictions provided by numerical analysis. Results of DDA simulations as plotted in Figure 6.9 provide a useful insight into this problem. While the demand for joint friction decreases with increasing joint spacing from $S_j = 25$ cm to $175$ cm, it increases when the block length increases from what appears to be an optimal scaled length of $S_j/S = 0.25$, where the required friction for stability is only $25°$, to what appears to be the worst case of $S_j/S = 1$.

The discovery of an optimal scaled block length of $S_j/S = 0.25$ is very worthwhile in itself, and can be applied to excavations through coal seams for example. In terms of optimal block dimension it is found to be $S_j/t = 3.5$.

We can now understand the failure of the immediate roof at Tel Beer-Sheva the first time the excavation was attempted by the skillful ancient engineers. With the average joint spacing in the field being only $25$ cm, the required friction angle for stability is $80°$, much higher than the available friction angle for the joints at the site (see Figure 6.9).

Note that the accuracy of first order DDA with simply deformable blocks assumption is compromised when the block size is large with respect to the modeled domain, because the blocks cannot bend when using first order approximation. With a single block forming the roof layer, as is the case in our example with joint spacing of $700$ cm, the layer will most likely bend before it shears along the abutment, and that complex failure mechanism cannot be modeled accurately with first order DDA. To demonstrate this consider Figure 6.10 where first order DDA is compared with UDEC with deformable blocks for the exact same problem. When UDEC is applied with a coarse grid the results are strikingly similar. When the UDEC mesh is refined, however, we see that block bending in the case of $700$ cm spacing allows roof stabilization even with a relatively low friction angle. Note also that when the block size decreases with respect to the size of the modeled domain the agreement between first order DDA and UDEC with deformable blocks greatly improves.

Interestingly, with both methods the existence of an optimal scaled block length of $S_j/S = 0.25$ is confirmed, suggesting that this could indeed be a rule of thumb for assessing the risk of roof collapse when excavating through horizontally layered and vertical jointed rocks. This would require the geological engineers to properly asses the layer thickness and the joint spacing at the site.

### 6.2.3 Excavation depth

The significance of the excavation span on underground opening stability has been realized long ago (e.g. Lauffer, 1958; Bieniawski, 1970) and has been incorporated in modern empirical rock mass classifications (e.g. Barton et al., 1974; Bieniawski, 1976) as a parameter of major importance. We have seen in the previous section that the relationship between excavation span and joint spacing is also important, as it controls the scaled block length $(S_j/S)$ which as shown has a strong influence on the stability of the immediate roof. When discussing shallow excavations, however, the issue of the available overburden must also be considered. A well-known rule of thumb is that at least two tunnel diameters should be allowed as overburden to ensure short term stability, particularly in portal areas (Brekke, personal communication). But what if we do not have two tunnel diameters? Could we expect the opening to remain stable when excavated through discontinuous rock masses while violating this rule of thumb? In the case of Tel Beer Sheva discussed in the previous section an average room diameter is 7 meters, therefore by this convention 14 meters of overburden would be required. The existing cover depth at that site is much less than that, and indeed, as we have seen, the opening collapsed immediately after it was attempted. On the other hand, in the case of Zedekiah cave discussed earlier in this chapter, the span of the main chamber is nearly 40 meters. Do we really need 80 meters of overburden to ensure the stability of this underground opening? The amount of overburden above Zedekiah cave is 25 meters only, and yet it has stood unsupported for two millennia, save some local slab detachments from the immediate roof. So clearly, some probing into this rule of thumb is called for, particularly now that we have adequate analytical tools to address such questions.

Our discussion so far has been restricted to rock masses consisting of horizontally layered and vertically jointed rock masses, and we will stay with this assumption here as well. One reason for this is that we have a lot of experience with this type of rock mass structure, another is that this is the most dangerous kind of structural setting with the least amount of interlocking between blocks. In this structural configuration almost every block surrounding the opening is free to move into the opening from a kinematic stand point and may thus be referred to as a "key block" following block theory terminology (Goodman and Shi, 1985).

A characteristic mesh for a horizontally layered and vertically jointed rock mass exhibiting mechanical layering is shown in Figure 6.11. We study the deformation of the roof using computational data output recorded at four measurement positioned in the roof as shown in Figure 6.12. In our simulations we vary the tunnel span (here labeled $B$) and overburden thickness $(h)$, scaling the horizontal distance to the computational domain boundary $(b)$ so as to maintain $b = 3B$ thus ensuring minimum stress

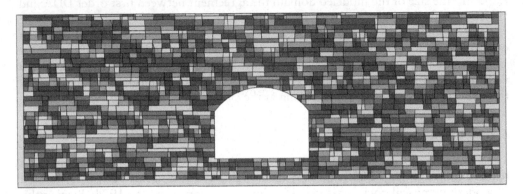

*Figure 6.11*   DDA mesh of a rock mass consisting of horizontal layers and vertical joints with mechanical layering.

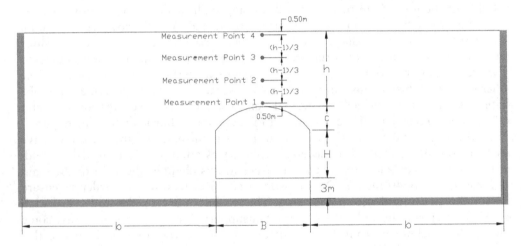

*Figure 6.12*   Position of measurement points for recording computational output data.

concentrations due to the opening near the boundaries. The relationships between tunnel span $B$, tunnel height $H$, and crown height $C$ remain constant in all simulations.

19 DDA models are analyzed with varying span vs. depth ratios as listed in Table 6.1. Because the mean joint spacing in the modeled rock mass remains the same, the total number of blocks in the mesh increases with increasing tunnel depth and tunnel diameter. The greatest number of blocks occurs at the model with the deepest tunnel of the widest span, namely Model 10 in Table 6.1 (11,104 blocks). Model 1 with the smallest tunnel diameter and shallowest cover results in the smallest number of blocks in the mesh (871). Depending on the kind of processor, it is evident that a significant demand of CPU time will would be needed to complete these simulations.

The performance of the modeled opening can be assessed by the deflection and horizontal stress evolution at mid-section of the immediate roof. As was seen in the case study of Zedekiah cave, the mere fact that the immediate roof exhibits vertical

*Table 6.1* DDA models used to study the span vs. overburden problem in horizontally layered and vertically jointed rock masses. Legend: B is tunnel span, H is tunnel height, c is crown height, b is distance to modeled domain boundary, and h is overburden thickness.

| Model | B (m) | H (m) | c (m) | b (m) | h (m) | Blocks |
|---|---|---|---|---|---|---|
| 1 | 10 | 5 | 2.5 | 20 | 6 | 871 |
| 2 | 14 | 7 | 3.5 | 28 | 6 | 1325 |
| 3 | 14 | 7 | 3.5 | 28 | 11 | 1780 |
| 4 | 17 | 8.5 | 4.25 | 34 | 6 | 2123 |
| 5 | 20 | 10 | 5 | 40 | 6 | 2221 |
| 6 | 20 | 10 | 5 | 40 | 10 | 2715 |
| 7 | 20 | 10 | 5 | 40 | 14 | 3197 |
| 8 | 40 | 20 | 10 | 80 | 15 | 8638 |
| 9 | 40 | 20 | 10 | 80 | 20 | 9948 |
| 10 | 40 | 20 | 10 | 80 | 25 | 11104 |
| 11 | 30 | 15 | 7.5 | 60 | 15 | 5720 |
| 12 | 30 | 15 | 7.5 | 60 | 22 | 6947 |
| 13 | 30 | 15 | 7.5 | 60 | 30 | 8157 |
| 14 | 25 | 12.5 | 6.25 | 50 | 12.5 | 4287 |
| 15 | 25 | 12.5 | 6.25 | 50 | 25 | 5925 |
| 16 | 25 | 12.5 | 6.25 | 50 | 19 | 5086 |
| 17 | 25 | 12.5 | 6.25 | 50 | 30 | 6591 |
| 18 | 22 | 11 | 5.5 | 44 | 11 | 3240 |
| 19 | 22 | 11 | 5.5 | 44 | 22 | 4683 |

deformation from the first time step still does not necessarily mean that the roof is doomed. Even in the most stable configuration some vertical shear along the abutments will always precede the arching mechanism. Once the arching mechanism is developed in the roof two observations will be made: 1) the vertical deflection will cease, and 2) horizontal stresses will reach a constant value the magnitude of which would be sufficient to keep the blocks in place given sufficient joint friction.

Using these stability guidelines we can group the simulation results of the 19 models into three main categories: 1) Unstable, 2) Marginally Stable, 3) Stable. In the "Unstable" category vertical deflection will never be arrested and horizontal stresses may never be developed, thus, ultimately, the entire roof may collapse and in very shallow opening the ground surface may break (Figure 6.13). In the "Marginally Stable" configuration we may loosen the immediate roof, however stable arching may develop from the middle roof and above (Figure 6.14). Finally, in the "Stable" category effective arching will arrest any further vertical deflection of the entire roof (Figure 6.15).

When plotting the results of these 19 simulations in span ($B$) vs. cover height ($h$) space an interesting and surprising conclusion is obtained. Instead of confirming the famous rule of thumb proposed by Terzaghi (1946) predicting that in blocky rock masses the height of the rock requiring support will be $0.5B$, here we get a nonlinear function. In relatively small span tunnels the demand for cover remains constant on a low value, much lower than the two tunnel diameters rule of thumb. Once the tunnel diameter exceeds a certain size, here found to be 18 m, the demand for cover increases very steeply. Although it never reaches the rule of thumb $h = 2B$, it does reach $h = 1B$

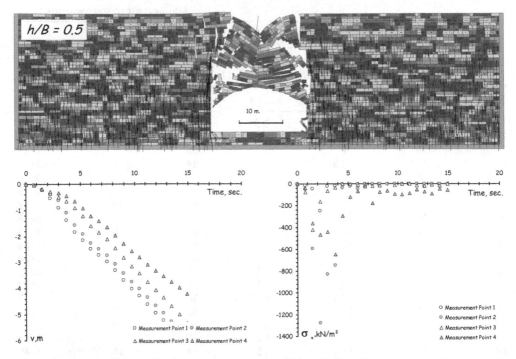

*Figure 6.13* Graphical illustration of "Unstable" configuration.

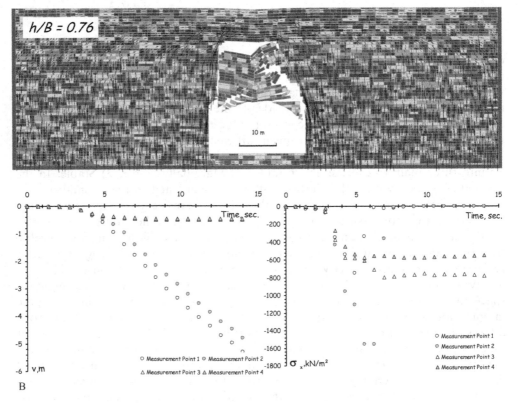

B

*Figure 6.14* Graphical illustration of "Marginally Stable" configuration.

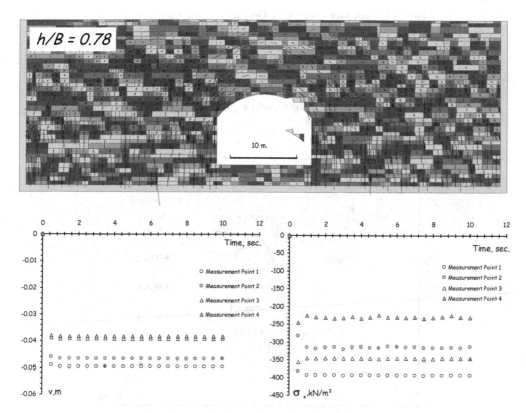

*Figure 6.15* Graphical illustration of "Stable" configuration.

*Table 6.2* Geometrical characteristics of the rock mass used in the 19 models.

| Joint set | Dip/Direction | Trace length | Mean spacing | Degree of randomness | Rock bridge length |
|-----------|---------------|--------------|--------------|----------------------|--------------------|
| 1 | 0/0 | $\infty$ | 0.70 m | 1.0 | 0 m |
| 2 | 88/182 | 5 m | 0.96 m | 0.5 | 2.5 m |
| 3 | 88/102 | 5 m | 0.78 m | 0.5 | 2.5 m |

at the most extreme case, here for tunnel diameter of 30 m. When the opening diameter continues to increase beyond that point the rate of change of demand for cover with respect to tunnel diameter becomes very small, namely, once we are deep enough, we are quite safe.

Of course these results are assumed to be valid only for the rock mass simulated here, the geometrical and mechanical properties of which are listed in Table 6.2 and Table 6.3. However, when plotting results from several other case studies in horizontally layered and vertically jointed rock masses but with different mechanical and geometrical properties (Hatzor and Benary, 1998; Bakun-Mazor *et al.*, 2009; Hatzor *et al.*, 2010), the results fall nicely along the nonlinear trend presented in Figure 6.16,

Table 6.3 Input parameters for DDA used in the 19 simulations.

| | |
|---|---|
| Unit weight | 22.54 kN/m³ |
| Young's Modulus | 15.32 GPa |
| Poisson's ratio | 0.21 |
| Friction angle of discontinuities | 30° |
| Cohesion of discontinuities | 0 MPa |
| Tensile strength of discontinuities | 0 MPa |
| Normal spring stiffness | 500 MN/m |
| Initial time step size | 0.0005 sec |
| Kinetic damping | 1% |

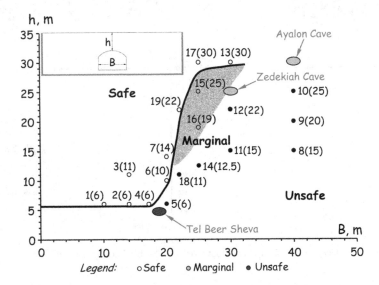

Figure 6.16 Overburden vs. span relationships in blocky rock masses as predicted by DDA. The solid line separates between unsafe and safe configurations. Three case studies in blocky rock masses in three different lithologies plot near the numerically predicted function possibly suggesting it's universality for blocky rock masses.

suggesting that this predicted trend indeed may be true for a large range of blocky rock masses.

## 6.3 DEEP UNDERGROUND EXCAVATIONS

### 6.3.1 Boundary conditions

In deep underground excavations the issue of initial *in-situ* stresses is of paramount importance. While in shallow excavations the deformation is driven and controlled primarily by gravitational loading, in deep excavations tectonic stresses may completely alter the stress field, and therefore cannot be ignored.

In order to properly simulate the mechanical deformation under high *in-situ* stresses the assumed initial stresses must be imposed on the modeled domain, <u>before</u> the opening is created in the model. It has been shown that if the opening exists in the

*Figure 6.17* The canyon in Jinsha river where Baihetan hydropower project is being constructed. The left bank consists of columnar basalts with basal planes dipping gently to the right (east) whereas in the right bank a thick sequence of southerly dipping sandstones overlies the columnar basalt sequence.

mesh from the beginning of the forward modeling stage, excessive displacements will be computed by DDA because it takes a relatively large number of time steps until the imposed initial stresses are fully developed and the available shear strength of all joints is fully mobilized (Tal *et al.*, 2014). The number of time steps required to achieve this equilibrium increases with the number of blocks comprising the DDA mesh, and therefore some experience is required in determination of the time step at which the opening should be removed in the model. In lack of such experience, the user can find by trial and error the time step at which the recorded stresses in some strategically located measurement points reach the imposed values, and then re-run the simulation with the opening being removed after that time step. Clearly, a modification to the original DDA code is necessary, where essentially the modeled domain first equilibrates under the imposed initial stresses in a "static" deformation stage, followed by "dynamic" forward modeling stage at the beginning of which the opening is removed.

Another important issue to realize when modeling deep underground excavations with DDA is the distance between the opening and the boundaries of the modeled domain. Clearly, when the average block diameter is in the order of several meters and the depth of the excavation is in the order of several thousand meters it is not feasible to model the entire problem with a DDA mesh comprised of so many discrete blocks. Since we are interested primarily in the deformation near the tunnel, the modeled domain boundaries may be set close to the opening, to a distance of say 3 to 5 tunnel diameters, to allow for complete dissipation of opening induced stress concentrations. However, if we do this, the boundaries of the modeled domain should be viscous, so as not to reflect waves that were supposed to propagate to infinity through the rock mass back into the analyzed domain, an artifact that will distort the computation. Applying viscous boundaries close to the opening therefore enables reducing the total number of blocks in the mesh, the interactions of which must be computed and results stored, in every time steps. This will save enormous computation time without compromising solution accuracy. Several methods for applying viscous, or non-reflective, boundaries specifically for DDA have recently been proposed (Jiao *et al.*, 2007; Bao *et al.*, 2012). In the discussion that follows extensive use is made of both sequential excavation and non-reflective boundaries enhancements for DDA.

### 6.3.2 Excavation damage zone

There has been a lot of discussion of the "excavation damage zone" in the rock mechanics literature in recent years (e.g. Cai and Kaiser, 2005; Hudson *et al.*, 2009; Li *et al.*, 2011; Li *et al.*, 2012), yet most of the published studies apply continuum based approaches to assess the failure mechanisms and extent of the EDZ. In hard crystalline rocks such as granites, a continuum mechanics approach may be absolutely justified. However when the rock mass is stratified, or jointed, or both, discontinuous approaches must be employed to delineate the boundaries of the excavation damage zone and to correctly assess the failure modes. While in continuous rock masses the EDZ will be generated by fracture of intact rock elements, in discontinuous rock masses the EDZ extent will be controlled by the geometry of the rock mass, specifically the orientation and spacing of discontinuities.

A case in point is the excavation of underground tunnels and shafts in Baihetan hydropower station, currently under construction in southwest China (Figure 6.17). The geology of this project is fascinating. The Jinsha river which flows from south to north at the project site, separates between Sichuan province on the left (west) bank and Yunnan province on the right (east) bank. In the left bank a thick sequence of columnar basalts that gently dip eastwards is exposed, whereas in right abutment a thick layer of southerly dipping sandstones overlies the columnar basalt sequence which forms both dam abutments.

An illustration of the structure of the columnar basalt is shown in Figure 6.18. In simulating this structure some distribution of spacing values about the mean were allowed for both joint sets and the bridge length in the columnar joints was scaled to allow for mechanical layering. The resulting mesh portrays well the geological structure in the field.

The complete mesh used for the analysis of a diversion tunnel in the left abutment is shown in Figure 6.19. On the left panel the location of the measurement points used to record the deformation in the forward modeling stage is shown, and on the right panel the principal stress trajectories are shown at the end of the static stage, just before opening removal and start of the forward modeling stage.

The removal of the tunnel was done in three stages as shown in Figure 6.20 to simulate as closely as possible the actual excavation methodology that was applied in the field.

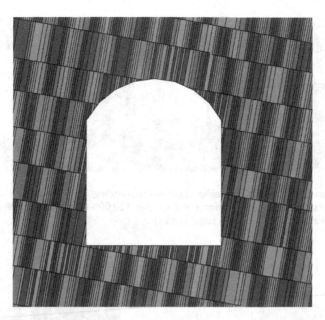

*Figure 6.18* Cross section through a 20 m span diversion tunnel in the left abutment of Baihetan Dam showing the eastwardly dipping basal planes (10/90) and the steeply dipping columnar joints (80/270). The mean spacing of the basal planes is 5 m, and of the columnar joints 0.3 m.

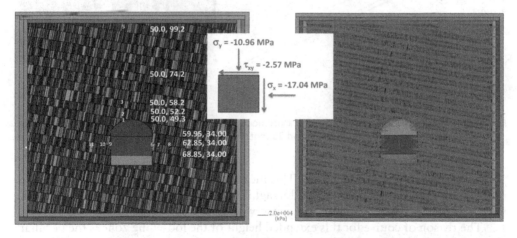

*Figure 6.19* Initial configuration of the mesh used to find the extent of the EDZ in the diversion tunnel (left) with the imposed initial in situ stresses (right). The assumed initial in-situ stress field is shown in inset.

The vertical displacement of the measurement points in the roof is shown in Figure 6.21. Note that the three lowermost measurement points never stop their vertical motion, meaning the vertical displacement is not arrested by an effective arching mechanism, which must develop higher in the roof. Judging from the output displayed in Figure 6.21 maximum arching stresses must develop somewhere between 10 and

*Figure 6.20* Sequence excavation modeling. Left – top heading removal after 17,000 time steps (0.109 s), Center – mid section removal after 120,000 time steps (0.798 s), Right – bench removal after 230,000 time steps (1.49 s).

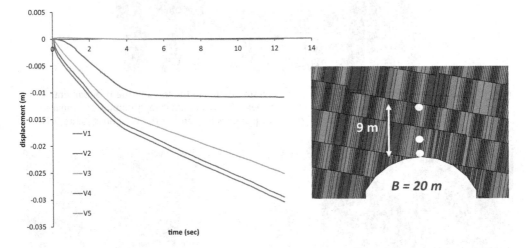

*Figure 6.21* Vertical displacement of five measurement points in the roof. The height of the arching stresses must be between 10 and 25 meters above the immediate roof.

25 meters about the immediate roof. This means that the height of the loosening zone in this rock mass extends more than Terzaghi's rule of thumb of 0.5B for blocky rock masses, which is typically considered conservative.

The reason of course for this extended height of the loosening zone is the peculiar structure of the rock mass. The steep inclination of the columnar joints restricts the ability of the arching stresses to be transferred properly from one block to another and thus an extremely large section of the immediate roof is loosened once the opening is created. In the sidewalls, however, the extent of the loosening zone is much more restricted, with sliding indicated in the right sidewall (Figure 6.22) and block toppling, as expressed by the measurement forward rotation in point 9, in the left sidewall Figure 6.23. In both side walls the loosening zone does not extend beyond 3 meters, in stark contrast to the situation in the roof where the loosening zone extends beyond 10 meters.

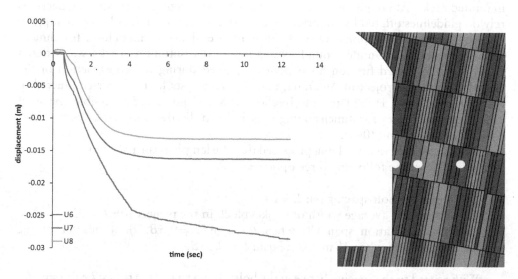

*Figure 6.22* Horizontal displacement vs. time in right sidewall. Measurement point 6 outputs suggest sliding of the wedges in the immediate sidewall into the opening, but measurement points 7 and 8 output suggests that this deformation is arrested less than 3 meters from the sidewall.

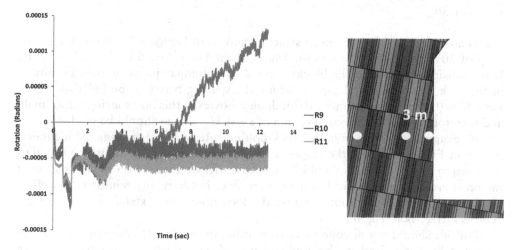

*Figure 6.23* Deformation in the left sidewall is dominated by forward rotation of the immediate blocks on the face as indicated by the plotted rotations in the figure. The deformation does not extend to more than 3 meters from the face.

We discuss constraints on rock bolting design imposed by this anisotropic nature of the rock mass in the next section.

## 6.3.3 Rockbolting

Rockbolting design in discontinuous rock masses is based largely on experience. Most empirical rock mass classifications provide some guidelines for dimensioning rockbolts in jointed rocks. Although they are not intended to be design tools, many engineers do rely on guidelines offered by empirical classification methods in the design of support of underground excavations. One of the most useful design guidelines for dimensioning rockbolts for underground excavation in discontinuous rocks was offered by T. A. Lang, based first on his experience gained during the construction of the Snowy Mountain project in Australia, and on many tests he has later conducted in the US (Lang, 1961, 1972; Lang and Bischoff, 1982). A good and accessible review of T. A. Lang guidelines for dimensioning rock bolts in discontinuous rock is provided by Brady and Brown (2004).

For blocky rock masses, Lang proposed that the length of the rock bolts ($L$) should be the longest of the following three options:

- Two times the bolt spacing ($s$): $L = 2\,s$
- Three times the average width of the keyblock in the project ($b$): $L = 3b$
- Half the excavation span ($B$) when $B < 6$ m: $L = 0.5B$, or a quarter of the excavation span when $18$ m $< B < 30$ m: $L = 0.25B$.

With regard to the spacing between the bolts, Lang suggested that it be the smaller of the following two options:

- $s = 0.5L$
- $s = 1.5b$

Let us now consider the rock mass structure shown in Figure 6.18. With an opening span of 20 m by Lang guidelines the length of the bolts should be $L = 0.25B = 5$ m. If we consider the size of the blocks formed by columnar joints as the "keyblock" in the rock mass, then by Lang guidelines the spacing between the bolts should be $s = 1.5b = 0.45$ m. From a practical standpoint, however, this bolt spacing is too small, and therefore the other suggested option of $s = 0.5L = 2.5$ m should be used.

An example of a "conservative" rock-bolting pattern based on Lang's guidelines is shown in Figure 6.24 and the response of the bolted roof as computed with DDA is shown in Figure 6.25. As would have been anticipated based on the discussion in the previous section, this rockbolting pattern does not have any reinforcement effect on the rock mass in the roof, because the loosening zone extends far beyond the "conservative" bolt length of 6 m.

But this should not discourage us from using the well tested empirical guidelines proposed by Lang. Rather, the application of these guidelines must be done with particular attention paid to the particular geometry of the rock mass at the site. The key issue here is to realize that the "average keyblock width", and important design parameter in Lang's guidelines, is not the same in this case for the roof and the sidewalls. In the sidewalls rockbolts are installed in direction roughly normal to the columnar joints, therefore the relevant block size for the sidewalls is given by the spacing between the columnar joints, namely 0.3 m. Since using the $L = 3b$ rule will obviously lead to too short length, we can use the $L = 0.25B$ rule ($18$ m $< B < 30$ m) to obtain a design rockbolt length of $L = 5$ m for the sidewalls. This length will supply ample reinforcement to the loosened zone in the sidewalls and predicted by DDA. In the roof rockbolts are installed in direction roughly normal to the basal planes, therefore the

*Figure 6.24* A "conservative" rockbolting design pattern based on T. A. Lang's guidelines for the diversion tunnel used as an example in the previous section. Bolt length in roof is 6 m and bolt spacing is 1.2 m.

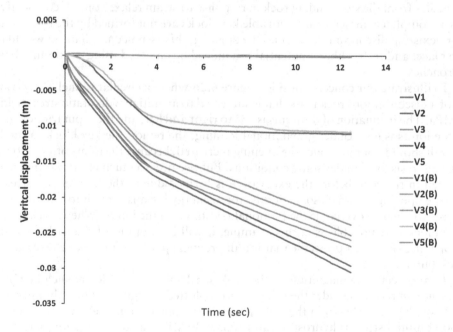

*Figure 6.25* Roof deformation with (B) and without the rockbolts as inferred from vertical displacement of five measurement points in the roof. Note that application of rockbolts according to the layout shown in Figure 6.24 hardly makes any difference and the same height of loosening zone is obtained above the immediate roof, with or without, the bolts.

relevant block width is determined by the mean spacing of the basal planes. Using the rule of $L = 3b$ for the roof here would require a bolt length of 15 m, which according to our modeling results should be sufficient to arrest any downward displacement in the roof.

We have seen in these two sections how the geometry of the rock mass structure controls both the extent of the loosening zone behind the free faces of the excavation, as well as the correct way to apply safe rockbolting support. In the next section we will examine how pre-existing blocks in the rock mass may be ejected as hazardous rockbursts when the excavation is created in a rock mass that is originally discontinuous and subjected to a high level of in situ stresses.

### 6.3.4 Rockbursts

It is widely accepted that rockbursts are generated in deep underground openings by fracture of initially intact rock. Therefore, analytical approaches based on fracture mechanics have been employed to discuss the governing mechanism (e.g. Cook, 1966; Fairhurst and Cook, 1966; Cook, 1976), to assess the extent of the excavation damage zone (e.g. Perras and Diederichs, 2016), and to propose effective support measures (e.g. Kaiser and Cai, 2012). Here we propose a different outlook on rockbursts. We argue that when a deep underground opening is excavated in an originally discontinuous rock mass that is subjected to very high initial stresses, before fracture of intact rock takes place, removable blocks (as defined by Goodman and Shi, 1985) may be ejected as projectiles from the surrounding rock in response to strain relaxation. Of course, if the rock is completely intact, or if removable keyblocks are not formed by the intersection of pre-existing discontinuities around the opening, this assumed mechanism would not take place, and the rockburst potential can indeed be analyzed using continuum based approaches.

To illustrate our concept consider Figure 6.26 where a circular tunnel is excavated out of a discontinuous rock mass that is subjected to an initial hydrostatic stress field of 50 MPa. The inclination of the joints is 45° to right and left, and the input friction angle for the joints is 65°. Under gravitational loading, the removable key block in the left sidewall should not move once the opening is created, because the joints offer sufficient frictional resistance under static conditions. But because so much elastic strain energy is stored in the rock before the excavation is formed due to the high in-situ stresses, once the opening is formed some of this stored energy is transferred into kinetic energy that is sufficient to overcome the frictional resistance of the joints. When this happens, if the block is removable as in this example, it will be shot out of the rock mass into the opening space as a projectile that for all practical purposes may be referred to as a "rock burst".

To appreciate the magnitude of the peak acceleration and velocity such key block ejections can reach, consider the time histories plotted in Figure 6.27 as obtained with DDA for the mesh shown in the right panel. The tunnel is removed at time 0.3 s, the imposed initial stress is hydrostatic and equals 30 MPa, the dip of the joints is 45° to both right and left, the input friction angle for the joints is 65°, the Young's modulus of the rock is 20 GPa, and Poison's ratio is 0.2. Note that an extremely high peak

*Figure 6.26* Ejection of a key block as a projectile in response to strain relaxation.

acceleration is obtained the first instant the key block is ejected from the host rock, but it quickly drops to zero while the velocity remains constant (note that this simulation is done with no gravity). The peak velocity obtained in this DDA simulation is similar to values of 3 m/s reported by Kaiser and Cai (2012), 10 m/s reported by Ortlepp and Stacey (1994), and to values between 0.6 to 2.5 m/s measured by Milev *et al.* (2001) using high speed video camera in situ.

The energy balance associated with opening the excavation in an initially continuous, homogeneous, linear-elastic (CHILE) rock has been calculated analytically by He *et al.* (2016) who have also showed that the zone of influence in terms of energy increase extends to a distance of three diameters from the tunnel center. Once the opening is formed, the energy increase in that annulus around the excavation must be balanced by the following three energy components: 1) the elastic strain energy that goes into intact rock elements, 2) the kinetic energy that moves removable blocks in the affected zone, and 3) the energy that dissipates by shear displacement of blocks along joints. Energy components 1 and 2 are readily available from DDA output, provided that a measurement point is positioned at the center of each block in the affected annulus. Since the energy increase in the affected annulus is found analytically (He *et al.*, 2016), the shear component of the total energy budget can be found by simple subtraction.

The evolution of the kinetic energy of the entire block system in the affected annulus as computed with DDA is shown in Figure 6.28 for several joint friction coefficients. The restraining effect of joint friction is readily apparent, where an increase of friction

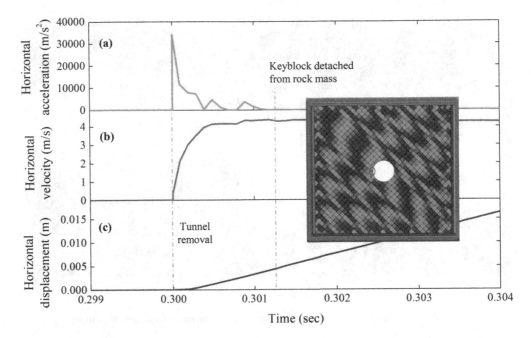

*Figure 6.27* Acceleration, velocity, and displacement of the ejected keyblock shown in inset as computed with DDA.

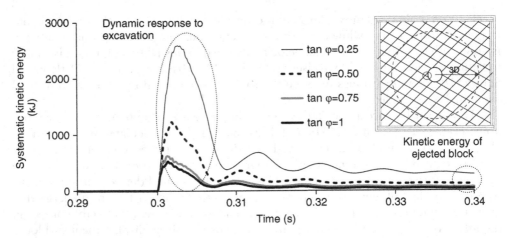

*Figure 6.28* Evolution of kinetic energy of the block system in the affected zone of the opening as a function of joint friction.

coefficient from 0.25 to 0.5 is sufficient to reduce the total kinetic energy of the block system in the affected domain roughly by a factor of 2.

The influence of initial stress and joint friction on the kinetic energy of the ejected key block, or the rockburst, is shown in Figure 6.29. With increasing initial in situ stress the kinetic energy of the rockburst naturally increases when the frictional resistance

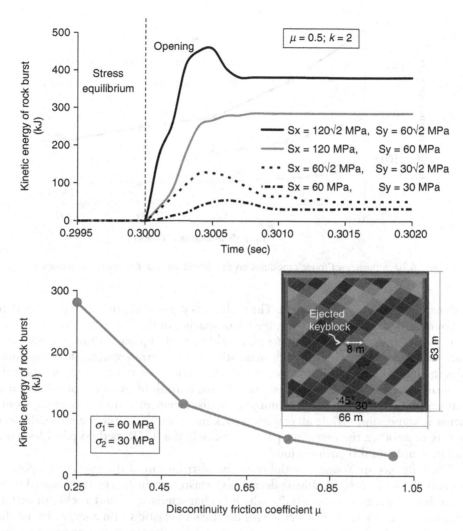

*Figure 6.29*  Kinetic energy of ejected keyblock as a function of initial stress (top) and joint friction (bottom). DDA mesh used in simulation shown in inset.

offered by the joints is kept constant (top panel). Similarly, for a given initial in situ stress the kinetic energy of the rockburst decreases with increasing frictional resistance of the joints (lower panel), as would be expected.

In order to probe further into the influence of rock mass characteristics on rockburst intensity, we will decrease somewhat the size of the studied annulus to reduce the total number of monitored blocks in the DDA mesh, from a distance of three to a distance of 1.5 tunnel diameters measured from tunnel center. It can be shown that the energy density concentration is highest in this restricted domain (He *et al.*, 2016).

We can investigate the influence of various rock mass parameters on the distribution of the energy components by changing the input parameters in DDA and checking

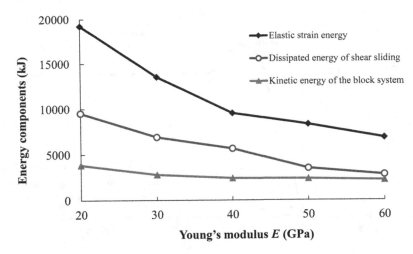

*Figure 6.30* Influence of Young's modulus on the distribution of the energy components.

the dynamic response of the system. The rock mass parameters that are considered are the Young's modulus of intact rock, the mean spacing of the joints, friction coefficient of the joints, and the inclination of unfavorably oriented joints. Young's modulus of intact rock elements scales the rock mass stiffness, the mean spacing between joints scales the mean block size in the rock mass, the friction coefficient of the joints scales the shear resistance of the rock mass, and the inclination of the unfavorable oriented joints scales the potential for obtaining block displacements into the opening from a kinematic standpoint. While all empirical rock mass classification systems address, in one way or another, the first three parameters, only the RMR method provides a way to address unfavorably inclined joints.

The influence of Young's modulus on the distribution of the energy components is shown in Figure 6.30. A drastic decrease in elastic strain energy is indicated when the modulus increases from 20 GPa which is characteristic of soft rocks, to 60 GPa which is characteristic of stiff rocks. This decrease in elastic strain energy also implies that much less elastic strain energy was stored in the rock mass before the opening was created and therefore that there is much less energy to be balanced after the excavation is created by the aforementioned three mechanisms. This means that with increasing rock mass stiffness the rockburst potential in fact decreases.

The influence of the mean block size in the rock mass on the energy distributions after the excavation is made is shown in Figure 6.31, where two important trends can be observed: 1) the dissipated energy by shear sliding increases with increasing block size, and 2) the kinetic energy of the block system decreases with increasing block size. Both results are intuitive, suggesting that with increasing block size in the rock mass the energy of rockburst is expected to decrease.

The influence of joint friction on the distribution of energy components after the excavation is made is shown in Figure 6.32. As would be expected, the dissipated energy by shear sliding increases with increasing joint friction, and the kinetic energy of the block system is reduced with increasing joint friction. This result implies that

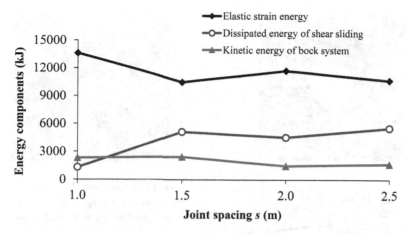

Figure 6.31 Influence of mean joint spacing on the distribution of the energy components.

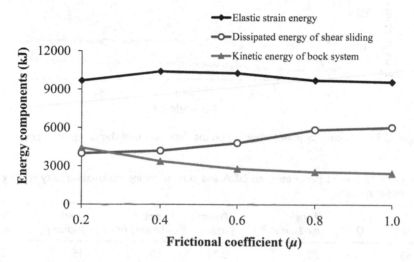

Figure 6.32 Influence of joint friction on the distribution of the energy components.

with increasing shear resistance of the rock mass the energy of rockbursts is expected to decrease.

Finally, the influence of unfavorably oriented joints is shown in Figure 6.33. Two joint sets are considered, one horizontal and the other inclined, with the dip angle $\alpha$ measured as shown in the upper left panel. The obtained result is not necessarily intuitive. We see that the highest kinetic energy for the block system is obtained when the dip of the inclined joint set is vertical, and it decreases with decreasing dip of the inclined joint set. This can be explained by the degree of interlocking in the block system which appears to decrease with increasing $\alpha$ as indicated by the red and yellow lines delineating the displacement vectors of the removable blocks in the modeled annulus. Note that in the left upper panel showing a block system with gently inclined joints

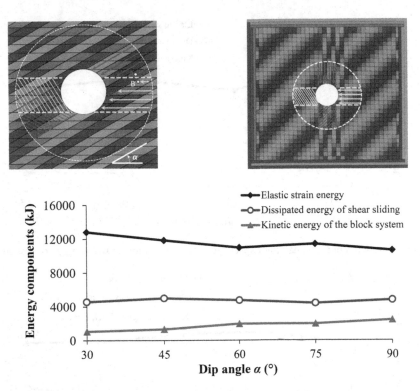

*Figure 6.33* Influence of joint inclination on the distribution of the energy components.

*Table 6.4* Input rock mass parameters to DDA and corresponding rockmass quality ranking by the major systems.

| GSI | RMR | Q | Young's modulus (GPa) | Poisson's ratio $\nu$ | Joint spacing (m) | Joint friction (°) | Monitored blocks |
|-----|-----|-----|-----|-----|-----|-----|-----|
| 60 | 65 | 10 | 30 | 0.23 | 1.0 | 35 | 567 |
| 70 | 75 | 31 | 50 | 0.22 | 1.5 | 40 | 254 |
| 80 | 85 | 95 | 70 | 0.21 | 2.5 | 45 | 97 |
| 90 | 95 | 289 | 90 | 0.20 | 5.0 | 50 | 27 |

the freedom of point B to move towards the opening is constrained by point A that must move first. Such constraints on the motion of blocks are absent in the case of the vertical joints (upper right panel) where indeed all the blocks delineated in red and yellow arrows are free to move simultaneously into the opening once the excavation is created.

These sensitivity analyses naturally lead to the discussion of the relationship between rock mass quality and the energy of rockbursts. As mentioned earlier, only the RMR method allows for consideration of unfavorably oriented joints and therefore here we shall use the RMR method for ranking rock mass quality. Of course,

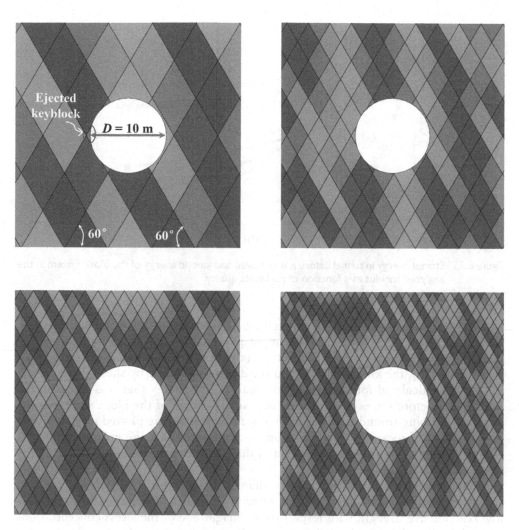

*Figure 6.34*  Four block systems used for DDA modeling of fours rock mass qualities listed in Table 6.4, decreasing in quality from upper left to lower right.

correlations between all major rock mass classifications are readily available (e.g. Hoek *et al.*, 1995).

In Table 6.4 input parameters for DDA simulations that represent different rock mass quality rankings according to the different empirical rock mass classification systems are listed. Note that the inclination of the unfavorable joint set is not included here, and in all simulations a symmetric joint inclination of 60 degrees is assumed. The number of monitored blocks in the analyzed annulus increases with decreasing rock mass quality because of the smaller joint spacing that is assumed for poor quality rocks. The DDA block systems that correspond to each of the four rock mass categories listed in Table 6.4 are shown in Figure 6.34.

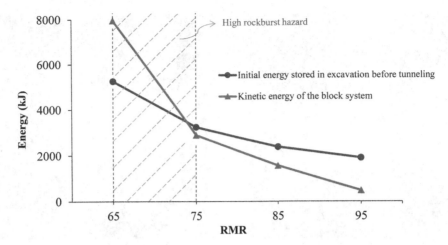

*Figure 6.35* Stored energy in tunnel before it is removed and kinetic energy of the block system in the analyzed annulus as a function of rockmass quality.

The four block systems shown in Figure 6.34 where subjected to an initial hydrostatic stress of 55 MPa with all other input parameters as listed in Table 6.4. The deformation of all monitored blocks in the analyzed annulus was studied and the elastic strain energy, the kinetic energy, and the dissipated energy by shear sliding along joints were calculated for each rockmass quality. The energy that was stored in the tunnel area before it was removed and the kinetic energy of the block system in the analyzed annulus around the excavation once it is formed are plotted in Figure 6.35 as a function of rockmass quality. It can be appreciated that the greatest rockburst energy is obtained with rockmass qualities that are between fair and good, in RMR terms between RMR = 65 and 75.

When the rock mass quality is lower than 65 it is feasible that tunneling induced stress concentrations will indeed fracture intact rock elements because of their relatively low compressive strength with respect to the magnitude of the stress concentrations around the tunnel. When the rock becomes sufficiently strong so that it can sustain the stress concentrations, the risk of rockbursts in the form of block ejections increases. Here we show that when rockbursts in the form of block ejections are feasible, their kinetic energy will be highest for fair rock (RMR = 65) and it will decrease drastically when the RMR value is greater than 75. Clearly, it would have been impossible to arrive at these conclusions without performing dynamic discontinuous deformation analysis.

# Chapter 7

# Rock slopes

## 7.1 INTRODUCTION

The discipline of slope stability in soil like materials is very developed with advanced analytical and numerical solutions for both static and dynamic problems. These approaches universally assume continuity of the medium where separation between the elements comprising the sliding mass is not allowed and all deformation is assumed to be concentrated along a well defend sliding surface. When the analyzed mass remains continuous throughout the deformation process, it becomes much easier to consider complex constitutive laws for the material, and to address pore pressure evolution with ongoing deformation. Moreover, addressing time dependency and creep deformation are easier when assuming continuity. Thus, the deformation and mechanism of failure of landslides have typically been analyzed assuming continuity, where the main questions were the triggering mechanism, the critical state at which the material failed, the amount of displacement, and possibly also the velocity evolution of the sliding mass. Famous landslides have been analyzed this way, for example the Vajont landslide (sometimes referred to as the Vaiont, or Mount Toc, landslide) in the Italian Alps (e.g. Mencl, 1966; Skempton, 1966; Jager, 1979; Trollope, 1980; Veveakis et al., 2007) which was in fact a rock slide (see Barla and Paronuzzi, 2013). Furthermore, continuity has been assumed to predict relationships between earthquake magnitude and expected displacement of the sliding mass (e.g. Jibson, 2007) based on analytical solutions which consider a single two-dimensional slice, such as those proposed by Newmark (1965).

As we have seen in this book, rock masses are rarely continuous. Even if a rock slope starts as a continuous body, once sliding begins it may quickly disintegrate into discrete blocks, either along pre-existing discontinuities or along deformation induced fractures. During sliding these blocks will interact with one another, while the entire discontinuous mass will move together along some well-defined shear surface, either pre-existing, or evolving. Clearly, analytical closed – form solutions do not exist for such complex dynamic processes. With the introduction of numerical discrete element methods, however, it became possible to compute the deformation of landslides without ignoring the interaction between the blocks comprising the rockmass, and even to allow for sliding induced disintegration of the mass to smaller blocks during sliding. Sitar, MacLaughlin and coworkers (Sitar et al., 2005) applied DDA to the case of the Vajont slide by discretizing the sliding mass in advance, before sliding initiated, into increasing number of blocks, and studying the influence of block kinematics on the velocity of the sliding mass, an issue of intensive debate among Vajont slide researchers. They found that peak velocity increased by up to 50% as the number of blocks increased, indicating that internal disintegration of the landslide mass results in increasing acceleration and higher peak velocity. The have also simulated pore pressure rise due to shear heating using their DDA velocity results, and showed that this mechanism could also explain peak velocity rise as the number of blocks increases. They concluded that the increase in peak velocity due to disintegration suggests that as much attention should be paid to the geometry of discontinuities, as is paid to shear strength and pore pressure in conventional slope stability studies.

In addition to the strong influence of the initial geometry of the sliding mass, the dynamic loading input is also very important in determining the final configuration of the sliding mass. Wu (2010) tested various methods of seismic record input into DDA using the Chiufenerhshan landslide that was triggered by the 1999 $M = 7.3$ Chi-Chi earthquake in Taiwan as a case study. By discretizing the enormous rock avalanche into many blocks, he was able to obtain quite accurately the post landslide topography as mapped in the field.

Clearly, both the initial geometry of the discontinuous sliding mass as well as the dynamic loading input should be considered simultaneously, to best simulate the deformation mode and the final configuration of a sliding mass. Zhang *et al.* (2015) applied DDA to the case of the Donghekou landslide which was triggered by the 2008 $M = 8$ Wenchuan earthquake that affected an area of 100,000 km$^2$ with nearly 90,000 fatalities in Sichuan province, China. They studied the effect of seismic loading, the number of blocks, and the block size on the kinematics of the deformation and on the runout, namely on the final position of the slide after the event. Similarly, the formation mechanism of the Tangjiashan landslide, also triggered by the same earthquake, was studied with DDA by Wu *et al.* (2009). They were able to constrain the duration of sliding (35 s), the peak velocity (30 m/s), to discuss the stress distribution along the body of the slide during the event, and to assess the degradation of the shear resistance during sliding.

The strong kinematic theory of DDA enables dynamic analysis of landslides that also involve, in addition to sliding, rigid block rotations such as in rock falls and toppling. Rock falls have been studied extensively by Ohnishi and coworkers with DDA (Yang *et al.*, 2004; Sasaki *et al.*, 2005; Wu *et al.*, 2005) and the results applied to real rock slopes in Japan and elsewhere. The classical analytical solution for block toppling (Goodman and Bray, 1976) has been checked and validated in 2D (Yeung, 1991; Yeung and Goodman, 1992) and extended to 3D (Yagoda Biran and Hatzor, 2013) with two dimensional and three dimensional DDA, respectively. The intriguing back-slumping mode where the blocks rotate backwards while sliding forward was initially discussed by Prof. Wittke (Wittke, 1965), and has been further explored with DDA (Kieffer, 1998; Goodman and Kieffer, 2000).

Existing rock slopes are stable under static conditions, naturally, and the major stability concern is what kind of external effects will destabilize the slope and will prompt motion. External effects are typically driven by environmental changes. Changes in the hydrology of the slope can induce pore pressure rise in the joints that may decrease the frictional resistance of the siding planes (Hoek and Bray, 1981) and may even change the failure mode of rock wedges from single plane to double plane sliding, or vice versa (Hatzor and Goodman, 1997). Seasonal climatic changes may cause thermal expansion of wedges in the tension crack that may drive the block forward by a wedging – ratcheting mechanism (Bakun-Mazor et al., 2013). Slow degradation of shear strength along the sliding plane due to changing humidity conditions and water chemistry may induce creep deformation along a potential sliding plane that can culminate in rupture of rock bridges and runout (Kemeny, 2005).

These effects are time dependent with a period of days to years. Earthquakes, in contrast, subject the slope to cyclic vibrations at much shorter periods of a second, or less. The only way to consider such rapid loading changes are through a dynamic analysis in the time domain, and DDA proves to be a capable method for doing just that, as shown in the verification examples presented in Chapter 5 using sinusoidal input motions. Seismically induced ground motions are never sinusoidal, but rather contain a rich frequency content, but this does not constrain the analysis in any way as it is performed in DDA in the time domain.

Outstanding issues pertinent to numerical rock slope analysis such as application of viscous damping at the domain boundaries and at contact points, dynamic shear strength degradation, dynamic block fracturing and pore pressure evolution during sliding, are actively being studied presently by the DDA research community. Here we shall focus on three issues that help illustrate the applicability of DDA to rock slope engineering: rotational failure modes, dynamic input motion, and reinforcement.

## 7.2 ROTATIONAL FAILURE MODES

Rock slopes exhibit several, frequently occurring, failure modes including single plane sliding, double plane sliding, forward rotation or "toppling", and backward rotation or "back-slumping". These failure modes involve the displacement or rotation of a single block and typically limit equilibrium analysis can be applied to find the factor of safety against failure. When the failing rock slope consists of multiple, interacting blocks numerical, discrete element, approaches must be adopted and a factor of safety cannot be clearly defined. Rather, the stability of the slope is discussed in terms of some displacement tolerance. When the failure mechanism involves further decomposition of blocks into smaller blocks during sliding as in large landslides or rock avalanches, fracture mechanics failure criteria must be employed in conjunction with the robust kinematical analysis performed with DDA.

Block sliding modes, either single plane or double plane, have been studied extensively in the general literature, and DDA verifications for sliding modes were discussed in Chapter 5. These modes, therefore, are not discussed here again, although in the discussion of dynamic input motion later in this chapter single plane sliding is used to illustrate some concepts. Large rockslides during which the sliding mass breakups into multiple blocks, have also been studied extensively in the DDA literature as reviewed in Chapter 1 and in the introduction to this chapter, and therefore are not discussed here as well. We therefore limit the discussion in this section to failure modes involving in-plane rotation: toppling, back-slumping, and overhanging slopes.

### 7.2.1 Toppling

The classical rock slope problem of block toppling as discussed by Goodman and Bray (1976) involves a set of slender parallel blocks that are resting on a stepped base, with

*Figure 7.1* Examples of toppling failures (after Goodman, 2013). A) Block Toppling, B) Flexural Toppling, C) Block – Flexural Toppling. Illustrations courtesy of Prof. R. E. Goodman.

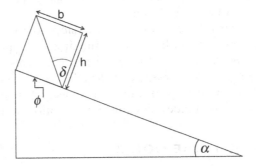

*Figure 7.2* Definition of terms for limit equilibrium analysis of the block on an incline problem.

the dips of the boundary joints of the blocks pointing at high angles into the slope. Recently, Goodman (2013) published a general classification of toppling failures as found in the field, some examples of which are reproduced in Figure 7.1.

Ashby (1971) and later Hoek and Bray (1981) discussed the static limiting conditions for a single block that rests on an inclined plane to undergo sliding, toppling, or both sliding and toppling, using the parameters as shown in Figure 7.2. They presented their results in a kinematic chart showing the boundaries between the failure modes as a function of slope inclination $\alpha$, friction angle $\phi$, and the aspect angle of the block $\delta$ (Figure 7.3A). Bray and Goodman (1981) who expanded the study to multiple blocks, treated boundary 3 as a "dynamic" boundary, and modified it as shown in Figure 7.3B.

With the publication of DDA the problem of block toppling was revisited by Yeung (1991) who found agreement between DDA and boundaries 1 and 2 and the modified boundary 3 as proposed by Bray and Goodman (1981), but found discrepancies with regard to boundary 4. He found that in some cases, within the "sliding and toppling" region as delineated in Figure 7.3B, only toppling is obtained with DDA. It has already been pointed out (Sagaseta, 1986) that the state of equilibrium at boundary 4 is dynamic, rather than static. Yeung (1991) treated boundary 4 as such and derived a complete analytical solution for it, following Sagaseta (1986). This evolution of the research demonstrates that when the numerical approach is accurate it can assist us in finding the correct solution even to problems that actually have an analytical solution; it becomes a guide of sorts, as the failure mode in DDA is not an assumption, but rather a result, of the analysis.

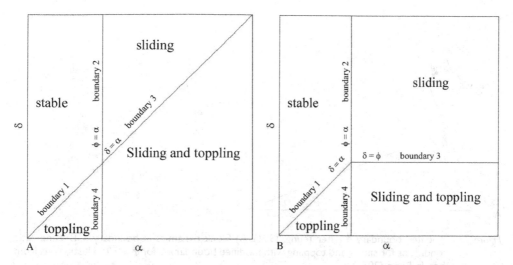

*Figure 7.3* A) Boundaries between failure modes for block toppling as proposed by Hoek and Bray (1981), B) Modified boundary 3 as proposed by Bray and Goodman (1981).

Consider Figure 7.4A below where a dynamic approach is taken to obtain boundary 4. When toppling is imminent, the center of rotation (the "hinge") tends to move upslope. This may prevent sliding even if sliding is theoretically permissible when $\phi < \alpha$. Boundary 4 separates between toppling with and without sliding, therefore when deriving the analytical solution for boundary 4 Yeung (1991) assumed limiting friction, namely $\phi = \alpha$. When the block is under pure rotation, the angular acceleration $\ddot{\theta}$ at the hinge and at the centroid must be identical. Writing dynamic force balance parallel and perpendicular to the sliding plane and taking moments about the block centroid, provide three equations with four variables ($\ddot{\theta}$, $\ddot{u}$, $\phi$, and $N$):

$$mg \sin \alpha - N \tan \phi = m\ddot{u} \cos \delta \tag{7.1}$$

$$N - mg \cos \alpha = m\ddot{u} \sin \delta \tag{7.2}$$

$$N \tan \phi \frac{h}{2} - N \frac{b}{2} = \frac{1}{12} m(h^2 + b^2)\ddot{\theta} \tag{7.3}$$

Equation 7.4 relates between $\ddot{\theta}$ and $\ddot{u}$:

$$\ddot{u} = \frac{1}{2}\ddot{\theta}\sqrt{h^2 + b^2} \tag{7.4}$$

The friction angle that will satisfy boundary 4 for every combination of $\alpha$ and $\delta$ is found by solving the four equations (for complete derivation see Yagoda-Biran (2013) or Yagoda Biran and Hatzor (2013)):

$$\tan \phi = \frac{3 \sin \delta \cos(\alpha - \delta) + \sin \alpha}{3 \cos \delta \cos(\alpha - \delta) + \cos \alpha} \tag{7.5}$$

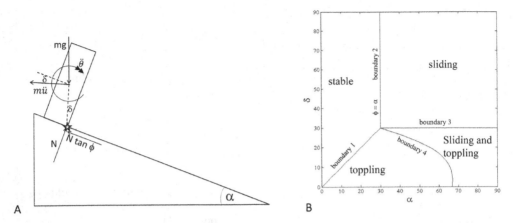

*Figure 7.4* Modified boundary 4 (after Yeung, 1991). A) Block dynamics at boundary 4, B) Kinematic conditions for sliding and toppling with modified boundary 4, for $\phi = 30°$. Illustrations from Yagoda-Biran (2013).

or:

$$\tan\alpha = \frac{3\cos^2\delta\tan\phi - 3\sin\delta\cos\delta + \tan\phi}{3\sin^2\delta - 3\sin\delta\cos\delta\tan\phi + 1} \tag{7.6}$$

The kinematic chart with the modified boundary 4 is shown in Figure 7.4B for the case of $\phi = 30°$. Indeed, Yeung (1991) obtained good agreement between 2D-DDA and the modified boundary 4 as shown in Figure 7.4B.

Note that in order to present the kinematic chart in two dimensions one of the three parameters $\alpha$, $\delta$, $\phi$ must be set at a fixed value, which limits the usefulness of the presentation. Yagoda-Biran (2013) showed an original way to present the chart in a three dimensional ($\alpha$, $\delta$, $\phi$) space so that no single value of these three independent parameters needs to be fixed, and her chart is reproduced in Figure 7.5A. The *Stable* mode is above the red ($\alpha = \delta$) surface and to the left of the blue ($\alpha = \phi$) surface. The *Sliding* mode is above the green ($\phi = \delta$) surface and to the right of the blue ($\alpha = \phi$) surface. The *Sliding and Toppling* mode is below the green ($\phi = \delta$) surface and in front of the curved surface representing Equation 7.6 (note that in this view the curved surface is actually behind the green surface). The *Toppling* mode is below the red ($\alpha = \delta$) surface and behind the curved surface representing Equation 7.6

In Figure 7.5B the boundaries are mapped as viewed from vector $(-1, -1, -1)$ which points from the object to the eye. When using this viewing direction, vector $(1, 1, 1)$ is reduced to a point, and the surfaces separating the different modes are reduced to lines. In this mapping the 3D space appears as a 2D space where it is easier to perceive the boundaries between the four modes.

Yagoda-Biran (2013) also considered pseudo-static earthquake inertia forces and how they might affect the kinematic boundaries. Typically, the peak ground acceleration (*PGA*) of the earthquake record is converted into a pseudo-static horizontal force

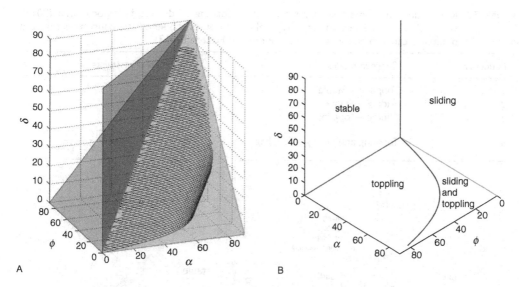

*Figure 7.5* The kinematic chart proposed by Yagoda-Biran (2013) for the four regions. A) Point of view similar to Figure 7.4B, B) Isometric point of view, viewing vector (−1, −1, −1). See text for explanation.

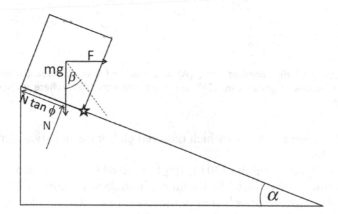

*Figure 7.6* Force diagram for a block on an incline with pseudo- static force *F*. The hinge of rotation is marked by a star (Yagoda-Biran, 2013).

$F$ acting at the centroid where the pseudo-static coefficient $k$ is defined as $F = kW$. By adding a new force $F$ a new angle $\beta$ can be defined (see Figure 7.6), between the block weight vector $W$ and the resultant of forces $F$ and $W$, namely:

$$\tan \beta = \frac{F}{W} = k \tag{7.7}$$

Yagoda-Biran (2013) modified the kinematic chart using the angle $\beta$ and defined the kinematic boundaries for the case of pseudo-static loading $F$ applied at the centroid

Table 7.1 Kinematic constraints assuming pseudo-static loading as defined by Yagoda-Biran (2013), where $\psi = \alpha + \beta$. The complete analytical derivation of the kinematic constraints for each boundary can also be found in Yagoda-Biran and Hatzor (2013).

| Boundary | Between modes | Kinematic constraints |
|---|---|---|
| 1 | Toppling – Stable | $\delta = \alpha + \beta$ |
| 2 | Sliding – Stable | $\phi = \alpha + \beta$ |
| 3 | Sliding – Toppling | $\delta = \phi$ |
| 4 | Toppling and Sliding – Toppling | $\tan \phi = \dfrac{3 \sin \delta \cos(\delta - \psi) + \sin \psi}{3 \cos \delta \cos(\delta - \psi) + \cos \psi}$ |

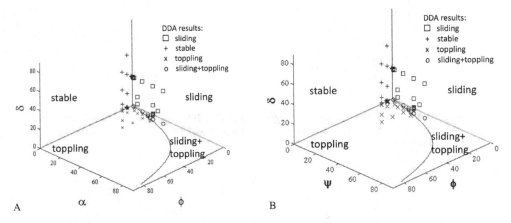

Figure 7.7 Verification of the modified static (A) and modified pseudo-static (B) kinematic charts with 3D-DDA (Yagoda-Biran, 2013) using the isometric view where surfaces are reduced to lines.

of the block. Her results, some of which required elaborate analytical derivations, are summarized in Table 7.1.

The results of Yagoda-Biran (2013) imply that when a horizontal force $F = kW$ acts on the centroid of the block, the kinematic boundaries become a function of three angles: $\phi$, $\delta$ and $\psi$, instead of only $\alpha$ as in the case of gravitational loading alone. The new angle $\psi$ can also be expressed in terms of $k$ instead of $\beta$:

$$\psi = \tan^{-1} \frac{k + \tan \alpha}{1 - k \tan \alpha} \tag{7.8}$$

Yagoda-Biran (2013) verified her modified kinematical boundaries for pseudo-static loading with both 2D as well as 3D DDA and found excellent agreement. In order to show the agreement between the analytical and numerical solutions the isometric view as shown in Figure 7.5B is used in Figure 7.7 below.

Finally, using the modified kinematic boundaries for the case of pseudo-static loading Yagoda-Biran and Hatzor (2013) showed how the failure mode might change with increasing $PGA$ (Figure 7.8), similar in essence to Hatzor and Goodman (1997) study on Pacoima Dam that showed how increasing water pressure in the joints could

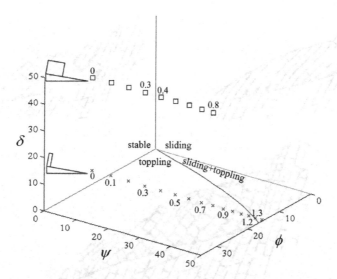

*Figure 7.8* Changing failure modes with increasing input *PGA* as computed with 2D-DDA for two block geometries: $\delta = 50°$ (square symbols) and $\delta = 15°$ (x symbols). For both geometries $\alpha = 10°$ and $\phi = 30°$. The mode of the block changes with changing value of $k$, as indicated near the symbols (Yagoda Biran and Hatzor, 2013).

change the failure mode of tetrahedral wedges from single to double plane sliding, or vice versa.

## 7.2.2 Block slumping

This unique failure mode in rock slopes was first discussed by Wittke (1965) and later revisited by Kieffer (1998) who extended the analysis to multiple blocks and studied it with 2D-DDA. Kieffer (1998) distinguished between "rock slumping" which involves backward rotational modes and "toppling" which involves forward rotational modes. Within the rock-slumping group, Kieffer (1998) distinguished between "flexural slumping", "block slumping", and "block flexural slumping" (Figure 7.9).

In all three modes of rock slumping the layers are restrained from sliding on the steeply dipping beds alone, and end up slipping simultaneously on both the bedding and the gently dipping basal crossing joint. This assigns to the layers the mechanical action of beams, which fracture and deform as sliding continues (Goodman and Kieffer, 2000).

Consider the tall and slender geometry of the block in Figure 7.10 which makes it theoretically susceptible to the block slumping failure mode. Because the resultant weight vector trajectory acts on the steeply inclined plane, sliding will commence by mobilizing shear resistance along both the steep and the shallow inclined planes simultaneously, equivalent to the steeply dipping bedding planes and the gently dipping basal cross-joints in Goodman and Kieffer (2000) classification. Here the gently dipping basal plane would more likely be a bedding plane whereas the steeply dipping plane could be a tension crack in the rock mass at some distance behind the face.

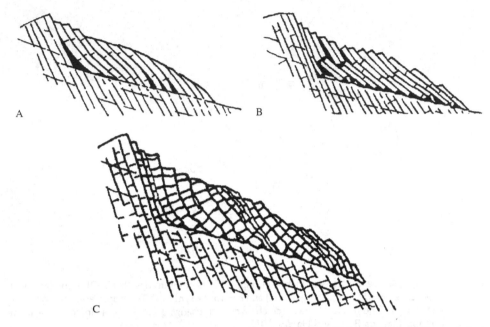

*Figure 7.9* Three categories of rock slumping modes as defined by Kieffer (1998). A) Flexural slumping, B) Block slumping, C) Block flexure slumping (after Goodman and Kieffer, 2000). Here the "bedding planes" are steeply dipping and the "cross-joints" are gently dipping. Reproduced with permission from ASCE.

In any case, rotation around a center located outside of the block may take place in this geometrical setting.

The forces acting on the block are shown in Figure 7.10. Assuming the friction angles on the two sliding planes are equal ($\phi_1 = \phi_2$) three equilibrium equations are necessary for solution of the contact forces $N_1$ and $N_2$ and the mobilized friction angle $\phi$, where $\alpha_1$ and $\alpha_2$ are the inclinations of the basal plane and the steep joint, respectively (Kieffer, 1998):

$$\sum F_v = 0 : W = N_1 \cos \alpha_1 + N_1 \tan \phi_1 \sin \alpha_1 + N_2 \cos \alpha_2 + N_2 \tan \phi_2 \sin \alpha_2 \quad (7.9)$$

$$\sum M_0 = 0 : W d_w + N_2 \tan \phi_2 d_2' = N_2 d_2 \quad (7.10)$$

$$\sum M_C = 0 : W x = N_2 \tan \phi_2 AC + N_1 \tan \phi_1 OC \quad (7.11)$$

Consider now the block geometry shown in Figure 7.11A which is susceptible to the block slumping mode as the weight vector of the block plots on the steeply inclined plane. Simultaneous solution of the three equilibrium equations for the geometry of the block in Figure 7.11A yields a mobilized friction angle value of $\phi = 22°$. A 2D-DDA model for the same block is shown in Figure 7.11B. Forward modeling of the block is performed with 2D-DDA with input friction angles for the joints decreasing from 43°

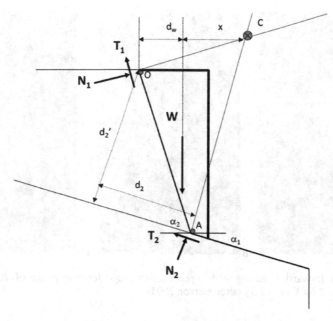

*Figure 7.10* Free body diagram used for static limit equilibrium analysis of the block slumping mode.

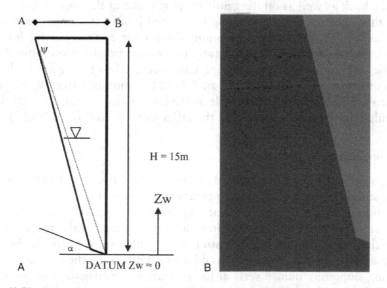

*Figure 7.11* A) Block geometry used for limit equilibrium analysis, B) DDA model used for verification.

to 22°. In all these simulations the block remains stable under gravitational load. However, once the input friction angle for the joints is reduced to 21°, block motion is initiated, exactly as predicted by the limit equilibrium analysis for the modeled block geometry. The motion of the block as computed with DDA clearly exhibits the block

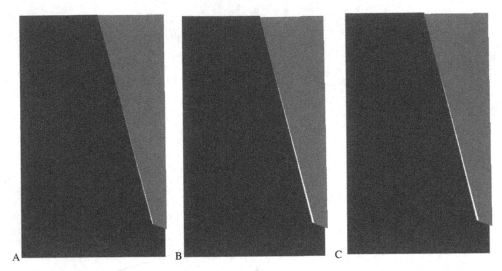

*Figure 7.12* DDA forward modeling with input friction angle for the joints of 20°. A) $t = 0.8\,s$, B) $t = 1.6\,s$, C) $t = 2.5\,s$ (after Hatzor, 2003).

slumping mode where friction is mobilized simultaneously on both the steeply dipping plane at the back as well as on the gently dipping plane at the base (Figure 7.12).

Note that once backward slumping is initiated joint water pressures will rapidly dissipate, as a joint with a wide base and sharp edge at the top will form behind the block at onset of motion. Consideration of water pressures and how they might affect static stability in backward slumping is therefore less important in the analysis of this failure mode. The formation of an *A* joint (terminology from R. E. Goodman, personal communication) which is wide at the base and narrows towards the top, is clearly evident in the final snap shot of the DDA simulation (Figure 7.12C).

### 7.2.3 Overhanging slopes

Overhanging slopes are seldom treated in the standard rock mechanics literature on rock slope stability, but they may pose a great risk because of possible forward rotation of the extruding portion of the slope that may ensue if a separation of the hanging part of the slope from the rock mass behind and below should take place. If the rock mass is initially discontinuous, pre-existing discontinuities may provide the necessary planes across which such separation could occur. Of course, the extruding tip of the overhanging slope nay induce vertical tensile fracture propagation at some distance from the face from the surface downwards, but strength failure in general is beyond the scope of this book, as we assume in DDA simply deformable, rigid, blocks.

Tsesarsky and Hatzor (2009) studied the stability of overhanging slopes with 2D-DDA in an attempt to find the critical relationships between the distance between the tip of the slope and the assumed tension crack behind the slope *L*, the distance between the base of the block and the tension crack *B*, the slope height *h*, and the slope angle $\alpha$ (Figure 7.13).

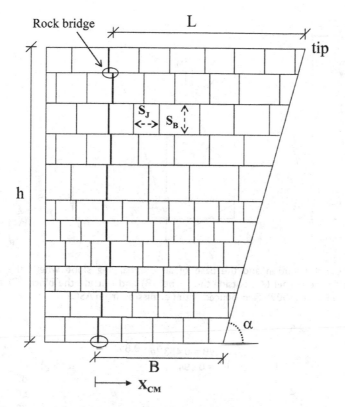

*Figure 7.13* A simple example of an overhanging slope in a horizontally layered and vertically jointed rock mass where the tension crack is formed by coalescence of cross joints at a distance L and B from the tip and the base of the block, respectively (after Tsesarsky and Hatzor, 2009). Reproduced with permission from ASCE.

The extruding slope and the vertical tension crack at the back form a polygon for the geometry of which a structural kernel lies between $B/3$ and $2B/3$ (Figure 7.14). When the weight vector plots within the structural kernel the stress distribution at the base of the block is uniformly compressive (Figure 7.14A). When the resultant plots outside the kernel (Figure 7.14B) part of the base experiences tensile stresses. When the resultant vector plots outside the polygon the block will overturn, provided that the base is disconnected (Figure 7.14C).

The $X$ coordinate of the center of mass ($X_{CM}$) of the polygon is:

$$X_{CM} = \frac{3B \cdot \tan \alpha + \frac{2}{3}h}{h(B+h)} \qquad (7.12)$$

The relationship between the location of the center of mass ($X_{CM}/B$) and the eccentricity ration ($B/L$) and how this affects the expected failure mode of the slope can be found analytically and the results are plotted in Figure 7.15. It can be seen from

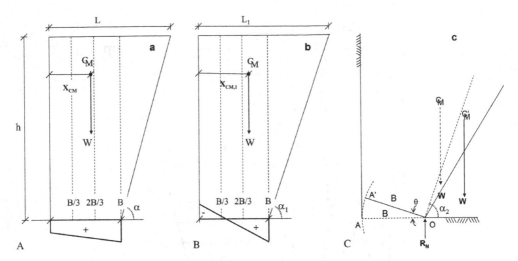

*Figure 7.14* Stress distribution and the base of an overhanging slope where the resultant plots inside the kernel (A), outside the kernel (B) and outside the block (C), after Tsesarsky and Hatzor (2009). Reproduced with permission from ASCE.

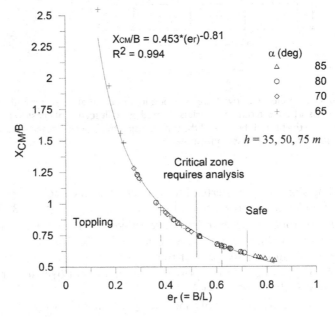

*Figure 7.15* Analytical relationship between the location of the center of mass with respect to the base of the slope ($X_{cm}/B$) and the eccentricity ratio ($e_r = B/L$), after Tsesarsky and Hatzor (2009). Reproduced with permission from ASCE.

inspection of Figure 7.15 that when the eccentricity ratio $e_r$ is smaller than 0.38 the slope is prone to forward rotation, and when $e_r > 0.62$ the slope is safe. The critical geometry that requires further analysis is when the eccentricity ratio is between these two bounds.

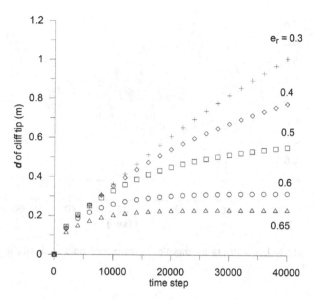

*Figure 7.16* Results of DDA for an overhanging slope with eccentricities ratio between 0.3 and 0.65 showing the displacement of the tip of the slope vs. number of time steps. Friction angle of discontinuities 41°. After Tsesarsky and Hatzor (2009). Reproduced with permission from ASCE.

DDA can be employed to perform such an analysis where the eccentricity ratio is being changed between models. The results of such analyses are shown in Figure 7.16. Indeed the expected relationship between the eccentricity ratio and the failure mode of the overhanging slope is confirmed with DDA. The role of reinforcement in stabilizing such a slope is discussed in the final section of this chapter.

## 7.3   DYNAMIC ROCK SLOPE STABILITY ANALYSIS

In the previous section we used static and pseudo-static approaches to determine block stability by means of limit equilibrium analysis. But life is really dynamic…and since DDA is a dynamic method there is no reason to restrict ourselves to static or pseudo-static approaches that only provide us with the state of equilibrium. With a truly dynamic approach we can find in addition to the state of limiting equilibrium also the rate and amount of displacement as well as the dynamic deformation pattern in cases of multi-block problems. In order to perform a true dynamic analysis however, a time dependent loading function must be incorporated correctly in the loading matrix instead of the constant gravitational acceleration. Methods to input the dynamic motion into the analyzed domain, and what kind of input motions should be selected, are explored in this section.

### 7.3.1   First introduction of time dependent accelerations to DDA

Inserting time dependent accelerations into 2D-DDA was first demonstrated by Shi (1999) when studying the stability of the East portal of Yurba Buena tunnel along the

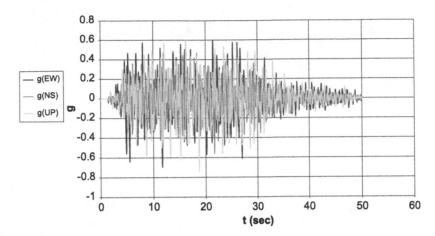

*Figure 7.17* Record of the Loma Prieta earthquake as measured on a rock site in the San Francisco Bay area.

track of the San Francisco Bay bridge in California. Shi used a modified record of the Loma Prieta earthquake, which struck that region in 1989, the three components of which are shown in Figure 7.17.

Seismometers that are attached to the ground typically provide motions in three directions, one vertical and two horizontal, typically aligned in $E–W$ and $N–S$ directions. For a two dimensional analysis only one horizontal component is sufficient, and therefore it would be required to find the resultant of the two horizontal components in the direction of the analyzed cross section which need not necessarily be aligned in one of those two directions. When running the dynamic simulation with direct input of horizontal and vertical accelerations it becomes possible to assess the significance of the vertical motions which are typically ignored.

## 7.3.2 Loading methods

An important issue to consider when running dynamic analysis with DDA is the point of application of the input accelerations in the modeled domain. One method of application, which has been referred to as "Method 1" (Wu, 2010), would be to introduce the time-dependent accelerations simultaneously at the centroids of all the blocks in the DDA block system. When using this approach it is assumed that there are no attenuations or amplifications within the modeled domain throughout the duration of the dynamic loading. This is the approach adopted by Shi (1999) when he first introduced time dependent accelerations in DDA and this method was used by Hatzor and Feintuch (2001) in the first verification study of dynamic input in DDA using the block on an incline problem. This approach would be perfectly valid when analyzing dynamic response of deep underground excavations to earthquake vibrations for example. In such cases, it would be sufficient to apply viscous boundaries at the edges of the modeled domain, to avoid artificial reflections back into the modeled domain, and to assume the entire rock mass in which the underground opening is embedded responds to the earthquake in the same manner.

In the case of surface structures that are resting on a shaking foundation such as masonry structures, it would be more accurate to introduce the loading function, either time dependent displacements, velocities, or accelerations, into a moving foundation block that represents the shaking ground. This way the waves can propagate through the block structure allowing for possible amplifications and ground-structure interactions. This method which was used by Kamai and Hatzor (2008) in a verification study of frictional block response to shaking foundation (see discussion in Chapter 5), has been referred to as "Method 2" (Wu, 2010).

Kamai and Hatzor (2008) in their verification study used a flat responding block so as to avoid possible rotations that are not considered in the analytical solution they developed for this problem (see Figure 5.16). Therefore, wave propagations were not studied in their research. Indeed, when they attempted to apply this loading method to existing historic masonry structures that have experienced shaking in the past (the Nabatean *Mamshit* town in the Negev and the Crusaders *Nimrod* castle in the Golan) they encountered difficulties in obtaining with DDA the same structural deformation pattern as mapped in the field. The way in which the connection between the block representing the shaking ground and the overlying blocky structure is modeled with DDA will certainly control the wave propagation pattern from the base to the top of the masonry structure and consequently the resulting structural deformation. Therefore, this issue of how to best represent a masonry structure with DDA for true dynamic analysis requires further research.

In the case of landslides that are comprised of many blocks, it is not clear which of the two methods is preferable, and indeed this issue has been a subject of some debate in the literature. Clearly if one seeks to perform an analysis similar to the Newmark method where the final displacement of a block resting on a pre-existing plane is sought, then Method 1 as used in DDA would be the exact equivalent to Newmark's method. Then by extension in case of multiple blocks where an analytical solution does not exist, Method 1 could be assumed to provide sufficiently accurate results.

This however is challenged by Wu (2010) who argues otherwise. Consider for example the two-block assembly shown in Figure 7.18, similar in essence to the block geometry used in the verification study of block response to cyclic motion of a frictional interface in Chapter 5 (Figure 5.16). The initial configuration is shown in Figure 7.18A where Point C in Block B and Point C' in Block A are in contact. Assume a constant acceleration is applied horizontally to the right. In Method 1, where the input acceleration goes into the centroid of Block B (Figure 7.18B), the final position of Point C will be to the right of Point C'. In Method 2, however, where loading is applied to the foundation block A and Block B is allowed to slide in response to the induced motion in Block A, the final position of Point C will be to the left of Point C' (Figure 7.18C). Wu (2010) argued that this is in conflict with d'Alembert's principle of mechanics, and therefore Method 1 should not be used in such cases.

Ning and Zhao (2012) studied extensively dynamic displacement of blocks on frictional planes with DDA and provided analytical solutions to a variety of cases including consideration of the role of vertical motions. With regard to the conflict with d'Alembert's principle of mechanics when applying Methods 1 as argued by Wu (2010), Ning and Zhao (2012) note that the relative displacement between the two

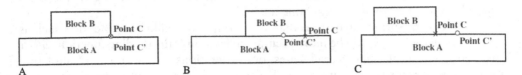

*Figure 7.18* Comparison between loading methods 1 and 2 under constant acceleration acting to the right. (A) Initial block configuration, (B) Final position of point C in Method 1, (C) Final position of point C in Method 2 (after Wu (2010)).

blocks, after all, is the same in the two methods. Therefore, if the input motion is applied to the base block in the opposite direction in Method 2, then the resulting position of Block 1 would be the same as if it was loaded in Method 1. They conclude therefore that the two methods are equivalent for single block sliding, but care has to be taken with regard to the direction of application of the dynamic load. If the existing record was measured on bedrock, equivalent in our analogy to Block A, then when the record is applied to the sliding mass directly (Method 1) as a volume force, it should be applied in the opposite direction. If however the record is applied to the base (Method 2), then it should be applied as constraint displacement time histories and it can retain the same directions as measured in the ground during the earthquake.

### 7.3.3 Consideration of local site effects

The discussion of dynamic loading methods in the previous section leads us to the very intriguing issue of local site effects. We have seen in Chapter 5 that DDA is capable of performing site response analysis just as well as designated software packages available for that purpose such as *SHAKE* (see Figure 5.36), provided the model is set up properly and the percent damping used in *SHAKE* is implemented correctly in DDA. This verification of site response capabilities with DDA calls for further consideration of *topographic* site effects which are very relevant to dynamic rock slope stability studies.

Theoretical considerations of topographic site effect and its influence on surface ground motion were discussed by many authors (e.g. Davis and West, 1973; Celebi, 1987; Bouchon and Barker, 1996; Chavez-Garcia *et al.*, 1996; Ashford *et al.*, 1997) and simulations of topographic amplifications have been performed using various theoretical methods (e.g. Bard and Tucker, 1985; Sanchez-Sesma and Campillo, 1991). These studies show that amplification of up to factor ten and more can be expected at ridge tops, and this has natural consequences to dynamic rock slope stability analysis.

The best way to assess if an elevated rock slope is prone to topographic site effect is to perform geophysical measurements in the field. An example of such an experimental campaign was performed by Zaslavsky *et al.* (1998) at Mount Masada in the context of dynamic rock slope stability analysis performed there (Hatzor, 2003; Hatzor *et al.*, 2004) and the main results are reviewed here.

A topographic map of the site and a cross section with the location of the measurement stations are shown in Figure 7.19. The recorded data consisted of several windows of micro-tremors and one earthquake. Ground motion amplification was estimated by Zaslavsky *et al.* (1998) using three spectral ratio methods: a) the Nakamura method

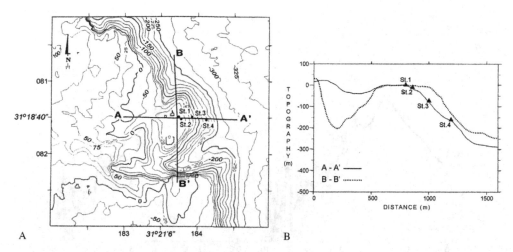

*Figure 7.19* Experimental topographic effect measurement campaign at Masada. A) Topographic map and location of geophones, B) Cross section A-A' (Zaslavsky *et al.*, 1998).

using ambient seismic noise, b) reference station technique, and c) receiver function estimates based on earthquake data.

Individual and average horizontal-to-vertical spectral ratios for Sites 1 and 2 as obtained from micro-tremors are shown in Figure 7.20. The dominant feature of all spectral ratios is the high spectral ratio level at a frequency of about 1.4 Hz. At this frequency we also observe differences between the EW and NS components. Such differences are characteristic of topographic effects. At the summit of Mt. Masada, the average spectral ratios reach maxima of about 2.5 in the EW direction and about 2 in the NS direction. It should be pointed out, however, that the Nakamura method provides in general a relatively reliable estimate of the predominant frequency of the site (resonance frequency) but it is less reliable for estimating the amplification level, especially at other frequencies.

The magnitude of spectral amplification is best estimated when comparing between sites at the top of the mountain and a reference site at the base (the reference station technique). In our case this would be between stations 1 and 2 at the top and station no. 4 at the base (see Figure 7.19B). The comparison between the top and base of the mountain was performed by Zaslavsky and Shapira (1998) utilizing a recorded earthquake of magnitude $M_L = 2.9$ and at a distance of 545 km from the site that struck Southeast Cyprus during the time of the survey at Masada. Only small variations in the site response of the two sites are indicted (Figure 7.21). The spectral amplification ratios show a prominent peak at about 1.3 Hz. The horizontal ground motion oriented EW is amplified by a factor of about 3.5, while it is about 2.0 in the NS direction. These results clearly indicate that Mt. Masada exhibits a preferential direction of resonance motion.

Finally, the horizontal-to-vertical spectral ratios for Sites 1 and 2 for the S wave window during the earthquake are plotted in Figure 7.22. The receiver function clearly exhibits resonance mode in frequency range 1.2–1.4 Hz, with amplification values of about 3.5.

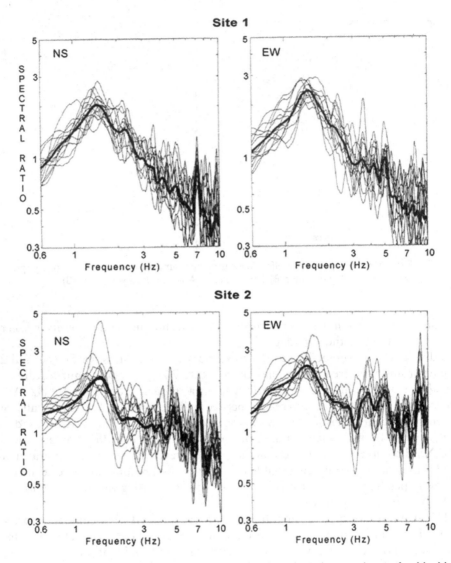

*Figure 7.20*  Individual and average (heavy lines) horizontal-to-vertical spectral ratio for Mt. Masada obtained from microtremors recorded at Sites 1 and 2 shown in Figure 7.19 (Zaslavsky *et al.*, 1998).

The final empirical response function for Masada as developed by Zaslavsky *et al.* (1998) is shown in Figure 7.23. Three characteristic modes are found, at frequencies of 1.06, 3.8, and 6.5 Hz.

We have shown the pronounced topographic site effect measured experimentally at Masada to demonstrate that such local effects may be rather significant and should therefore be incorporated in dynamic analysis if realistic results are sought.

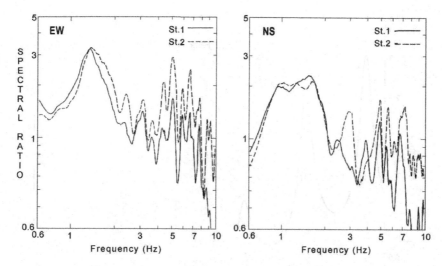

*Figure 7.21* Spectral ratios for Sites 1 and 2 with respect to Site 4 computed from an earthquake the struck Cypress during the survey and was recorded at the site (Zaslavsky *et al.*, 1998).

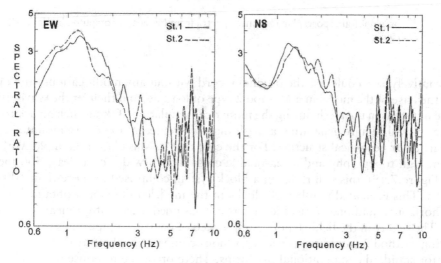

*Figure 7.22* Horizontal-to-vertical spectral ratios obtained from earthquake data for Sites 1 and 2 at Mt. Masada (Zaslavsky *et al.*, 1998).

Assuming a seismic record for the region is available (sometime referred to as a design earthquake) but with no local site effects, given a measured topographic effect such as the one shown in Figure 7.23, the record will have to be modified so as to account for the topographic site effect in a mathematical procedure known as "convolution". Such a procedure, if performed, will allow dynamic analysis using Method 1 where the time dependent accelerations are inserted directly into the centroid of the blocks at the top of the mountain, and this will be very efficient computationally.

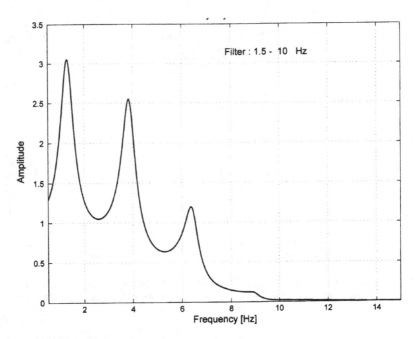

*Figure 7.23* The empirical topographic response spectra from Masada as computed by Zaslavsky *et al.* (1998).

Alternatively, one could use the original record without any modifications, and apply it to the base of the model in a Method 2 type of loading, and then let the waves propagate upwards in the mesh during the numerical simulation. Taking such an approach, however, will require generating a mesh that will represent the entire mountain with its particular geological structure. For the case of Mt. Masada for example, with the 250 m high topography and a reasonable cross section width of at least 500 meters (see Figure 7.19), this will result in a block system comprised of more than 100,000 blocks. This estimated number of blocks in the modeled domain is obtained if only one horizontal and one vertical joint sets are assumed, each with a mean spacing of 1 m. The reality in the field is more complicated of course. Computing such a mesh in loading method 2 would require an enormous computational effort that is not available for standard computational platforms. Therefore if the response function of the site is known, it would be much more efficient to apply the loading function in Mode 1 directly at the top of the mountain, without compromising accuracy.

To demonstrate this approach consider the record of the $M = 7.1$ Nuweiba earthquake which struck the Gulf of Aqaba in 1995. The motion was recorded in the city of Eilat, some 100 km north of the epicenter in a measurement station that was located on 50 meters of alluvial sediments (Figure 7.24). Zaslavsky *et al.* (2000) performed "deconvolution" of the measured record to obtain an analytical rock response function using information they collected from nearby boreholes with regard to the dynamic properties of the strata which are required for this analytical procedure. The rock record thus obtained is shown in Figure 7.25.

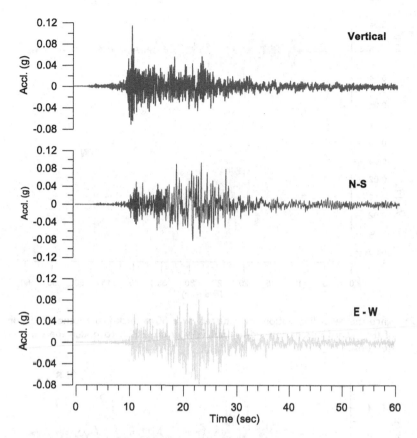

*Figure 7.24* Record of the *Nuweiba* M = 7.1 Earthquake of Nov. 1995 as recorded in the city of Eilat, some 100 km north of the epicenter.

In a dynamic analysis of Masada rock slopes, Hatzor *et al.* (2004) used the rock record developed by Zaslavsky *et al.* (2000) for the Nuweiba earthquake as measured in Eilat, and performed convolution with the experimentally measured topographic response spectrum at Masada (Figure 7.23). They have thus obtained the expected ground motion at the top of the mountain, should a similar earthquake hit the region in the nearby, seismically active, Dead Sea rift valley assuming such an earthquake will have similar characteristics as the Nuweiba earthquake (Figure 7.26).

With the available dynamic motion for the top of Masada it is now possible to perform dynamic DDA analysis for block at the top of the mountain in Method 1, without having to simulate the dynamic deformation of the whole mountain in Method 2, without compromising accuracy.

### 7.3.4 Application to single plane sliding

Let us now examine how such a procedure can be applied in practice, using the top of Mt. Masada as a case study. The approach can be used in dynamic analysis of

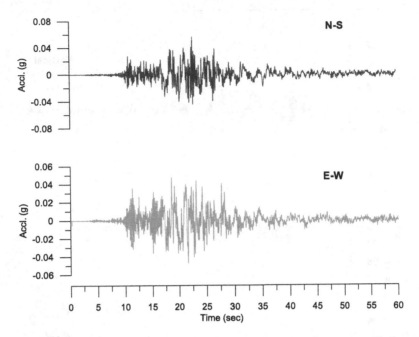

*Figure 7.25* Since the recording station was situated on a 50 m thick Pleistocene fill, deconvolution of the record was performed by Zaslavsky *et al.* (2000) to obtain the expected "Rock Response".

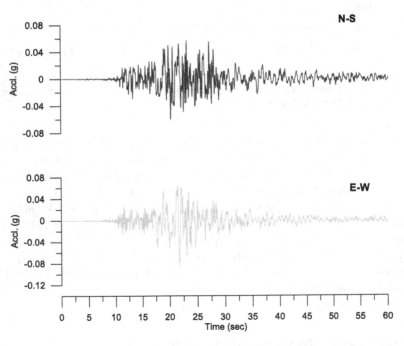

*Figure 7.26* Convolution of the Nuweiba "rock response" record with the experimentally measured Masada topographic response spectrum for obtaining the input motion to be used at the top of the mountain.

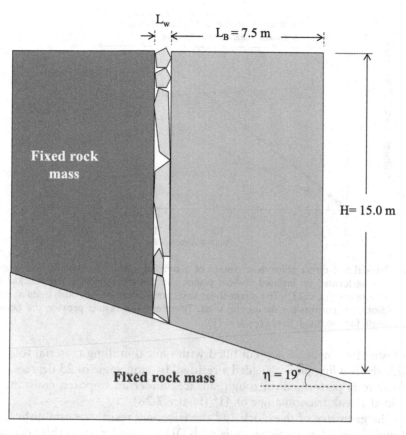

*Figure 7.27* Example of a block that is separated from the rock mass by a tension crack filled with debris and resting on an inclined base at the top of the Snake-Path cliff, Mount Masada (Bakun-Mazor, 2011). The block is held in place by virtue of friction alone.

any rock slope expected to experience amplifications due to topography during strong earthquakes. Consider the prismatic block that was mapped at the top of the East face of Mt. Masada in a location locally known as the "Snake Path" Cliff (Figure 7.27). We shall refer to this block as *Block 1* in the discussion that follows. The Block rests on a bedding plane dipping 19 degrees to ESE and has separated over time from the rock mass by opening of a vertical joint with accumulated siding to a distance of 200 mm over time. In a different study we explore the role of thermal fluctuations on the opening of the joint and sliding of the bock in a thermally induced wedging – ratcheting mechanism (Bakun-Mazor *et al.*, 2013). Here we shall only consider motion of this block due to seismic vibrations. Our goal would be to try to constrain the earthquake magnitude in the seismically active Dead-Sea rift valley below Masada that could have caused the mapped block displacement of 200 mm.

Because the block has separated from the rock mass, it is held in place by virtue of friction alone. In order to establish the factor of safety against sliding in static conditions the available friction angle of the bedding plane has to be determined.

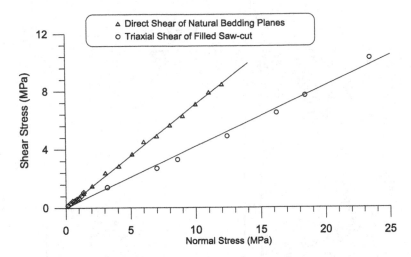

*Figure 7.28* Triaxial and direct shear test results of a bedding planes at Masada. The triaxial tests are performed on inclined saw-cut planes and therefore provide the residual frictional resistance ($\phi_{res} = 23°$). The direct shear tests are performed or rough bedding planes and dilation is permissible during the tests. Therefore the results provide the peak friction angle for the rough joints ($\phi_{peak} = 41°$).

Triaxial tests of an inclined saw-cut filled with typical infilling material found in the field inside the bedding planes yielded a residual friction angle of 23 degrees, whereas direct shear tests performed on rough surfaces under an imposed constant normal stress yielded a peak friction angle of 41° (Figure 7.28).

With the geometry of the block and the frictional resistance established, we can proceed with forward dynamic analysis with DDA. In order to do this accurately we need to first select a good value for the contact spring stiffness to be used in the dynamic DDA simulations. This was done by Bakun-Mazor (2011) by subjecting the block to sinusoidal input motions in loading method 1, namely the time-dependent accelerations were applied at the centroid of the block. Two input frequencies were used, corresponding to the two resonance modes determined experimentally for Masada, 1.3 Hz and 3.8 Hz, because the optimal penalty value is sensitive also to the frequency of the input vibrations. The analysis was repeated with changing contact spring stiffness values until the minimum error was obtained for the two frequencies. A peak friction angle of 41 degrees was assumed for the sliding interface. The results are shown in Figure 7.29. A contact spring stiffness of $k = 500$ GN/m was used as input for forward modeling with DDA, with a time step of 0.005 s. The method for choosing the optimal stiffness value using the relative errors obtained with the two frequencies is discussed by Bakun-Mazor *et al.* (2013).

The horizontal yield acceleration ($a_{yield}$) for an inclined bedding plane dipping 19° with peak friction angle of 41° as in the case of Block 1 is readily obtained using pseudo-static analysis (see Goodman and Seed, 1966) at $a_{yield} = 0.404\,g$. The $a_{yield}$ value thus obtained constrains the epicenter location of the maximum expected earthquake ($M_w = 7.5$) at the Dead Sea rift capable of triggering sliding of Block 1, to a distance of up to 20 km from Masada (see Figure 7.30).

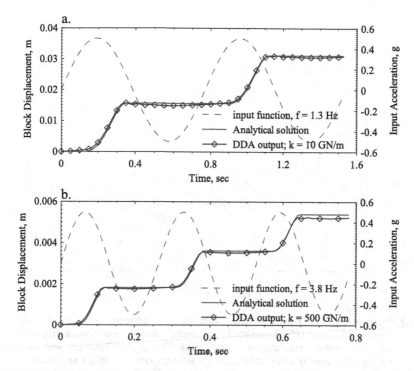

*Figure 7.29*  Finding the optimal contact spring stiffness for forward analysis by comparing DDA results with the analytical (Newmark's) solution for the two resonance modes measured for the top of Masada (Figure 7.23). A friction angle of 41° is assumed for the base plane. The yield acceleration is 0.404 g (Bakun-Mazor, 2011).

DDA results for seismic analysis of Block 1 subjected to amplified Nuweiba records corresponding to $M_w = 6.0, 6.5, 7.0, 7.5$ Dead Sea rift earthquakes at an epicenter distance of 1 km from Masada are shown in Figure 7.31 (Bakun-Mazor *et al.*, 2013). For moderate earthquakes ($M_w \leq 6.5$) the block displacement per single event is expected to be lower than 42 mm, whereas for strong earthquakes ($M_w \geq 7.0$) the block is expected to slide more than 447 mm along the inclined bedding plane in a single event. The mapped opening of the tension crack in the field is only 200 mm (see Figure 7.27), and this value constrains feasible earthquake scenarios that could have struck the region since this block was situated in its current geomorphological setting.

## 7.4  ROCKBOLT REINFORCEMENT

In the original 2D-DDA version rockbolts are implemented as springs connecting between two points in the mesh ($X_1 Y_1 - X_2 Y_2$) with a known stiffness. The bolt connection is linear elastic with infinite yield strength. No interaction between the bolt and rock mass is modeled and perfect bonding is assumed throughout the bolt length. Although these assumptions may seem overly simplified, this approach allows effective examination of the role of rockbolt reinforcement in discontinuous rock masses,

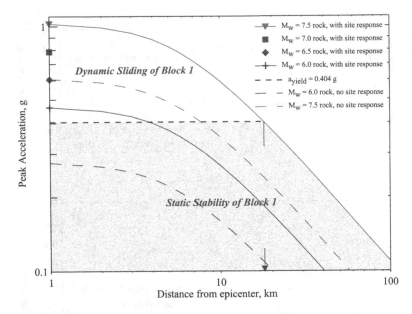

*Figure 7.30* Attenuation relationships for the Dead Sea fault adjacent to Mount Masada based on Boore and Joyner (1997) empirical relationship between earthquake magnitude, distance, and peak ground acceleration in rock sites. The dashed lines are without considering topographic amplification. Solid lines are with topographic effect and are relevant to the top of Masada. Shaded region indicates static stability of the modeled block at the top of Masada (Block 1). After Bakun-Mazor et al. (2013).

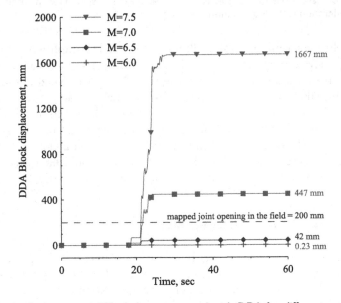

*Figure 7.31* Plastic displacement of Block 1 as computed with DDA for different earthquake magnitudes assumed to take place at a distance of 1 km from Masada. This chart indicates that slope could not have experienced an earthquake magnitude greater than 6.5 in its current geometrical configuration. After Bakun-Mazor et al. (2013).

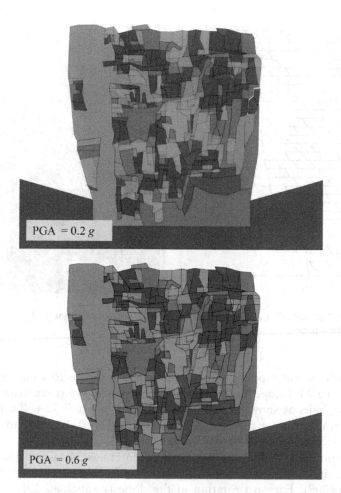

*Figure 7.32* Dynamic deformation of the Upper terrace of the North face of Masada. Top – response to the record shown in Figure 7.26 scaled to *PGA* = 0.2 g with no rockbolt reinforcement, Bottom – response to the record shown in Figure 7.26 scaled up to *PGA* = 0.6 g with spot bolting applied only in the West slope.

provided that the bolts are loaded within their yield strength and good anchoring is ensured when installed in the field. The dimensioning of the bolt in the field, in terms of bolt length and diameter, can be represented in DDA by the corresponding bolt stiffness that goes into the code as an input parameter, where typically a Young's modulus of steel is assumed ($k_{bolt}$ = AE/L).

The bolt action to restrain block motions due to dynamic loading is well illustrated by the case of upper terrace in the North face of Masada discussed, the geological structure of this wad discussed in Chapter 4. The cross section shown in Figure 4.24B was subjected to the earthquake record shown in Figure 7.26 in Mode 1, and the response of the cross section to the input record when scaled to *PGA* = 0.2 g is shown in upper panel of Figure 7.32. It can be clearly seen that given such an earthquake the

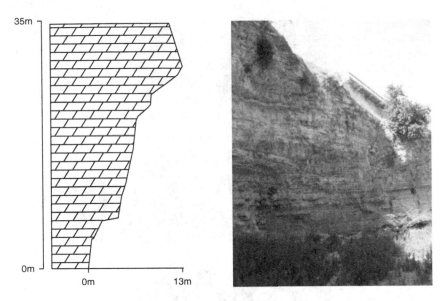

*Figure 7.33* The "Gibborim" overhang in Haifa near the Carmel tunnels. Left – cross section, Right – photo.

West slope will exhibit toppling failure, endangering the visitors on the path that is located along the West slope just below the upper terrace. However, installing a simple pattern of rockbolts as shown in the lower panel of Figure 7.32 will arrest any such toppling, even if the input record is up scaled to $PGA = 0.6\,g$, due to the expected topographic effect in Masada (Figure 7.23).

Consider now the "Gibborim" overhang in Haifa near the Carmel tunnels project with cross section and photo shown in Figure 7.33, that was studied by Tsesarsky and Hatzor (2009). Forward rotation of the slope is anticipated if a tension crack is assumed at a distance of 5 m from the foot of the slope ($B = 5$). Indeed, forward rotation of the slope is indicated with DDA, as shown in Figure 7.34. A friction angle of 37° is assumed both for the horizontal and vertical joints, and therefore initially the overhanging polygon rotates as a single rigid body under gravitational loading, as no connection is provided at the base of the slope. This is a realistic assumption as rock joints in general possess very little, if any, tensile strength.

The forward rotation of the overhang can be arrested if a rockbolting pattern that extends beyond the assumed location of the tensile crack is installed. This is demonstrated in Figure 7.35 where the deformation of the slope with the installed rockbolts after 20 seconds of gravitational loading is shown.

A very useful byproduct of DDA with rockbolts is the tension force evolution in every bolt as the analysis progresses in the time domain. The developed tension in the bolt will depend on the assumed stiffness of the bolt, which relates to the installed bolt diameter. The tension evolution in the bolts when a bolt diameter of 2 inch is assumed is plotted in Figure 7.36 where the bolts are numbered from No. 1 at the foot of the slope to No. 9 at the slope tip. The extension of the bolts can be appreciated from this

Figure 7.34 Deformation of the Gibborim overhang as computed with 2D-DDA if no rockbolt reinforcement is applied (assumed joint friction angle is 37°).

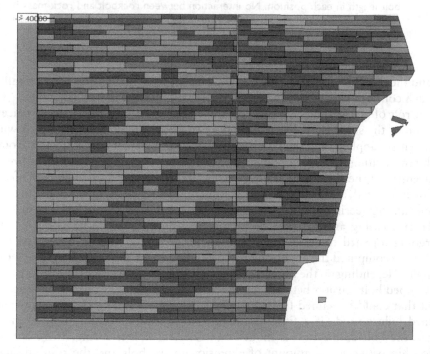

Figure 7.35 Designed rockbolting reinforcement pattern for the Gibborim overhang with an assumed tension crack at a distance from the base of B = 5 m.

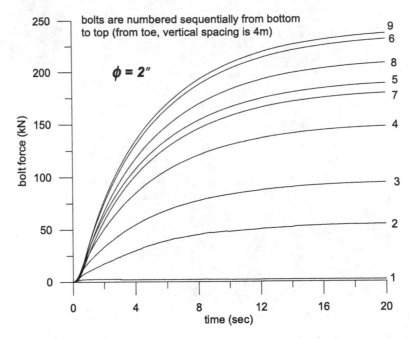

*Figure 7.36* Bolt tension evolution computed with DDA for the Gibborim overhang. Bolt stiffness is calculated for steel Young's modulus of 200 GPa, bolt diameter of 2 inch, and respective bolt length in each position. No interaction between rockbolt and rock mass is considered, namely the bolt is fully bonded throughout its entire length (after Tsesarsky and Hatzor, 2009). Reproduced with permission from ASCE.

plot and their stabilizing effect that can be inferred once the tension forces in all bolts approach constant values, depending on the location of the bolt in the slope.

The role of bolt stiffness is clearly portrayed in Figure 7.37 where the displacement evolution of the slope tip is plotted for the two bounding eccentricity ratios with no reinforcement applied, and then with three values of installed bolt diameters from 1 to 3 inch representing increasing bolt stiffness. Clearly, with increasing bolt stiffness the displacement of the slope is restrained. With all three modeled diameters the forward rotation is arrested, and therefore the decision of which bolt diameter to use in the field becomes an engineering decision involving both economic and practical considerations.

The increasing anchoring force with increasing elevation of the bolt in a block undergoing forward rotation has also been observed by Yagoda Biran and Hatzor (2013) who computed the anchoring force with 2D-DDA for a block that either slides or topples, depending on the input friction angle (Figure 7.38). The observed difference in developed bolt tension between sliding and toppling modes reveals once again the insight that could be gained from the results of proper numerical analysis where the obtained failure mode is a final result, and not an initial assumption, as in DDA. Because the bolt connection is modeled in DDA as a stiff spring, there is a linear relationship between the amount of extension in the bolt and the resulting tension force that will be developed through the assumed input stiffness of the bolt in the

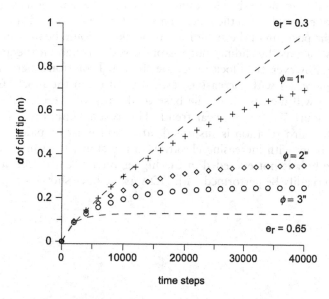

Figure 7.37 The role of increasing bolt stiffness (via increasing bolt diameter in reality) on slope deformation for the Gibborim overhang, as expressed by the displacement evolution of the slope tip with time steps. The dashed lines are the computed displacement of the tip without rockbolt reinforcement for the two boundary eccentricity ratios (after Tsesarsky and Hatzor, 2009). Reproduced with permission from ASCE.

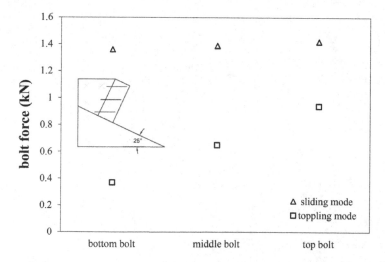

Figure 7.38 Comparison between the anchoring force required to restrain motion in sliding and toppling modes using 2D-DDA in static loading. The same block geometry is used in both simulations, the change in mode is induced by changing the input friction angle from $\phi = 15°$ (sliding) to $\phi = 30°$ (toppling) (after Yagoda Biran and Hatzor, 2013).

analysis. In the sliding mode the block undergoes displacement, as a single rigid body, of equal amount everywhere in the block. This is why the developed anchoring force is not related to the position of the anchor in the block; It would be the same everywhere in the block as long as the sliding mode is preserved throughout the analysis. In the toppling mode, however, the block undergoes forward rotation where the horizontal component $u$ increases with increasing vertical position in the block. Consequently, the bolt tension will increase from the base to the top of the block, and as can be inferred from Figure 7.38, in a linear trend. This result implies that in overhanging slopes where forward rotation is suspected, attention must be paid to the expected loading of the bolts with increasing elevation in the slope, to ensure that the design capacity of the bolt is not exceeded in the higher most areas of the slope. Such an assessment can readily be performed with DDA as we have shown here.

# Chapter 8

# Shi's new contact theory

**EDITORIAL NOTE**

This chapter presents the new contact theory developed by Dr. Gen-hua Shi and has been written exclusively by Dr. Shi especially for this book.

## FOREWORD

Contacts between two general blocks are the fundamental problem for discontinuous analysis. There are different contact points in different block positions, and there may be infinite contact point pairs in the same block position. A new concept of "entrance block" is introduced in this chapter for solving the contacts between two general blocks. The boundary of an entrance block is a contact cover system. Contact covers may consist of contact vectors, edges, angles or polygons. Each contact cover defines a contact point and all closed-contact points define the movements, rotations and deformations of all blocks as in real cases. Given a reference point, the concept of entrance block simplifies the contact computation in the following ways:

1  The shortest distance between two blocks can be computed by the shortest distance between the reference point and the surface of the entrance block.
2  As the reference point outside the entrance block moves onto the surface of entrance block, the first entrance takes place. This first entrance point on the entrance block surface defines the contact points and related contact locations.
3  If the reference point is already inside the entrance block, it will exit the entrance block along the shortest path. The corresponding shortest exit point on the entrance block surface defines the contact points and related contact locations.

All blocks and angles here are defined by inequality equations. Algebraic operations on blocks and angles are described here. Since the blocks and angles are point sets with infinite points, the geometric computations are difficult, and therefore the geometric computations are performed by related algebraic operations.

## 8.1  INTRODUCTION

A contact is ubiquitous in structural, mechanical, robotic, and geotechnical engineering. The difference between continuous computation and discontinuous computation is that discontinuous computation involves contacts (Goodman *et al.*, 1968). The basis of discontinuous computation is to detect where and when contacts happen, which are the major objectives of this chapter. This is one of the most difficult issues in numerical computation involving discontinuities and the simulation of sliding, holding, and impacting. As a result, contact computation has been drawing much attention from many scholars in the past decades.

A "distinct object" as used here is referred to as a "block" in the aspect of geometry. A block system (Poetsch, 2011) contains more than one block, but only when two blocks are close enough can they possibly contact and these two blocks are then defined as neighboring blocks. Before computing contacts, algorithms for finding neighboring blocks (e.g. Belytschko and Neal, 1991; Wu *et al.*, 2014) should be used in contact analysis.

For the contact problem of two neighboring blocks, several methods, including the common plane (Cundall 1988; Nezami *et al.*, 2006), the single surface algorithm (Benson and Hallquist 1990), hierarchy territory algorithm (Zhong and Nilsson 1996; He 2010), the "master-slave" approach (Jelenić and Crisfield 1996), and the block-particle algorithm (Li *et al.*, 2004) were proposed.

These methods have solved many contact problems (Konyukhov and Schweizerh 2013), such as the contact between rounded convex angles, balls, 2D convex angles, and 3D planes. However, it is more advantageous that a global contact theory, one that is topologically proven and applied to the objects with arbitrary shapes, should be developed. With the use of such global contact theory, difficult contact problems, such as the contact between symmetric angles, the rotation of parallel edges, the separation of two contacted polygons, the contact between two concave blocks, the contact of blocks with small edge length and under large penetration distance, and the destabilization process of a densely compacted block system, can be easily solved.

The difficulty of contact judgment is associated with the arbitrary shape of the boundary and the infinite interior points. In order to tackle contacts, appropriate mathematical tools are needed for transforming geometric and topological computations to algebraic computations. Blocks and the components of blocks, such as solid angles, polygons, and edges, are defined as point sets. The operation of simple point sets has been provided by Minkowski and Geometrie der Zahlen (1910). However, the mathematical tool for dealing with complex point sets is still lacking.

In this study, new mathematical tools are first proposed and used to operate point sets (Sections 8.2 and 8.3). A new concept "the entrance block" is proposed and associated with an operation of point sets (Section 8.3). With the use of the entrance block, the relationship between the two blocks reduces to the relationship between a reference point and the entrance block, thus considerably simplifying the complexity of contact computation (Section 8.4). The entrance block is defined and represented by its boundaries and conforms with many theorems (Section 8.4). With these theorems, the entrance blocks in different conditions are formed and discussed in Sections 8.5 to 8.12, including contacts between 2D angles, 2D blocks, 3D angles, and 3D blocks. These solids can be of any shape, i.e. convex or concave, with rounded or sharp angles. In Sections 8.5 to 8.12, the solutions of the entrance block both in algebraic and geometrical ways are proven. In fact, the solving of the entrance block is to build and solve inequality equations that employ point sets.

Using the entrance block and its contact covers, the possible contact positions and degrees of freedom can be identified. Thus the problem as to where to apply possible and suitable contact springs or penalties is solved.

To date, the idea of contact theory (Shi 2013a, 2013b) has been fully developed and employed. In this theory, the geometrical and topological computation is transferred into algebraic computation. Subsequently, the proposed contact theory becomes computational. Most of the computation methods, including discontinuous deformation analysis (Shi 1988), discrete element method (Cundall 1988; Li *et al.*, 2004; Nezami *et al.*, 2006), numerical manifold method (Shi 1991), and finite element method (Goodman *et al.*, 1968; Taylor 2004) can use this theory to compute contacts.

## 8.2   GEOMETRIC REPRESENTATIONS OF ANGLES AND BLOCKS

The difference between continuous computation and discontinuous computation is that the discontinuous computation involves contacts. In order to compute contacts, suitable mathematical representations of blocks are needed. Blocks and the components of blocks such as polygons, angles, edges or vectors are defined as point sets. These point sets are represented by inequality equations. The orientation of blocks, angles, polygons, edges and vectors is also defined for the algebraic operation. Since the geometric computation is difficult and different comparing with algebraic computation, the geometric computations are transferred here to algebraic computations.

### 8.2.1   Discontinuous computations

The basic essence of contacts can be described mathematically by inequality equations. Therefore, the basic tasks of discontinuous computation are to build inequality equations and to solve inequality equations.

Discontinuous computation follows time steps. If the displacements of each time step are smaller than a given limit, linear approximation is accurate enough within a time step. As with the regular equations, in each time step, inequality equations have to be transferred to linear inequality equations in order to be solved. This step-linear approach allows the contact theory to be used directly. The algorithms of solving inequality equations are well established by all major discontinuous computation methods; solving inequality equations is essentially transferring a subset of inequality equations to normal equations.

However, the major challenges of discontinuous computation stem from building the appropriate inequality equations. In this the most difficult task for building inequality equations is to find the contact position, where the possible springs, penalties or linear complementarities can be applied. The physical meaning of building inequality equations is where to apply contact springs, penalties or linear complementarities (Zheng and Li 2015). This chapter presents inequality equations for contacts by defining the entrance block and its contact covers.

### 8.2.2   Related mathematical symbols

Throughout this chapter, the conventional symbols are used:

$A, B$ are blocks.
$\partial A, \partial B$ are the boundaries of blocks $A, B$ respectively.
$int(A)$ is the set of inner points: $x \in A$, and $x \notin \partial A$.
$x, y, z$ are values or real numbers.

$a = (x_i, y_i, z_i), b = (x_j, y_j, z_j)$ are 3D points.

$a = (x_i, y_i, 0), b = (x_j, y_j, 0)$ are 2D points.

$a\,b$ is an edge which is from $a$ to $b$.

$(a\,b) = (x_j - x_i, y_j - y_i, z_j - z_i)$ is a vector.

$\vec{n}_i = (x_i, y_i, z_i)$ is a vector from $(0,0,0)$ to $(x_i, y_i, z_i)$ and often refers to an inner normal vector of a block or polygon, which points the inside of the block and polygon.

$\vec{n}_i \| \vec{n}_j$ means $\vec{n}_i = t\vec{n}_j$.

$\vec{n}_i \uparrow\uparrow \vec{n}_j$ means $\vec{n}_i = t\vec{n}_j$, $t > 0$.

The computation of $a$ and $b$ follows the vector operation rule (as shown in Figures 8.1 & 8.2),

$$a \pm b = (x_i \pm x_j, y_i \pm y_j, z_i \pm z_j)$$

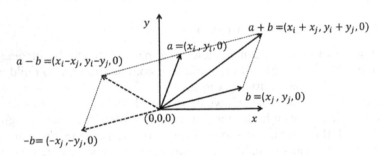

Figure 8.1  Addition and subtraction of 2D vectors and points.

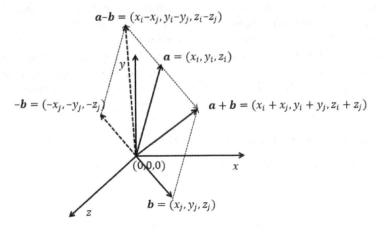

Figure 8.2  Addition and subtraction of 3D vectors and points.

### 8.2.3 Algebraic operations of blocks

Define:

$$A + B = \bigcup_{a \in A, b \in B} (a + b).$$

This is known as Minkowski Sum (1910).

Define:

$$-B = \bigcup_{b \in B} (-b),$$

$$A - B = A + (-B).$$

Therefore:

$$A - B = A + (-B) = \bigcup_{a \in A, b \in B} (a - b).$$

### 8.2.4 Representation of solid angles

Throughout this paper, an angle means a two dimensional angle. A three dimensional angle means a solid angle. It is convenient to define a one dimensional solid angle. A one-dimensional solid angle $A$ with a top $(0,0,0)$ along $\vec{e}_1$ is a ray and represented as:

$$\angle \vec{e}_1 = \bigcup_{t \geq 0} t\vec{e}_1.$$

A one-dimensional line $A$ along $\vec{e}_1$ passing $(0,0,0)$ is represented as:

$$\vec{e}_1 = \bigcup_{\forall t} t\vec{e}_1.$$

A 2D solid angle $\angle \vec{e}_1 \vec{e}_2$ means the swept area of a ray when rotating the ray from vector $\vec{e}_1$ to $\vec{e}_2$ along the direction from $ox$ to $oy$, i.e. anti-clockwise direction. Vectors $\vec{e}_1$ and $\vec{e}_2$ are the edge vectors of solid angle $\angle \vec{e}_1 \vec{e}_2$ (Figures 8.3 & 8.4).

A 2D convex solid angle $A$ with top $(0, 0, 0)$ is represented as:

$$\angle \vec{e}_1 \vec{e}_2 = \bigcup_{t_1 \geq 0, t_2 \geq 0} t_1 \vec{e}_1 + t_2 \vec{e}_2.$$

If $\angle \vec{e}_1 \vec{e}_2$ is a 2D convex angle (Figure 8.3),

$$\vec{e}_1 \times \vec{e}_2 \uparrow\uparrow (0, 0, +1).$$

If $\angle \vec{e}_1 \vec{e}_2$ is a concave angle (Figure 8.4),

$$\vec{e}_1 \times \vec{e}_2 \uparrow\uparrow (0, 0, -1).$$

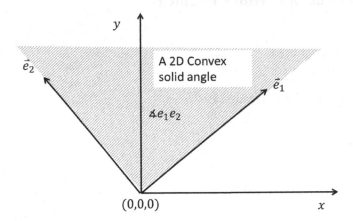

*Figure 8.3* A 2D convex angle $\angle \vec{e}_1 \vec{e}_2$.

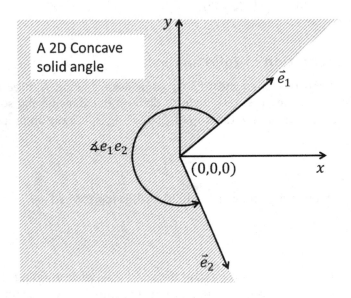

*Figure 8.4* A 2D concave angle $\angle \vec{e}_1 \vec{e}_2$.

The boundary of 2D solid angle $\angle \vec{e}_1 \vec{e}_2$ is 1D solid angles $\angle \vec{e}_1$ and $\angle \vec{e}_2$. From Section 8.2.3, it can be proved that:

$$-\angle \vec{e}_1 \vec{e}_2 = \angle (-\vec{e}_1)(-\vec{e}_2).$$

A 3D solid angle $A$ with top $(0, 0, 0)$ is represented as:

$$\angle \vec{e}_1 \vec{e}_2 \cdots \vec{e}_{u-1} \vec{e}_u.$$

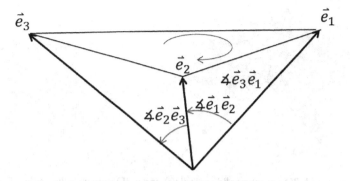

Figure 8.5 A 3D convex angle $\angle \vec{e}_1 \vec{e}_2 \vec{e}_3$.

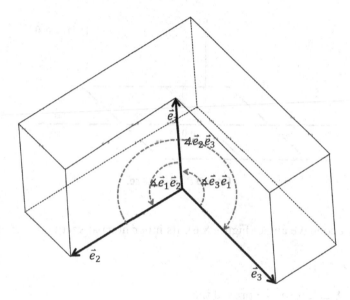

Figure 8.6 A 3D concave angle $\angle \vec{e}_1 \vec{e}_2 \vec{e}_3$.

Vectors $\vec{e}_1, \ldots, \vec{e}_{u-1} \vec{e}_u$ are the edge vectors of solid angle $\angle \vec{e}_1 \vec{e}_2 \cdots \vec{e}_{u-1} \vec{e}_u$ (Figures 8.5 & 8.6). The boundary of 3D solid angle $\angle \vec{e}_1 \vec{e}_2 \cdots \vec{e}_{u-1} \vec{e}_u$ is 2D solid angles $\angle \vec{e}_r \vec{e}_{r+1}$. Solid angle $\angle \vec{e}_r \vec{e}_{r+1}$ rotates from $\vec{e}_r$ to $\vec{e}_{r+1}$ in right-hand rule. The rotation axis is the outside normal of $\angle \vec{e}_r \vec{e}_{r+1}$.

If $\angle \vec{e}_r \vec{e}_{r+1}$ is a convex angle (Figure 8.5), its inner normal vector (Figure 8.7) is:

$$\vec{n}_{12} = \vec{e}_{r+1} \times \vec{e}_r.$$

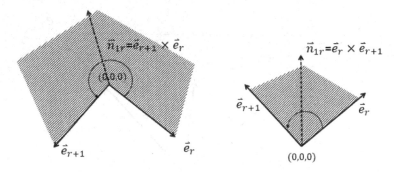

Figure 8.7 A 3D concave solid angle (left) and a 3D convex solid angle (right).

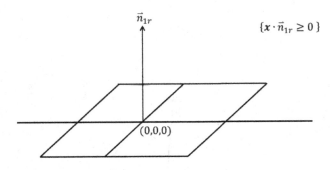

Figure 8.8 Half space.

If $\angle\vec{e}_r\vec{e}_{r+1}$ is a concave angle (Figure 8.6), its inner normal vector (Figure 8.7):

$$\vec{n}_{12} = \vec{e}_r \times \vec{e}_{r+1}.$$

From Section 8.2.3, it can be proved that

$$-\angle\vec{e}_1\vec{e}_2 \cdots \vec{e}_{u-1}\vec{e}_u$$

$$= \angle(-\vec{e}_u)(-\vec{e}_{u-1}) \cdots (-\vec{e}_2)(-\vec{e}_1).$$

## 8.2.5   Inequality equations of angles

A plane $A$ passing $(0, 0, 0)$ is defined as the set of $x$ satisfying the equation

$$\perp n_{1r} = \{x \cdot \vec{n}_{1r} = 0\}$$

where $\vec{n}_{1r}$ is the normal vector of the plan $A$ (Figure 8.8).

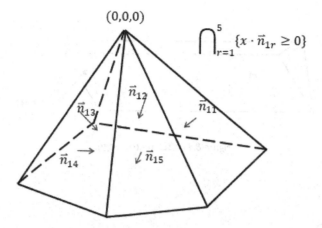

$$\bigcap_{r=1}^{5}\{x \cdot \vec{n}_{1r} \geq 0\}$$

*Figure 8.9* 3D convex angle.

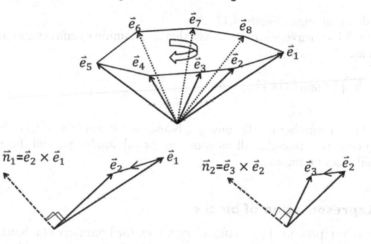

*Figure 8.10* Normal vectors of face angles.

Half space A passing $(0,0,0)$ can be represented as the set of $x$ satisfying the equation (Figure 8.8),

$$\uparrow n_{1r} = \{x \cdot \vec{n}_{1r} \geq 0\}.$$

Here $\vec{n}_{1r}$ is normal to plane $\perp n_{1r}$ and points the inside of A.

If a solid 3D angle A is convex, A is the intersection of half space (Figures 8.9 and 8.10),

$$A = \angle \vec{e}_1 \vec{e}_2 \cdots \vec{e}_{u-1} \vec{e}_u = \bigcap_{r=1}^{u}\{x \cdot \vec{n}_{1r} \geq 0\}$$

$$= \bigcap_{r=1}^{u} \uparrow n_{1r}.$$

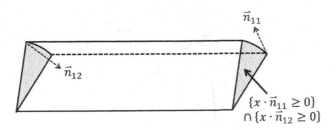

Figure 8.11   3D vector edge.

When $u = 2$,

$$A = \uparrow n_{11} \bigcap \uparrow n_{12}$$

is a solid dihedral angle (Figure 8.11).

If angle $A$ is concave, it can be proven that $A$ is a union of convex sub-angles which is defined as:

$$A = \bigcup_i \bigcap_r \uparrow n_{1i(r)}, i(r) \in \{1, \ldots, u\}.$$

Here $i$ is the number of 3D convex sub-angles; $r$ is number of faces belonging to the $i^{th}$ 3D convex sub-angle. All faces of convex sub-angles are still the extension of the original faces of angle $A$.

## 8.2.6   Representation of blocks

A block can be represented by its boundary. Given the boundary of a block, this block will be completely defined. The boundary of a 3D block is composed by polygons. A polygon which is a part of the boundary of a 3D block can be defined as

$$a_1 a_2 \cdots a_{p-1} a_p$$

where $a_{p+1} = a_1$ and $a_1 a_2 \cdots a_{p-1} a_p a_{p+1}$ rotates in the right-hand rule, *i.e.* the rotation axis points outward to block $A$ (Figures 8.12–14).

This polygon is also represented by its boundary

$$\partial a_1 a_2 \cdots a_{p-1} a_p = a_1 a_2 \cup a_2 a_3 \cup \cdots \cup a_{p-1} a_p \cup a_p a_1.$$

Denote the polygon of $a_i a_{i+1}$ as

$$P(a_i a_{i+1}) = a_1 a_2 \cdots a_{p-1} a_p.$$

$a_i a_{i+1}$ is an edge of polygon.

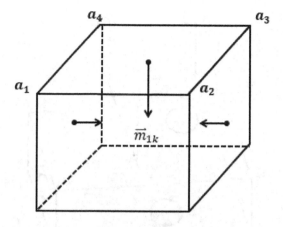

*Figure 8.12* A convex 3D block.

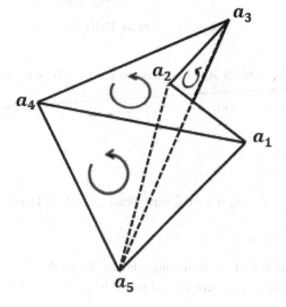

*Figure 8.13* Concave 3D block.

Denote $a_{1k}$ is any vertex on polygon face $k$, the inner normal vector $\vec{m}_{1k}$ of face $k$ points the inside of block $A$ (Figure 8.12) and

$$(a_{1k}a_i) = \vec{a}_i.$$

The inner vector $\vec{m}_{1k}$ can be defined as,

$$\vec{m}_{1k} = \sum_{i=1}^{p} \vec{a}_{i+1} \times \vec{a}_i.$$

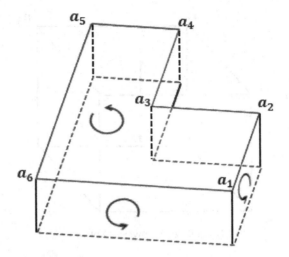

*Figure 8.14* Concave 3D block.

Denote $e$ as a vertex of block $A$, all vertices connecting with $e$ by edges are:

$$e_1 e_2 \cdots e_{u-1} e_u, \quad e_{u+1} = e_1.$$

Define:

$$(e e_r) = \vec{e}_r.$$

Therefore, $\angle \vec{e}_1 \vec{e}_2 \cdots \vec{e}_{u-1} \vec{e}_u$ is a solid angle passing $(0, 0, 0)$. Denote:

$$\angle e = \angle \vec{e}_1 \vec{e}_2 \cdots \vec{e}_{u-1} \vec{e}_u.$$

Similar to the condition of 3D solid angle (Figure 8.10), the inner normal vectors of boundary solid angle $\angle \vec{e}_r \vec{e}_{r+1}$ are $\vec{n}_{1r}$ and point the inside of block $A$.

### 8.2.7   Structure of blocks

Solid angles can still be recognized as one kind of infinite blocks.

Denote the following as the set of all boundary vertices of block $A$ and $B$, respectively

$$A(0), B(0).$$

Denote the following as the set of all boundary vectors or edges of block $A$ and $B$ respectively

$$A(1), B(1).$$

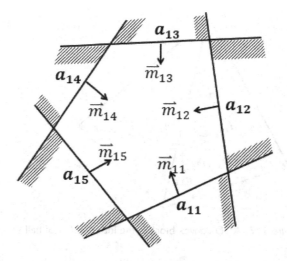

Figure 8.15  A 2D convex block is the intersection of half planes.

Denote the following as the set of all boundary 2D solid angles or polygons of block $A$ and $B$ respectively

$A(2), B(2)$.

A cover system of block $A$ means $A$ is the union of its closed subset $A_i$

$$A = \bigcup_i A_i.$$

Closed set $A_i$ means

$A_i \subset \partial A_i$.

This definition of cover system here is different from general topology.

## 8.2.8   Inequality equations of blocks

A block can also be represented by inequality equations. For a general block $A$, the face $k$ is a part of a plane passing $a_{1k}$ and corresponds to half space passing $a_{1k}$. This half space is represented by the following inequality equation,

$$\{(x - a_{1k}) \cdot \vec{m}_{1k} \geq 0\} = \uparrow m_{1k} + a_{1k}$$

where $\vec{m}_{1k}$ is the inner normal vector of face $k$ and points the inside of $A$ (Figures 8.15 & 8.16).

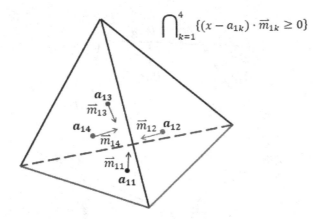

Figure 8.16 A 3D convex block is the intersection of half space.

If block $A$ is convex, $A$ is the intersection of half space (Figure 8.16):

$$A = \bigcap_{k=1,\ldots,f} (\uparrow m_{1k} + a_{1k}),$$

If block $A$ is concave, it can be proven that, $A$ is a union of convex sub-blocks, which is defined as:

$$A = \bigcup_i \bigcap_k (\uparrow m_{1i(k)} + a_{1i(k)}), \quad i(k) \in \{1,\ldots,u\}.$$

Here $i$ is the number of 3D convex sub-blocks; $k$ is the number of faces belonging to the $i$th 3D convex sub-blocks. All faces of convex sub-blocks are still the extension of the original faces of block $A$.

## 8.3 DEFINITION OF THE ENTRANCE BLOCK

The complicated contact conditions between two general blocks $A$ and $B$ can be simplified as the relations that a referring point is located inner, outer, or lies on the boundary of an entrance block. The entrance block is also represented by inequality equations.

The computation of an entrance block is geometric computation. The way of computing entrance block here is to transfer geometric computation to algebraic computation.

### 8.3.1 Relations of two blocks

The distance between block $A$ and block $B$ is defined and denoted as:

$$|A, B| = \min\{|b - a| \setminus \forall a \in A, \forall b \in B\}.$$

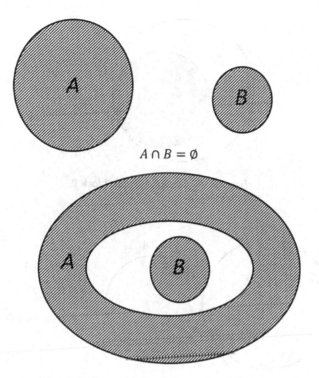

$$A \cap B = \emptyset$$

Figure 8.17  Blocks A and B have empty intersection.

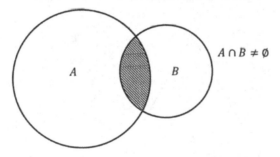

$$A \cap B \neq \emptyset$$

Figure 8.18  Blocks A and B have common points.

There are three different kinds of relation, i.e. contact conditions, between two blocks A and B.

Case 1: A and B are separated (Figure 8.17): $A \cap B = \emptyset$

In this condition, the distance between A and B is: $|A, B| = \varepsilon > 0$,

Case 2: A and B are overlapped, *i.e.* penetrated (Figures 8.18 & 8.19) $A \cap B \neq \emptyset$

Case 3: A and B contacted (Figures 8.20 & 8.21)

$$A \cap B \neq \emptyset, \quad A \cap B \subset (\partial A \cap \partial B) \Leftrightarrow$$

$$A \cap B \neq \emptyset, \quad int(A \cap B) = \emptyset.$$

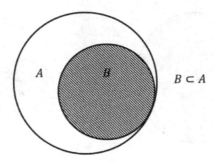

Figure 8.19   Block A includes block B.

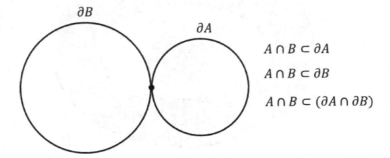

Figure 8.20   Blocks A and B contact at one point.

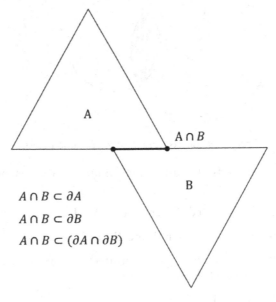

Figure 8.21   Blocks A and B contact on the boundaries of A and B.

The following equation is called contact condition:

$$A \cap B \subset (\partial A \cap \partial B) \neq \emptyset.$$

To know these different cases, the "entrance block" is defined in the next section.

### 8.3.2 Definition of entrance block

The use of entrance block can help to transfer the contact relations between two blocks to the entrance relations between one block and one point. Given a reference point $a_0$ of block $A$, the entrance block $E(A, B)$ is defined as:

$$E(A, B) = \bigcup_{a \in A, b \in B} (b - a + a_0) = B - A + a_0.$$

The reference point $a_0$ moves together with block $A$. Moving block $A$ without rotation, every point of $E(A, B)$ is a position of $a_0$ while block $A$ and block $B$ have common points.

### 8.3.3 Entrance block and parallel movement

As defined $A + \delta$ is parallel movement of $A$ along $\delta$.

#### Theorem of parallel movement

The entrance block $E(A, B)$ is irrelevant with the specific position of block $A$ (Figure 8.22).

#### Proof:

$$E(A + \delta, B) = B - (A + \delta) + (a_0 + \delta) = B - A + a_0 = E(A, B),$$
$$E(A + \delta, B) = E(A, B).$$

### 8.3.4 Entrance block of a block and a point

Assuming $B$ is a general block and $A = a_0$ is a point, the entrance block is $B$. (Figure 8.23)

$$E(A, B) = E(a_0, B) = B - a_0 + a_0 = B.$$

### 8.3.5 The entrance block of a block and a ball

Defining $A$ as a ball

$$A = \{|x - a_0| \leq r\}.$$

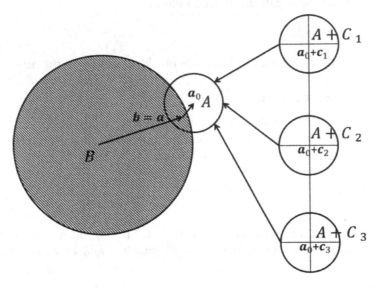

Figure 8.22  The entrance block $E(A,B)$ remains the same after parallel movements of block $A$.

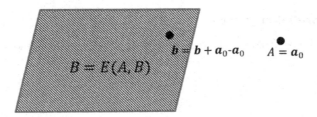

Figure 8.23  Point-block entrance.

### Theorem of point-block distance (Figure 8.24)

$$|a_0, B| \leq r \iff a_0 \in E(A, B)$$

### Proof:

$$|a_0, B| \leq r \iff$$
$$\exists b \in B, \quad |a_0 - b| \leq r \iff$$
$$\exists b \in B, \quad b \in A \iff$$
$$\exists b \in A \cap B \iff a_0 \in E(A, B).$$

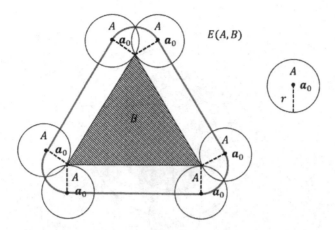

*Figure 8.24* The entrance block between a triangle and a point.

### 8.3.6  The entrance ball between two balls

Assuming $A$ and $B$ are 3D or 2D balls, $a_0$ and $b_0$ are the centers of $A$ and $B$ respectively, the equations of $A$ and $B$ are as follows,

$$A = \{|x - a_0| \leq r_1\},$$
$$B = \{|x - b_0| \leq r_2\}.$$

Denote ball

$$C = \{|x - b_0| \leq (r_1 + r_2)\}.$$

**Theorem of ball entrance**

$$E(A, B) = C.$$

**Proof:**

$$\forall\, a \in A, b \in B, E(a, b) = b - a + a_0,$$
$$|E(a, b) - b_0| = |(b - b_0) - (a - a_0)|$$
$$\leq |b - b_0| + |a - a_0| \leq r_2 + r_1$$
$$\Rightarrow E(A, B) \subset C.$$

If

$$|c - b_0| \leq (r_1 + r_2),$$
$$\exists\, a \in A, b \in B, b = a - a_0 + c \Rightarrow$$
$$E(A, B) \supset E(a, b) = b - a + a_0 = c$$
$$\Rightarrow E(A, B) \supset C.$$

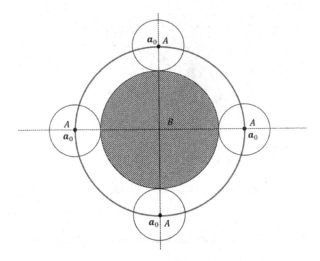

*Figure 8.25* The Entrance ball of the contact between two balls.

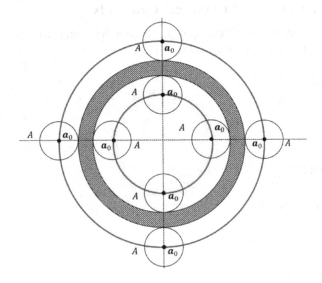

*Figure 8.26* Entrance ring of a disk and a ring.

Therefore

$$E(A, B) = C.$$

The entrance block $E(A, B)$ is a ball with radius $r_1 + r_2$ and centered on $b_0$.

$$E(A, B) = \{|x - b_0| \le (r_1 + r_2)\}.$$

This is a basic formula of the ball to ball contact (Figures 8.25 & 8.26).

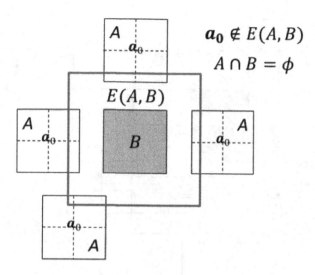

$$a_0 \notin E(A, B)$$
$$A \cap B = \phi$$

Figure 8.27   Entrance block of two squares.

### 8.3.7  Theorem of separation

*Theorem of separation (Figure 8.27)*

$$A \cap B = \emptyset \Leftrightarrow a_0 \notin E(A, B).$$

**Proof:**

$$A \cap B = \emptyset \Leftrightarrow \forall a \in A, \forall b \in B, a \neq b$$
$$\Leftrightarrow \forall a \in A, \forall b \in B, b - a + a_0 \neq a_0 \Leftrightarrow$$
$$a_0 \notin E(A, B).$$

The following theorems are about the block relations.

### 8.3.8  Theorem of entrance

$$A \cap B \neq \emptyset \Leftrightarrow a_0 \in E(A, B).$$

### 8.3.9  Theorem of distance

$$|A, B| = \varepsilon \Leftrightarrow |a_0, E(A, B)| = \varepsilon.$$

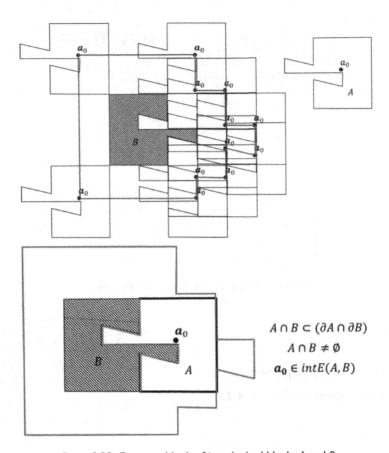

$$A \cap B \subset (\partial A \cap \partial B)$$
$$A \cap B \neq \emptyset$$
$$a_0 \in int E(A, B)$$

*Figure 8.28* Entrance block of interlocked blocks A and B.

**Proof:**

$$|A, B| = \varepsilon \Leftrightarrow \exists\, a \in A, b \in B, |a, b| = \varepsilon,$$
$$\forall a \in A, \forall b \in B, |a, b| \geq \varepsilon \Leftrightarrow$$
$$\exists\, a \in A, b \in B, |a_0, b - a + a_0| = \varepsilon,$$
$$\forall a \in A, b \in B, |a_0, b - a + a_0| \geq \varepsilon$$
$$\Leftrightarrow |a_0, E(A, B)| = \varepsilon.$$

## 8.3.10 Theorem of lock

See Figure 8.28.

$$(A \cap B) \subset (\partial A \cap \partial B) \neq \emptyset,$$
$$\exists\, \varepsilon > 0, \forall |\delta| < \varepsilon, (A + \delta) \cap B \neq \emptyset \Leftrightarrow$$
$$(A \cap B) \subset (\partial A \cap \partial B) \neq \emptyset, a_0 \in int(E(A, B)).$$

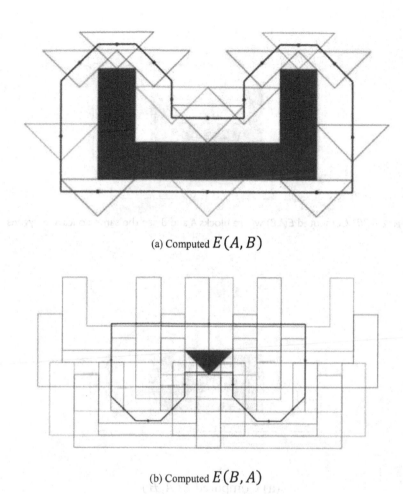

(a) Computed $E(A, B)$

(b) Computed $E(B, A)$

Figure 8.29   E(A,B) and E(B,A), where block A is a triangle and block B is a concave polygon.

**Proof:**

$$(A \cap B) \subset (\partial A \cap \partial B) \neq \emptyset,$$
$$\exists \varepsilon > 0, \forall |\delta| < \varepsilon, (A + \delta) \cap B \neq \emptyset \Leftrightarrow$$
$$(A \cap B) \subset (\partial A \cap \partial B) \neq \emptyset,$$
$$\exists \varepsilon > 0, \forall |\delta| < \varepsilon, \exists a \in A, b \in B, a + \delta = b,$$
$$b - a + a_0 = a_0 + \delta, a_0 + \delta \in E(A, B) \Leftrightarrow$$
$$(A \cap B) \subset (\partial A \cap \partial B) \neq \emptyset,$$
$$a_0 \in int(E(A, B)).$$

The definition of entrance block transfers the entrances and contacts of two blocks to the entrances and contacts of one point and one block.

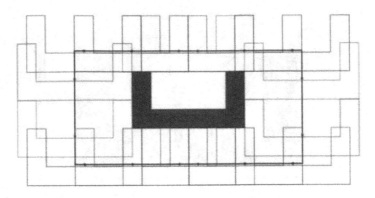

*Figure 8.30* Computed *E(A,B)* where blocks *A* and *B* are the same concave polygons.

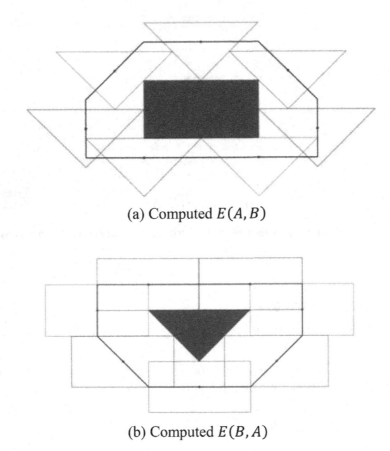

(a) Computed $E(A, B)$

(b) Computed $E(B, A)$

*Figure 8.31*  *E(A,B)* and *E(B,A)* where blocks *A* is a triangle and *B* is a rectangle.

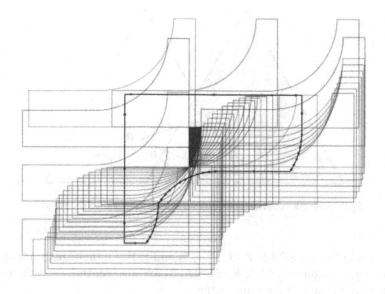

*Figure 8.32*  Computed *E(A,B)* where block *A* is a convex polygon and *B* is a concave polygon.

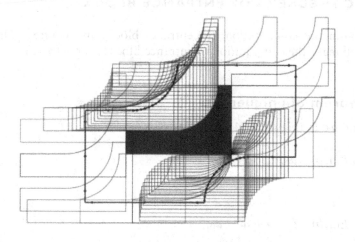

*Figure 8.33*  Computed *E(A,B)* where blocks *A* and *B* are the same concave polygons.

## 8.3.11  Examples of the entrance block

The proof of solving 2D entrance block is provided in Section 8.8. Here some complex entrance blocks are illustrated without proof.

Figures 8.29 to 8.33 are computed 2D $E(A, B)$, where block $B$ is a polygon indicated by the solid body, block $A$ is drawn by thin lines and $E(A, B)$ is drawn by thick lines.

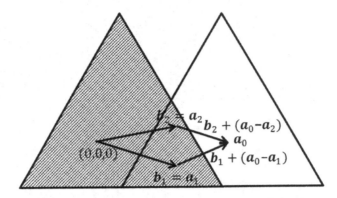

*Figure 8.34* All common point pairs $a = b$ correspond to the same $a_0$.

As shown in Figures 8.29 & 8.31, $E(A, B)$ and $E(B, A)$, they are centrally symmetrical. Similarly, in Figures 8.30 & 8.33, $E(A, B)$ are also centrally symmetrical in the conditions that $A, B$ are of the same shape.

## 8.4  BASIC THEOREMS OF ENTRANCE BLOCK

In the discontinuous computation, the entrance blocks are required. The following theorems will make the computation of entrance blocks more efficient.

### 8.4.1  Theorem of uniqueness

For $a \in A$ and $b \in B$, (Figure 8.34)

$$a = b \Leftrightarrow E(a, b) = a_0.$$

**Proof:**

$$a = b \Leftrightarrow E(a, b) = b - a + a_0 = a_0.$$

### 8.4.2  Theorem of finite covers of entrance blocks

Assuming $A$ and $B$ are $n$ dimensional blocks, $n = 1, 2, 3$.

$$E(A, B) = \bigcup_{m=0}^{n} E(A(m), B(n - m)),$$
$$E(A, B) = E(\partial A, B) \cup E(A, \partial B).$$

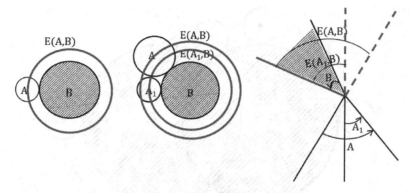

Figure 8.35 Examples for theory of including.

**Proof:**

$$\forall a \in A, \forall b \in B,$$
$$\exists \delta, a + \delta \in A(m) \subset A, b + \delta \in B(n - m) \subset B \Rightarrow$$
$$E(a, b) = E(a + \delta, b + \delta) \subset E(A(m), B(n - m)) \Rightarrow$$
$$E(A, B) = \bigcup_{m=0}^{n} E(A(m), B(n - m)) \Rightarrow$$
$$E(A, B) = E(\partial A, B) \cup E(A(n), B(0)) \Rightarrow$$
$$E(A, B) = E(\partial A, B) \cup E(A, \partial B).$$

This theorem indicates that $E(A(m), B(n - m))$ forms a finite cover system on $E(A, B)$.

### 8.4.3  Theorem of including

Assume block $A_1$ and $A$ have the same reference point $a_0$. (Figure 8.35)

If $A_1 \subset A, B_1 \subset B$ then
$E(A_1, B) \subset E(A, B),$
$E(A, B_1) \subset E(A, B).$

**Proof:**

$$A_1 \subset A \rightarrow E(A_1, B) = B - A_1 + a_0$$
$$\subset B - A + a_0 = E(A, B)$$
$$\Rightarrow E(A_1, B) \subset E(A, B).$$
$$B_1 \subset B \Rightarrow E(A, B_1) = B_1 - A + a_0$$
$$\subset B - A + a_0 = E(A, B)$$
$$\Rightarrow E(A, B_1) \subset E(A, B).$$

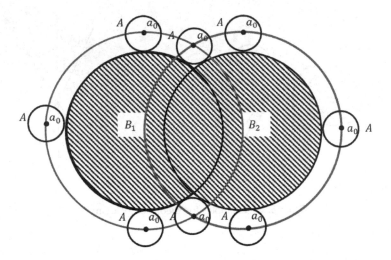

Figure 8.36 Example for theory of union.

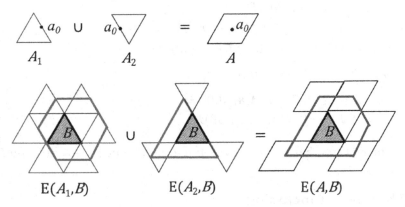

Figure 8.37 Another example for theory of union.

### 8.4.4 Theorem of union

Assuming blocks $A_1, A_2$ and $A$ have the same reference point $a_0$ (Figures 8.36 & 8.37), then

$$A = A_1 \cup A_2 \Rightarrow E(A_1, B) \cup E(A_2, B) = E(A, B).$$
$$B = B_1 \cup B_2 \Rightarrow E(A, B_1) \cup E(A, B_2) = E(A, B).$$

### 8.4.5 Theorem of symmetry

Using $a_0 = (0, 0, 0)$, $b_0 = (0, 0, 0)$, then

$$E(A, B) = -E(B, A).$$

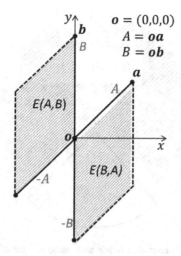

*Figure 8.38* Example for theory of symmetry.

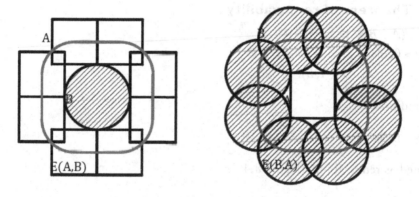

*Figure 8.39* Another example for theory of symmetry.

**Proof:**

$$E(A, B) = B - A + a_0$$
$$= -(A - B + b_0) + (a_0 + b_0)$$
$$= -E(B, A) + (a_0 + b_0).$$

The entrance block $E(A, B)$ and entrance block $E(B, A)$ are centrally symmetrical (Figures 8.38 & 8.39).

## 8.4.6 Theorem of contact

$$int(A \cap B) = \emptyset \Leftrightarrow (A \cap B) \subset (\partial A \cap \partial B).$$

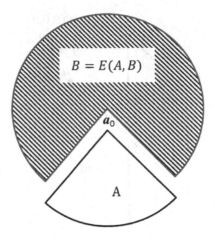

*Figure 8.40* Entrance block and removability of block A.

### 8.4.7 Theorem of removability

$$(A \cap B) \subset (\partial A \cap \partial B) \neq \emptyset,$$
$$\forall \varepsilon > 0, \exists |\delta| < \varepsilon, (A + \delta) \cap B = \emptyset \Leftrightarrow$$
$$a_0 \in \partial E(A, B).$$

The condition:

$$\forall \varepsilon > 0, \exists |\delta| < \varepsilon, \quad (A + \delta) \cap B = \emptyset$$

is defined as removability (Figure 8.40).

### 8.4.8 Theorem of inner points

$$int(A \cap B) \neq \emptyset \Rightarrow a_0 \in int(E(A, B)).$$

***Proof:***

$$int(A \cap B) \neq \emptyset \Rightarrow a = b,$$
$$\exists \varepsilon > 0, \forall |\delta| < \varepsilon, b + \delta \in B$$
$$\Rightarrow b + \delta - a + a_0 = a_0 + \delta \in E(A, B)$$
$$\Rightarrow a_0 \in int(E(A, B)).$$

### 8.4.9 Theorem of convex blocks

The majority of blocks are convex blocks. As non-convex blocks can also be divided into convex blocks, the following theorem of convex blocks is important.

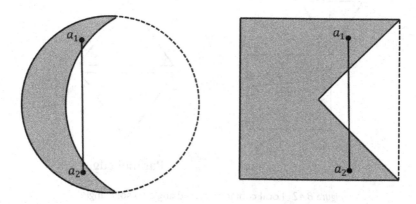

*Figure 8.41*  Non-convex blocks.

## Theorem of Convex Blocks

If block $A$ and $B$ are convex, the entrance block $E(A, B)$ is convex.

### Proof:

Assuming

$$E(a_1, b_1) \in E(A, B), E(a_2, b_2) \in E(A, B),$$
$$a_1 \in A, b_1 \in B, a_2 \in A, b_2 \in B.$$
$$\forall t, 0 \leq t \leq 1,$$
$$(1 - t)E(a_1, b_1) + tE(a_2, b_2)$$
$$= E((1 - t)a_1 + ta_2, (1 - t)b_1 + tb_2)$$
$$\in E(A, B)$$

$E(A, B)$ is convex (Figure 8.41).

In Figures 8.24, 8.25, 8.27, 8.31, 8.35, 8.37, 8.38 and 8.39, $A$ and $B$ are convex blocks, then $E(A, B)$ are convex blocks.

## 8.5   BOUNDARIES OF THE ENTRANCE SOLID ANGLES OF 2D SOLID ANGLES

The boundaries of entrance solid angle determine the entrance solid angles completely. This section will be the theoretical basis of the computation of the entrance solid angle boundary. The boundaries of 2D entrance solid angles are the subsets of the contact vectors.

In the previous codes of 2D discontinuous deformation analysis (DDA, Shi 1988) and 2D numerical manifold method (NMM, Shi 1991), the algorithms for solving the contacts are also based on computing these boundaries and consistent with the results

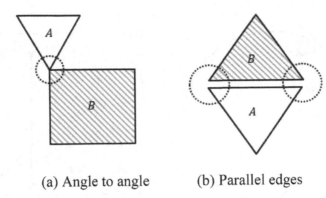

(a) Angle to angle        (b) Parallel edges

*Figure 8.42* Local contacts of solid angle to solid angle.

of this general entrance solid angle boundary. However, the new theory can simplify these contact algorithms of 2D DDA and NMM. In complicated cases the new theory can find the contact covers precisely.

### 8.5.1   Local entrance solid angle of 2D blocks

As the discontinuous computation follows time steps, the step displacements can be chosen small enough so that all contacts become independent contacts of local angle to angle contacts and angle to edge contacts. These local angle to angle contacts and angle to edge contacts are equivalent to solid angle to solid angle contacts and solid angle to edge contacts respectively (Figure 8.42).

### 8.5.2   Existence of entrance solid angle boundary of 2D solid angles

The following theorem is the necessary condition of the existence of the boundary of entrance solid angles of 2D solid angles. Considering the entrance of two 2D angles:

$$A = \angle \vec{e}_1 \vec{e}_2 + e,$$

$$B = \angle \vec{h}_1 \vec{h}_2 + h.$$

***Theorem of empty boundary of 2D entrance solid angle***

If $\angle e_1 e_2$ and $\angle h_1 h_2$ have a common inner vector, $\vec{v}_0$,

$$\vec{v}_0 \in int(\angle \vec{e}_1 \vec{e}_2 \cap \angle \vec{h}_1 \vec{h}_2).$$

then $E(A, B)$ is the whole plane (Figure 8.43),

$$\partial E(A, B) = \emptyset.$$

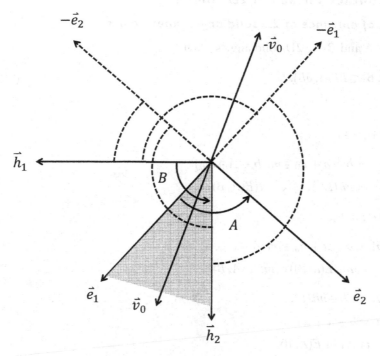

*Figure 8.43* The entrance angle of overlapped 2D solid angles.

**Proof:**

$$\exists \varepsilon > 0,$$

$$C = \{\vec{v}_0 \cdot x / |x| \le \varepsilon\},$$

$$C \subset (\angle \vec{e}_1 \vec{e}_2 \cap \angle \vec{h}_1 \vec{h}_2).$$

$$C - \angle \vec{v}_0$$

is the whole plane and

$$\partial(C - \angle \vec{v}_0) = \emptyset.$$

$$E(A, B) = B - A + a_0$$

$$= \angle \vec{h}_1 \vec{h}_2 - \angle \vec{e}_1 \vec{e}_2 + h - e + a_0$$

$$\supset C - \angle \vec{v}_0 + h - e + a_0.$$

$$\partial(C - \angle \vec{v}_0 + h - e + a_0) = \emptyset$$

$$\Rightarrow \partial E(A, B) = \emptyset.$$

Therefore, $E(A, B)$ is the whole plane.

### 8.5.3   Contact surface of 2D solid angles

*Theorem of entrance of 2D solid angle inner points*

Assuming $A$ and $B$ are 2D solid angles, then

$$\partial E(A, B) \subset E(\partial A, \partial B).$$

**Proof:**

If $a \in int(A), b \in B$,

$$E(a, b) = b - a + a_0 \in int(b - A + a_0)$$
$$= int(E(A, b)) \subset int(E(A, B)).$$

If $a \in A, b \in int(B)$,

$$E(a, b) = b - a + a_0 \in int(B - a + a_0)$$
$$= int(E(a, B)) \subset int(E(A, B)).$$

If $a \in int(A)$ or $b \in int(B)$,

$$E(a, b) = b - a + a_0$$

is an inner point of $E(A, B)$.
   Therefore

$$\partial E(A, B) \subset E(\partial A, \partial B).$$

### 8.5.4   Entrance of boundary vector to vector of 2D solid angles

Assuming $A$ and $B$ are 2D solid angles,

$$A = \angle \vec{e}_1 \vec{e}_2 + e,$$
$$B = \angle \vec{h}_1 \vec{h}_2 + h.$$

*Theorem of entrance of boundary vector-vector of 2D solid angles*

$$\partial E(A, B) \subset E(A(0), B(1)) \cup E(A(1), B(0)).$$

**Proof:**

From Section 5.3,

$$\partial E(A, B) \subset E(\partial A, \partial B) = E(A(1), B(1)).$$

If $\vec{e}_i$ and $\vec{h}_j$ are not parallel and

$$a \in \angle \vec{e}_i + e, \quad b \in \angle \vec{h}_j + h, \quad i, j = 1, 2,$$
$$a \neq e, \quad b \neq h$$

then

$$E(a, b) = b - a + a_0$$

$$\in int(\angle \vec{h}_j + h - \angle \vec{e}_i - e + a_0)$$

$$= int(E(\angle \vec{e}_i + e, \angle \vec{h}_j + h)) \subset int(E(A, B)).$$

$$E(\angle \vec{e}_i + e, \angle \vec{h}_j + h) \cap \partial E(A, B) \subset$$

$$(E(e, \angle \vec{h}_j + h) \cup^E (\angle \vec{e}_i + e, h)).$$

If $\vec{e}_i \parallel \vec{h}_j$, from the theorem of Section 8.4.2

$$E(\angle \vec{e}_i + e, \angle \vec{h}_j + h) = (E(e, \angle \vec{h}_j + h) \cup E(\angle \vec{e}_i + e, h)).$$

Therefore

$$\partial E(A, B) \subset E(A(0), B(1)) \cup E(A(1), B(0)).$$

### 8.5.5   Contact vectors of angle to angle contact

Assume $A$ and $B$ are the same 2D angles as in Section 8.5.4. Denote $\vec{n}_{1i}$ as the inner normal of $\angle e_i$,

$$\vec{n}_{11} = (0, 0, 1) \times \vec{e}_1,$$

$$\vec{n}_{12} = (0, 0, -1) \times \vec{e}_2.$$

Denote $\vec{n}_{2j}$ as the inner normal of $\angle h_j$,

$$\vec{n}_{21} = (0, 0, 1) \times \vec{h}_1,$$

$$\vec{n}_{22} = (0, 0, -1) \times \vec{h}_2.$$

#### Theorem of 2D vertex-vector contact

$$\exists b \in int(\angle \vec{h}_j + h), \quad E(e, b) \in \partial E(A, B)$$
$$\Rightarrow int(\angle \vec{e}_1 \vec{e}_2 \cap \uparrow n_{2j}) = \emptyset.$$

**Proof:**

If

$$int(\angle \vec{e}_1 \vec{e}_2 \cap \uparrow n_{2j}) \neq \emptyset,$$

$$\exists \, \vec{e}_0 \in \angle \vec{e}_1 \vec{e}_2, \quad \vec{e}_0 \cdot \vec{n}_{2j} > 0,$$

$$E(e, b) \in$$

$$int(\angle h_j - \angle \vec{e}_0 + h - e + a_0 \cup \angle \vec{h}_1 \vec{h}_2 + h - e + a_0)$$

$$= int(E(\angle \vec{e}_0 + e, \angle \vec{h}_j + h) \cup E(e, \angle \vec{h}_1 \vec{h}_2 + h))$$

$$\subset int(E(A, B) \cup E(A, B)) = int(E(A, B)).$$

$$\Rightarrow E(e, b) \in int(E(A, B)), \quad E(e, b) \notin \partial E(A, B).$$

The following theorem is similar.

### Theorem of 2D vector-vertex contact

$$\exists \, a \in int(e + \angle \vec{e}_i), \quad E(a, h) \in \partial E(A, B)$$

$$\Rightarrow int\left(\angle \vec{h}_1 \vec{h}_2 \cap \uparrow n_{1i}\right) = \emptyset.$$

**Proof:**

If

$$int(\angle \vec{h}_1 \vec{h}_2 \cap \uparrow n_{1i}) \neq \emptyset,$$

$$\exists \, \vec{h}_0 \in \angle \vec{h}_1 \vec{h}_2, \quad \vec{h}_0 \cdot \vec{n}_{1i} > 0,$$

$$E(a, h) \in$$

$$int(\angle h_0 - \angle e_i + h - e + a_0 \cup -\angle \vec{e}_1 \vec{e}_2 + h - e + a_0)$$

$$= int(E(\angle \vec{e}_i + e, \angle \vec{h}_0 + h) \cup E(\angle \vec{e}_1 \vec{e}_2 + e, h))$$

$$\subset int(E(A, B) \cup E(A, B)) = int(E(A, B)).$$

$$\Rightarrow E(a, h) \in int(E(A, B)), \quad E(a, h) \notin \partial E(A, B).$$

### Theorem of entrance of solid angle and half plane

$$int\left(\angle \vec{e}_1 \vec{e}_2 \cap \uparrow n_{2j}\right) = \emptyset \Leftrightarrow$$

$$\angle \vec{e}_1 \vec{e}_2 \cap \uparrow n_{2j} \subset \left(\partial \angle \vec{e}_1 \vec{e}_2 \cap \partial \uparrow n_{2j}\right) \Leftrightarrow$$

$$E\left(\angle \vec{e}_1 \vec{e}_2, \uparrow n_{2j}\right) = \uparrow n_{2j} + a_0 \Leftrightarrow$$

$$\vec{e}_1 \times \vec{e}_2 \uparrow\uparrow (0, 0, +1), \vec{e}_1 \cdot \vec{n}_{2j} \leq 0, \vec{e}_2 \cdot \vec{n}_{2j} \leq 0.$$

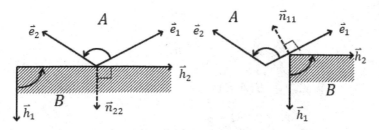

Figure 8.44  Contact of a solid angle and a boundary vector.

The equations of the first line are defined as contact condition and $E(e, \angle \vec{h}_j + h)$ is defined as one contact vector (Figure 8.44).

$$int(\angle \vec{h}_1 \vec{h}_2 \cap \uparrow n_{1i}) = \emptyset \Leftrightarrow$$
$$\angle \vec{h}_1 \vec{h}_2 \cap \uparrow n_{1i} \subset (\partial \vec{h}_1 \vec{h}_2 \cap \partial \uparrow n_{1i}) \Leftrightarrow$$
$$E(\angle \vec{h}_1 \vec{h}_2, \uparrow n_{1i}) = - \uparrow n_{1i} + a_0 \Leftrightarrow$$
$$\vec{h}_1 \times \vec{h}_2 \uparrow\uparrow (0, 0, +1), \vec{h}_1 \cdot \vec{n}_{1i} \leq 0, \vec{h}_2 \cdot \vec{n}_{1i} \leq 0.$$

The equations of the first line are defined as contact condition, and $E(\angle \vec{e}_i + e, h)$ is defined as another contact vector (Figure 8.44).

### 8.5.6  Finite covers of 2D parallel vector to vector entrance

Assume $A$ and $B$ are the same 2D solid angles as in Section 8.5.4. The inner normal vectors $\vec{n}_{ij}, i, j = 1, 2$ are defined in Section 8.5.5.

### Theorem of parallel vector contact

Assuming

$$\vec{e}_i \parallel \vec{h}_j,$$
$$\exists a \in int(e + \angle \vec{e}_i), \quad b \in int(h + \angle \vec{h}_j),$$
$$E(a, b) \in \partial E(A, B) \Rightarrow \vec{n}_{1i} \uparrow\uparrow -\vec{n}_{2j}.$$

**Proof:** (Figure 8.45)

$$\vec{n}_{1i} \uparrow\uparrow \vec{n}_{2j} \Rightarrow$$
$$\exists \vec{e}_0 \in \angle \vec{e}_1 \vec{e}_2, \quad \vec{e}_0 \cdot \vec{n}_{2j} > 0,$$
$$E(a, b) \in int(E(\angle \vec{e}_i + e, \angle \vec{h}_j + h)) \subset$$

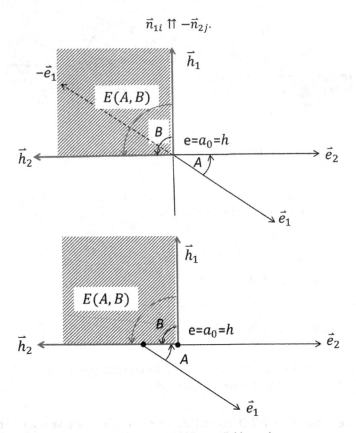

*Figure 8.45* Contact vectors of 2D parallel boundary vectors.

$$int(E(\angle \vec{e}_0 + \angle \vec{e}_i + e, \angle h_j + h) \cup E(\angle \vec{e}_i + e, \angle \vec{h}_1 \vec{h}_2 + h)) \subset$$
$$int(E(A, B) \cup E(A, B)) = int(E(A, B)).$$
$$\Rightarrow E(a, b) \in int(E(A, B)), \quad E(a, b) \notin \partial E(A, B).$$

Therefore

$$\vec{n}_{1i} \Uparrow -\vec{n}_{2j}.$$

### Theorem of contact vector covers

Under condition of the previous theorem, if

$$\vec{e}_i \Uparrow \vec{h}_j,$$
$$E(\angle \vec{e}_i + e, \angle \vec{h}_j + h)$$
$$= E(e, \angle \vec{h}_j + h) \cup E(\angle \vec{h}_j + e, h).$$

If

$$-\vec{e}_i \uparrow\uparrow \vec{h}_j,$$
$$E(\angle \vec{e}_i + e, \angle \vec{h}_j + h)$$
$$= E(e, \angle \vec{h}_j + h) = E(\angle \vec{e}_i + e, h).$$

**Proof:**

$$E(e, \angle \vec{h}_j + h) = \angle \vec{h}_j + h - e + a_0,$$
$$E(\angle \vec{e}_i + e, h) = -\angle \vec{e}_i + h - e + a_0.$$

If

$$\vec{e}_i \uparrow\uparrow \vec{h}_j,$$
$$E(\angle \vec{e}_i + e, \angle \vec{h}_j + h)$$
$$= \angle \vec{h}_j + h - \angle \vec{e}_i - e + a_0$$
$$= \angle \vec{h}_j + h - \angle \vec{h}_j - e + a_0$$
$$= \vec{h}_j + h - e + a_0$$
$$= E(e, \angle \vec{h}_j + h) \cup E(\angle \vec{h}_j + e, h).$$

Here the contact vectors $E(e, \angle \vec{h}_j + h)$ and $E(\angle \vec{h}_j + e, h)$ are two connected covers of $E(\angle \vec{e}_i + e, \angle \vec{h}_j + h)$.

If

$$\vec{h}_j \uparrow\uparrow -\vec{e}_i,$$
$$E(\angle \vec{e}_i + e, \angle \vec{h}_j + h)$$
$$= \angle \vec{h}_j + h - \angle \vec{e}_i - e + a_0$$
$$= \angle \vec{h}_j + h - e + a_0$$
$$= E(e, \angle \vec{h}_j + h) = E(\angle \vec{e}_i + e, h).$$

Therefore the contact of parallel vectors can be transferred to contacts of a vertex and a vector (Figure 8.46).

$$a_0 \in E(e, \angle \vec{h}_j + h) \Leftrightarrow e \in \angle \vec{h}_j + h,$$
$$a_0 \in E(\angle \vec{e}_i + e, h) \Leftrightarrow h \in \angle \vec{e}_i + e.$$

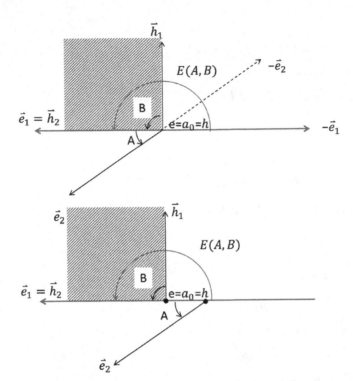

Figure 8.46  Contact vectors of 2D parallel boundary vectors.

## 8.5.7  Contact vectors of 2D entrance solid angle

Assume $A$ and $B$ are the same 2D angles as in Sections 8.5.4 and 8.5.5. From Section 8.5.4,

$$\partial E(A, B) \subset E(A(0), B(1)) \cup E(A(1), B(0))$$
$$= \left( \bigcup_{j=1,2} E(e, \measuredangle \vec{h}_j + h) \right) \cup \left( \bigcup_{i=1,2} E(\measuredangle \vec{e}_i + e, h) \right)$$

Denote $C(0, 1)$ as the union of all contact vectors of the form

$$E(e, \measuredangle \vec{h}_j + h),$$

$$C(0, 1) = \bigcup_{contact\ vector} E(e, \measuredangle \vec{h}_j + h)$$

From Section 8.5.5,

$$\vec{e}_1 \times \vec{e}_2 \upuparrows (0, 0, +1),$$
$$\vec{e}_1 \cdot \vec{n}_{2j} \leq 0, \quad \vec{e}_2 \cdot \vec{n}_{2j} \leq 0.$$

Denote $C(1,0)$ as the union of all contact vectors of the form

$$E(\angle \vec{e}_i + e, h),$$

$$C(1,0) = \bigcup_{contact\ vector} E(\angle \vec{e}_i + e, h)$$

From Section 8.5.5,

$$\vec{h}_1 \times \vec{h}_2 \upuparrows (0,0,+1),$$

$$\vec{h}_1 \cdot \vec{n}_{1i} \leq 0, \quad \vec{h}_2 \cdot \vec{n}_{1i} \leq 0.$$

The following theorem is the conclusion of Section 8.5.5.

### Theorem of 2D contact vectors

$$\partial E(A,B) \subset C(0,1) \cup C(1,0) \subset E(A,B).$$

## 8.6   CONTACT VECTORS OF 2D SOLID ANGLES

Based on Section 8.5.1, if the step movement $\rho$ is small enough, the entrances will be simplified as entrances of solid angles. The contact vectors or contact vector covers on the contact surface between 2D solid angles are computed in the following sections.

### 8.6.1   Contact vectors of a 2D convex solid angle and a 2D concave solid angle

The 2D convex solid angle $A$ and the concave solid angle $B$ are represented as (Figure 8.47)

$$A = e + \angle \vec{e}_1 \vec{e}_2,$$

$$B = h + \angle \vec{h}_1 \vec{h}_2,$$

$$\vec{e}_1 \times \vec{e}_2 \upuparrows (0,0,+1),$$

$$\vec{h}_1 \times \vec{h}_2 \upuparrows (0,0,-1).$$

The inner normal vectors of $\vec{e}_1, \vec{e}_2, \vec{h}_1, \vec{h}_2$ are

$$\vec{n}_{11} = \vec{e}_1 \times \vec{e}_2 \times \vec{e}_1,$$

$$\vec{n}_{12} = \vec{e}_2 \times \vec{e}_1 \times \vec{e}_2,$$

$$\vec{n}_{21} = \vec{h}_2 \times \vec{h}_1 \times \vec{h}_1,$$

$$\vec{n}_{22} = \vec{h}_1 \times \vec{h}_2 \times \vec{h}_2.$$

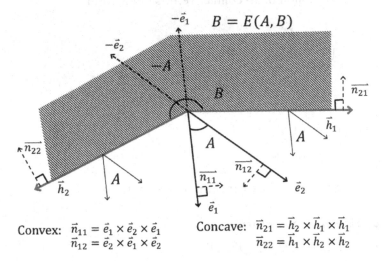

Convex: $\vec{n}_{11} = \vec{e}_1 \times \vec{e}_2 \times \vec{e}_1$
$\vec{n}_{12} = \vec{e}_2 \times \vec{e}_1 \times \vec{e}_2$

Concave: $\vec{n}_{21} = \vec{h}_2 \times \vec{h}_1 \times \vec{h}_1$
$\vec{n}_{22} = \vec{h}_1 \times \vec{h}_2 \times \vec{h}_2$

*Figure 8.47* Contact vectors of a 2D convex solid angle and a 2D concave solid angle.

$A$ is a 2D convex angle which is the intersection of two half planes. $B$ is a 2D concave angle which is the union of two half planes.

$$A = (\uparrow n_{11} + e) \cap (\uparrow n_{12} + e),$$
$$B = (\uparrow n_{21} + h) \cup (\uparrow n_{22} + h).$$

### Theorem of contact vectors of 2D concave-convex solid angle contact

Assuming

$$\vec{e}_1 \times \vec{e}_2 \uparrow\uparrow (0, 0, +1),$$
$$\vec{h}_1 \times \vec{h}_2 \uparrow\uparrow (0, 0, -1),$$

then

$$int(\angle \vec{e}_1 \vec{e}_2 \cap \angle \vec{h}_1 \vec{h}_2) = \emptyset \;\Leftrightarrow$$
$$E(A, B) = B - e + a_0,$$
$$\partial E(A, B) = \partial B - e + a_0.$$

### Proof:

$$int(\angle \vec{e}_1 \vec{e}_2 \cap \angle \vec{h}_1 \vec{h}_2) = \emptyset \;\Leftrightarrow$$
$$int(\angle \vec{e}_1 \vec{e}_2 \cap \uparrow n_{21}) = \emptyset,$$
$$int(\angle \vec{e}_1 \vec{e}_2 \cap \uparrow n_{22}) = \emptyset \;\Leftrightarrow$$

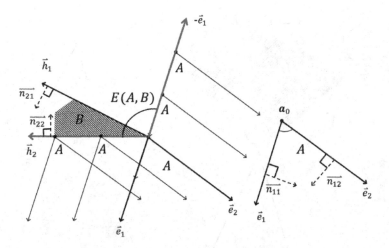

Figure 8.48 An example of contact vectors of two 2D solid angles.

$$E(\angle \vec{e}_1 \vec{e}_2, \uparrow n_{21}) = \uparrow n_{21} + a_0,$$

$$E(\angle \vec{e}_1 \vec{e}_2, \uparrow n_{22}) = \uparrow n_{22} + a_0,$$

$$E(A, B) = E(\angle \vec{e}_1 \vec{e}_2 + e, \uparrow n_{21} + h) \cup$$

$$E(\angle \vec{e}_1 \vec{e}_2 + e, \uparrow n_{22} + h)$$

$$= (E(\angle \vec{e}_1 \vec{e}_2, \uparrow n_{21}) + h - e) \cup$$

$$(E(\angle \vec{e}_1 \vec{e}_2, \uparrow n_{22}) + h - e)$$

$$= E(\angle \vec{e}_1 \vec{e}_2, \uparrow n_{21}) \cup E(\angle e_1 e_2, \uparrow n_{22}) + h - e$$

$$= \uparrow n_{21} \cup \uparrow n_{22} + h - e + a_0$$

$$= B - e + a_0,$$

$$\partial E(A, B) = \partial B - e + a_0.$$

The contact vectors and boundary vectors of $\partial E(A, B)$ are

$$E(e, \angle \vec{h}_1 + h) = h - e + a_0 + \angle \vec{h}_1,$$

$$E(e, \angle \vec{h}_2 + h) = h - e + a_0 + \angle \vec{h}_2.$$

## 8.6.2  Contact vectors of 2D convex solid angles

2D convex solid angles $A$ and $B$ (Figures 8.48 & 8.49) are represented as

$$A = e + \angle \vec{e}_1 \vec{e}_2,$$

$$B = h + \angle \vec{h}_1 \vec{h}_2,$$

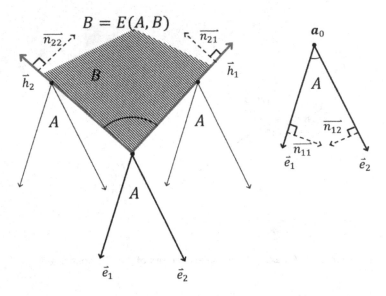

*Figure 8.49* Another example of contact vectors of two 2D solid angles.

$$\vec{e}_1 \times \vec{e}_2 \uparrow\uparrow (0,0,1),$$
$$\vec{h}_1 \times \vec{h}_2 \uparrow\uparrow (0,0,1).$$

The inner normal vectors of the edges are

$$\vec{n}_{11} = \vec{e}_1 \times \vec{e}_2 \times \vec{e}_1,$$
$$\vec{n}_{12} = \vec{e}_2 \times \vec{e}_1 \times \vec{e}_2,$$
$$\vec{n}_{21} = \vec{h}_1 \times \vec{h}_2 \times \vec{h}_1,$$
$$\vec{n}_{22} = \vec{h}_2 \times \vec{h}_1 \times \vec{h}_2.$$

A 2D convex angle is the intersection of two half planes:

$$A = \{(\boldsymbol{x} - \boldsymbol{e}) \cdot \vec{n}_{11} \geq 0\}$$
$$\cap \{(\boldsymbol{x} - \boldsymbol{e}) \cdot \vec{n}_{12} \geq 0\},$$
$$B = \{(\boldsymbol{x} - \boldsymbol{h}) \cdot \vec{n}_{21} \geq 0\}$$
$$\cap \{(\boldsymbol{x} - \boldsymbol{h}) \cdot \vec{n}_{22} \geq 0\}.$$

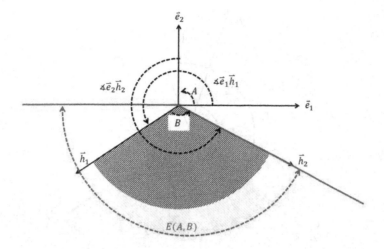

*Figure 8.50* Angle method for finding contact vectors of two 2D solid angles.

Also

$$A = (\uparrow n_{11} + e)$$
$$\cap (\uparrow n_{12} + e).$$
$$B = (\uparrow n_{21} + h)$$
$$\cap (\uparrow n_{22} + h).$$

Based on Section 8.5.2, if $\angle e_1 e_2$ and $\angle h_1 h_2$ have a common inner vector $\vec{v}_0$, $E(A, B)$ is the whole plane without boundary.

### Theorem of contact vectors of 2D solid angle

Assuming

$$int(\angle \vec{e}_1 \vec{e}_2 \cap \angle \vec{h}_1 \vec{h}_2) = \emptyset.$$

If

$$\vec{e}_1 \times \vec{h}_1 \uparrow\uparrow (0, 0, 1),$$
$$E(e, \angle \vec{h}_1 + h)$$

is a boundary vector of the entrance angle (Figures 8.50 & 8.51).
If

$$\vec{e}_2 \times \vec{h}_2 \uparrow\uparrow (0, 0, 1),$$
$$E(\angle \vec{e}_2 + e, h)$$

is another boundary vector of the entrance angle.

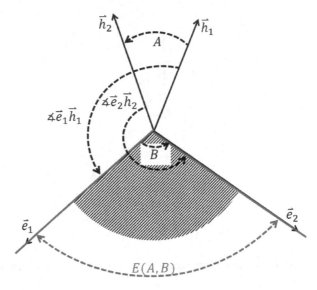

(a) Example of the contact between two convex angles

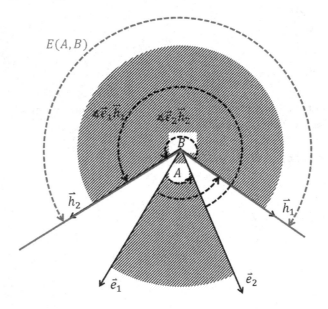

(b) Example of contact between a convex angle and a concave angle

*Figure 8.51* Rotation method for finding contact vectors of two 2D solid angles.

**Proof:**

$$int(\measuredangle \vec{e}_1 \vec{e}_2 \cap \measuredangle \vec{h}_1 \vec{h}_2) = \emptyset \Leftrightarrow$$

$\vec{e}_1, \vec{e}_2, \vec{h}_1, \vec{h}_2$ rotates from *ox* to *oy*.

If

$$\vec{e}_1 \times \vec{h}_1 \uparrow\uparrow (0,0,1) \Rightarrow$$

$$\vec{e}_1 \cdot \vec{n}_{21} \le 0, \quad \vec{e}_2 \cdot \vec{n}_{21} \le 0 \Rightarrow$$

$$E(\angle \vec{e}_1 \vec{e}_2, (\uparrow n_{21})) = \uparrow n_{21} + a_0 \Rightarrow$$

$$\partial E(\angle \vec{e}_1 \vec{e}_2 + e, (\uparrow n_{21}) + h) = h - e + a_0 + \vec{h}_1 \Rightarrow$$

$$E(e, \angle \vec{h}_1 + h) \subset E(A, B) \subset$$

$$E(\angle \vec{e}_1 \vec{e}_2 + e, (\uparrow n_{21}) + h),$$

$$E(e, \angle \vec{h}_1 + h) \subset \partial E(\angle \vec{e}_1 \vec{e}_2 + e, (\uparrow n_{21}) + h) \Rightarrow$$

$$E(e, \angle \vec{h}_1 + h) \subset \partial E(A, B)$$

If

$$\vec{e}_1 \times \vec{h}_1 \uparrow\uparrow (0,0,-1) \Rightarrow$$

$$\vec{h}_1 \cdot \vec{n}_{11} \le 0, \quad \vec{h}_2 \cdot \vec{n}_{11} \le 0 \Rightarrow$$

$$E((\uparrow n_{11}), \angle \vec{h}_1 \vec{h}_2) = \uparrow n_{11} + a_0 \Rightarrow$$

$$E(\angle \vec{e}_1 + e, h) \subset E(A, B) \subset$$

$$E((\uparrow n_{11}) + e, \angle \vec{h}_1 \vec{h}_2 + h),$$

$$E(\angle \vec{e}_1 + e, h) \subset \partial E((\uparrow n_{11}) + e, \angle \vec{h}_1 \vec{h}_2 + h) \Rightarrow$$

$$E(\angle \vec{e}_1 + e, h) \subset \partial E(A, B).$$

Figures 8.50 & 8.51 illustrate two methods for finding the contact vectors, which obey the upper theorem. The previous 2D DDA and NMM adopted the angle method. If

$$\vec{e}_2 \times \vec{h}_2 \uparrow\uparrow (0,0,1) \Rightarrow$$

$$\vec{h}_1 \cdot \vec{n}_{12} \le 0, \quad \vec{h}_2 \cdot \vec{n}_{12} \le 0 \Rightarrow$$

$$E((\uparrow n_{12}), \angle \vec{h}_1 \vec{h}_2) = \uparrow n_{12} + a_0 \Rightarrow$$

$$E(\angle \vec{e}_2 + e, h) \subset E(A, B) \subset$$

$$E((\uparrow n_{12}) + e, \angle \vec{h}_1 \vec{h}_2 + h),$$

$$E(\angle \vec{e}_2 + e, h) \subset \partial E((\uparrow n_{12}) + e, \angle \vec{h}_1 \vec{h}_2 + h) \Rightarrow$$

$$E(\angle \vec{e}_2 + e, h) \subset \partial E(A, B).$$

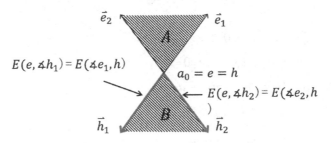

Figure 8.52 Symmetric 2D solid angle contact.

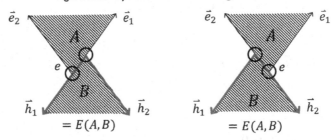

Figure 8.53 Finding contact vectors of symmetric 2D solid angle contact.

If

$$\vec{e}_2 \times \vec{h}_2 \upuparrows (0,0,-1) \Rightarrow$$

$$\vec{e}_1 \cdot \vec{n}_{22} \leq 0, \quad \vec{e}_2 \cdot \vec{n}_{22} \leq 0 \Rightarrow$$

$$E(\angle \vec{e}_1 \vec{e}_2, (\uparrow n_{22})) = \uparrow n_{22} + a_0 \Rightarrow$$

$$E(e, \angle \vec{h}_1 + h) \subset E(A, B) \subset$$

$$E(\angle \vec{e}_1 \vec{e}_2 + e, (\uparrow n_{22}) + h),$$

$$E(e, \angle \vec{h}_1 + h) \subset \partial E(\angle \vec{e}_1 \vec{e}_2 + e, (\uparrow n_{21}) + h) \Rightarrow$$

$$E(e, \angle \vec{h}_1 + h) \subset \partial E(A, B)$$

These two methods can be applied to the case of the contact vectors between two vertically opposite convex angles (Figures 8.52 & 8.53).

### 8.6.3 Entrance solid angle of 2D round corner convex solid angle and concave solid angle

Assuming $A$ is the round corner solid angle of a solid angle $A_0$,

$$A_0 = (\uparrow n_{11} + e)$$

$$\cap (\uparrow n_{12} + e),$$

$$\vec{n}_{11} \cdot \vec{n}_{11} = 1, \quad \vec{n}_{12} \cdot \vec{n}_{12} = 1,$$

$A_1$ is the circle of the round corner of $A$,

$$A_1 = \{|x - a_0| \leq r\},$$

### Theorem of round corner solid angle

Under the previous assumption in this section,

$$A = E(A_1, A_2),$$
$$A_2 = (\uparrow n_{11} + e_0)$$
$$\cap (\uparrow n_{12} + e_0),$$
$$e_0 = e + r(\vec{n}_{11} + \vec{n}_{12})/(1 + \vec{n}_{11} \cdot \vec{n}_{12}).$$

## Proof:

Define an inner angle $A_2$ by:

$$A_2 = \{(x - e - r\vec{n}_{11}) \cdot \vec{n}_{11} \geq 0\}$$
$$\cap \{(x - e - r\vec{n}_{12}) \cdot \vec{n}_{12} \geq 0\}.$$

The angle top $e_0$ of $A_2$ satisfies,

$$(e_0 - e - r\vec{n}_{11}) \cdot \vec{n}_{11} = 0,$$
$$(e_0 - e - r\vec{n}_{12}) \cdot \vec{n}_{12} = 0 \Rightarrow$$
$$(e_0 - e) \cdot \vec{n}_{11} - r = 0,$$
$$(e_0 - e) \cdot \vec{n}_{12} - r = 0 \Rightarrow$$
$$\exists t, \quad e_0 - e = t(\vec{n}_{11} + \vec{n}_{12}),$$
$$t(1 + \vec{n}_{11} \cdot \vec{n}_{12}) - r = 0,$$
$$e_0 = e + r(\vec{n}_{11} + \vec{n}_{12})/(1 + \vec{n}_{11} \cdot \vec{n}_{12}).$$

As illustrated in Figure 8.54, angle $A_2$ can be represented by

$$A_2 = (\uparrow n_{11} + e_0) \cap (\uparrow n_{12} + e_0).$$

The round corner angle $A$ is

$$A = E(A_1, A_2).$$

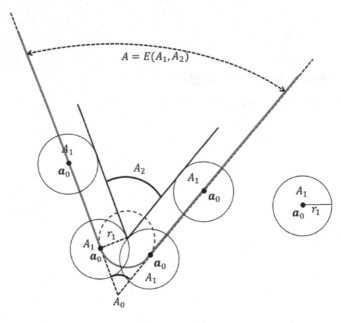

*Figure 8.54* Round corner angle is an entrance angle.

### Theorem of entrance of round corner solid angle and concave solid angle

Round corner solid angle $A$ is the same. Concave solid angle $B$ is the union of two half planes,

$$B = (\uparrow n_{21} + h) \cup (\uparrow n_{22} + h),$$

$$\vec{n}_{21} \cdot \vec{n}_{21} = 1, \quad \vec{n}_{22} \cdot \vec{n}_{22} = 1.$$

If

$$int((\uparrow n_{11} \cap \uparrow n_{12}) \cap (\uparrow n_{21} \cup \uparrow n_{22})) = \emptyset,$$

$E(A, B)$ is the union of the following two half planes,

$$E(A, B) = (\uparrow n_{21} + h_0) \cup (\uparrow n_{22} + h_0),$$

$$h_0 = h - r(\vec{n}_{21} + \vec{n}_{22})/(1 + \vec{n}_{21} \cdot \vec{n}_{22}).$$

### Proof:

$$A = E(A_1, A_2).$$

Let $e_0$ be the reference point of $A$ (Figure 8.55).

$$int((\uparrow n_{11} \cap \uparrow n_{12}) \cap (\uparrow n_{21} \cup \uparrow n_{22})) = \emptyset \Rightarrow$$

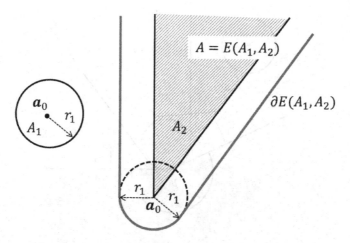

Figure 8.55 Round corner angle is the entrance angle of a 2D solid angle and a disk.

$$int(A_2 \cap (\uparrow n_{21})) = \emptyset, int(A_2 \cap (\uparrow n_{22})) = \emptyset,$$
$$E(A,B) = E(A, (\uparrow n_{21} + h)) \cup E(A, (\uparrow n_{22} + h)).$$
$$E(A, (\uparrow n_{21} + h))$$
$$= (\uparrow n_{21} + h) + e_0 - (A_2 - A_1 + a_0)$$
$$= (\uparrow n_{21} + h) - (A_2 - e_0) - (-A_1 + a_0)$$
$$= (\uparrow n_{21} + h) - (A_1 - a_0)$$
$$= (\uparrow n_{21} + h - r\vec{n}_{21}),$$
$$E(A, (\uparrow n_{22} + h))$$
$$= (\uparrow n_{22} + h) + e_0 - (A_2 - A_1 + a_0)$$
$$= (\uparrow n_{22} + h) - (A_2 - e_0) - (-A_1 + a_0)$$
$$= (\uparrow n_{22} + h) - (A_1 - a_0)$$
$$= (\uparrow n_{22} + h - r\vec{n}_{22}).$$
$$E(A,B) = (\uparrow n_{21} + h - r\vec{n}_{21})$$
$$\cup (\uparrow n_{22} + h - r\vec{n}_{22}).$$
$$E(A,B) = \{(x - h + r\vec{n}_{21}) \cdot \vec{n}_{21} \geq 0\}$$
$$\cup \{(x - h + r\vec{n}_{22}) \cdot \vec{n}_{22} \geq 0\}.$$

The angle top $h_0$ of $E(A,B)$ (Figure 8.56) satisfies the following equations

$$(h_0 - h + r\vec{n}_{21}) \cdot \vec{n}_{21} = 0,$$
$$(h_0 - h + r\vec{n}_{22}) \cdot \vec{n}_{22} = 0 \Rightarrow$$

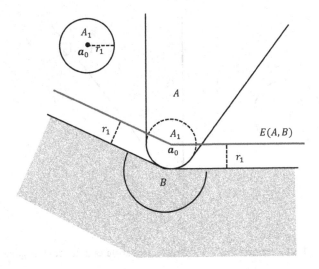

*Figure 8.56* Contact vectors of a round corner convex solid angle and a concave solid angle.

$$(b_0 - b) \cdot \vec{n}_{21} + r = 0,$$

$$(b_0 - b) \cdot \vec{n}_{22} + r = 0$$

$$\Rightarrow \exists t, \quad b_0 - b = t(\vec{n}_{21} + \vec{n}_{22}),$$

$$t(1 + \vec{n}_{21} \cdot \vec{n}_{22}) + r = 0,$$

$$b_0 = b - r(\vec{n}_{21} + \vec{n}_{22})/(1 + \vec{n}_{21} \cdot \vec{n}_{22}).$$

$$E(A, B) = (\uparrow n_{21} + b_0)$$

$$\cup (\uparrow n_{22} + b_0).$$

### 8.6.4 Entrance round corner solid angle of 2D round corner convex solid angles

$A$ is the round corner angle of solid angle $A_0$. $B$ is the round corner angle of solid angle $B_0$. The normal vector $\vec{n}_{ij}$ points the inside of a half plane.

$$A_0 = \{(x - e) \cdot \vec{n}_{11} \geq 0\}$$

$$\cap \{(x - e) \cdot \vec{n}_{12} \geq 0\}$$

$$B_0 = \{(x - b) \cdot \vec{n}_{21} \geq 0\}$$

$$\cap \{(x - b) \cdot \vec{n}_{22} \geq 0\}$$

Assuming the normal vectors are unit vectors:

$$\vec{n}_{11} \cdot \vec{n}_{11} = 1, \quad \vec{n}_{12} \cdot \vec{n}_{12} = 1,$$

$$\vec{n}_{21} \cdot \vec{n}_{21} = 1, \quad \vec{n}_{22} \cdot \vec{n}_{22} = 1.$$

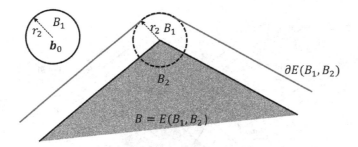

*Figure 8.57* Entrance angle of a 2D solid angle and a disk.

$A_1$ and $B_1$ are discs:

$$A_1 = \{|x - a_0| \leq r_1\},$$
$$B_1 = \{|x - b_0| \leq r_2\}.$$

Let

$$e_0 = e + r_1(\vec{n}_{11} + \vec{\cdot}n_{12})/(1 + \vec{n}_{11} \cdot \vec{n}_{12}),$$
$$b_0 = b + r_2(\vec{n}_{21} + \vec{n}_{22})/(1 + \vec{n}_{21} \cdot \vec{n}_{22}),$$

Define two inner angles $A_2$ and $B_2$ by the following equations: (Figure 8.57)

$$A_2 = \{(x - e_0) \cdot \vec{n}_{11} \geq 0\},$$
$$\cap \{(x - e_0) \cdot \vec{n}_{12} \geq 0\}.$$
$$B_2 = \{(x - b_0) \cdot \vec{n}_{12} \geq 0\},$$
$$\cap \{(x - b_0) \cdot \vec{n}_{22} \geq 0\}.$$

The round corner angles $A$ and $B$ are defined as follows:

$$A = E(A_1, A_2) = A_2 - A_1 + a_0,$$
$$B = E(B_1, B_2) = B_2 - B_1 + b_0,$$

where $b_0$ is the reference point of $B_1$. Based on Section 8.3.6, $E(A_1, B_1)$ is disc

$$E(A_1, B_1) = \{|x - b_0| \leq r_1 + r_2\}.$$

**Theorem of entrance round corner solid angle**

Under the previous assumption of this section,

$$E(A, B) = E(E(A_1, B_1), \quad E(A_2, B_2)).$$

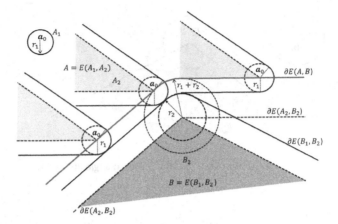

*Figure 8.58* The entrance angle of two 2D round corner convex solid angles is also a round corner convex solid angle.

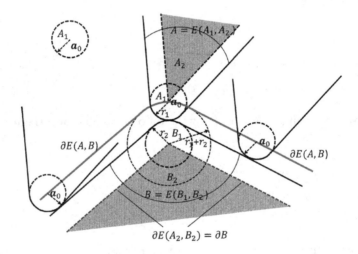

*Figure 8.59* Another example of round corner convex entrance solid angle.

## Proof:

Let $e_0$ be the reference point of $A$.

$$E(A, B) = (B_2 - B_1 + b_0) + e_0 - (A_2 - A_1 + a_0),$$
$$E(A, B) = (B_2 - A_2 + e_0) + (b_0 - B_1) + (A_1 - a_0),$$
$$E(A, B) = E(A_2, B_2) - B_1 + b_0 + A_1 - a_0,$$
$$E(A, B) = E(A_2, B_2) - E(A_1, B_1) + b_0,$$

If let $b_0$ be the reference point of the disk $E(A_1, B_1)$,

$$E(A, B) = E(E(A_1, B_1), \quad E(A_2, B_2)),$$

$E(A, B)$ is the entrance block of the disk $E(A_1, B_1)$ and the solid angle $E(A_2, B_2)$, i.e. also a round corner convex solid angle (Figures 8.58 & 8.59).

## 8.7   BOUNDARIES OF AN ENTRANCE BLOCK OF 2D BLOCKS

An entrance block is determined by its boundary. The following theorems are for understanding and finding the boundaries of 2D entrance blocks. The boundaries of 2D entrance blocks are the subsets of the contact edges.

### 8.7.1   Entrance of 2D block inner points

*Theorem of entrance of 2D block inner points*

Assuming $A$ and $B$ are 2D blocks, then

$$\partial E(A, B) \subset E(\partial A, \partial B).$$

**Proof:**

Assuming

$$a \in A, \quad b \in B.$$

If $a \in int(A)$

$$E(a, b) = b - a + a_0 \in int(b - A + a_0)$$
$$= int(E(A, b)) \subset int(E(A, B)).$$

If $b \in int(B)$ (Figure 8.60)

$$E(a, b) = b - a + a_0 \in int(B - a + a_0)$$
$$= int(E(a, B)) \subset int(E(A, B)).$$
$$E(a, b) = b - a + a_0 \in int(E(A, B)).$$

Therefore

$$\partial E(A, B) \subset E(\partial A, \partial B).$$

### 8.7.2   Entrance of 2D edge to edges

Assuming $A$ and $B$ are 2D blocks,

$$A = a_1 a_2 \cdots a_{p-1} a_p, \quad a_{p+1} = a_1,$$
$$B = b_1 b_2 \cdots b_{q-1} b_q, \quad b_{q+1} = b_1.$$

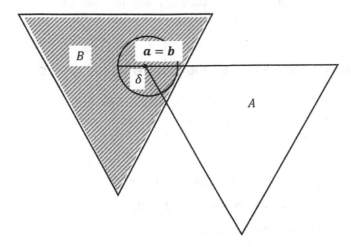

*Figure 8.60* The entrance point of a vertex and an inner point is still an inner point of entrance block.

### Theorem of entrance of 2D edge-edge

$$\partial E(A, B) \subset E(A(0), B(1)) \cup E(A(1), B(0)).$$

### Proof:

From Section 8.6.1,

$$\partial E(A, B) \subset E(\partial A, \partial B) = E(A(1), B(1)).$$

If $a_i a_{i+1}$ and $b_j b_{j+1}$ are not parallel and,

$$a \in int(a_i a_{i+1}), \quad i = 1, \ldots, p,$$
$$b \in int(b_j b_{j+1}), \quad j = 1, \ldots, q,$$

Referring to Figure 8.61,

$$E(a, b) = b - a + a_0$$
$$\in int(b_j b_{j+1} - a_i a_{i+1} + a_0)$$
$$= int(E(a_i a_{i+1}, b_j b_{j+1})) \subset int(E(A, B)).$$
$$E(a_i a_{i+1}, b_j b_{j+1}) \cap \partial E(A, B) \subset$$
$$(E(\partial a_i a_{i+1}, b_j b_{j+1}) \cup E(a_i a_{i+1}, \partial b_j b_{j+1})).$$

If $(a_i a_{i+1}) \parallel (b_j b_{j+1})$, from the theorem of Section 8.4.2

$$E(a_i a_{i+1}, b_j b_{j+1}) =$$
$$E(\partial a_i a_{i+1}, b_j b_{j+1}) \cup E(a_i a_{i+1}, \partial b_j b_{j+1}).$$

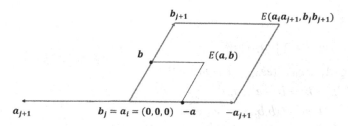

Figure 8.61   The entrance point of inner points of edges is an inner point of the entrance block.

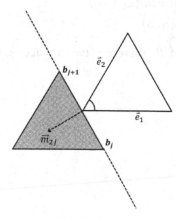

Figure 8.62   Contact edge of a vertex and an edge.

Therefore

$$\partial E(A, B) \subset E(A(0), B(1)) \cup E(A(1), B(0)).$$

### 8.7.3   Contact edges of 2D block to block contact

Assume $A$ and $B$ are the same 2D blocks as in Section 8.7.2. Given a vertex $e = a_i$ of $A$, there are two vertices connecting with $e$ by edges which are

$$e_1 = a_{i+1}, \quad e_2 = a_{i-1}.$$

Denote (Figure 8.62)

$$\vec{e}_1 = (ee_1), \vec{e}_2 = (ee_2),$$

$$\angle a_i = \angle \vec{e}_1 \vec{e}_2.$$

$$\vec{m}_{2j} = (0, 0, 1) \times (b_j b_{j+1}).$$

**Theorem of 2D vertex-edge contact**

$$\exists b \in int(b_j b_{j+1}), E(e, b) \in \partial E(A, B)$$
$$\Rightarrow int\left(\angle a_i \cap \uparrow m_{2j}\right) = \emptyset.$$

*Proof:*

If

$$int(\angle a_i \cap \uparrow m_{2j}) \neq \emptyset.$$

$$\exists e_0, ee_0 \subset A, \quad \vec{e}_0 = (ee_0), \quad \vec{e}_0 \cdot \vec{m}_{2j} > 0,$$
$$E(e,b) = b - e + a_0 \in$$
$$int((B - e + a_0) \cup (b_j b_{j+1} - ee_0 + a_0))$$
$$= int(E(e,B) \cup E(ee_0, b_j b_{j+1}))$$
$$\subset int(E(A,B) \cup E(A,B)) = int(E(A,B))$$
$$\Rightarrow E(e,b) \notin \partial E(A,B).$$

Given a vertex $h = b_j$ of $B$, there are two vertices connecting with $h$ by edges which are

$$h_1 = b_{j+1}, \quad h_2 = b_{j-1}.$$

Denote

$$\vec{h}_1 = (hh_1), \quad \vec{h}_2 = (hh_2),$$

$$\angle b_j = \angle \vec{h}_1 \vec{h}_2.$$

Inner normal of $a_i a_{i+1}$ is

$$\vec{m}_{1i} = (0, 0, 1) \times (a_i a_{i+1}).$$

**Theorem of 2D edge-vertex contact**

If

$$\exists a \in int(a_i a_{i+1}), \quad E(a, h) \in \partial E(A, B)$$
$$\Rightarrow int(b_j \cap \uparrow m_{1i}) = \emptyset.$$

*Proof:*

Referring to Figure 8.63, if

$$int(\angle b_j \cap \uparrow m_{1i}) \neq \emptyset,$$

$$\exists h_0, \quad hh_0 \subset B, \quad \vec{h}_0 = (hh_0), \quad \vec{h}_0 \cdot \vec{m}_{1i} > 0,$$
$$E(a, h) = h - a + a_0 \in$$
$$int((h - A + a_0) \cup (hh_0 - a_i a_{i+1} + a_0))$$
$$= int(E(A, h) \cup E(a_i a_{i+1}, hh_0))$$
$$\subset int(E(A,B) \cup E(A,B)) = int(E(A,B)),$$
$$\Rightarrow E(a, h) \notin \partial E(A, B).$$

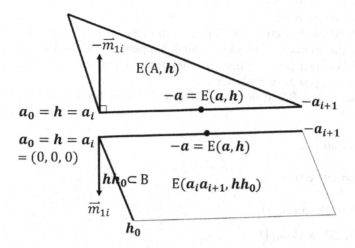

*Figure 8.63* 2D vertex to edge contacts.

## Theorem of entrance of solid angle and half plane

$$int(\angle a_i \cap \uparrow m_{2j}) =$$

$$int(\angle \vec{e}_1 \vec{e}_2 \cap \uparrow m_{2j}) = \emptyset \Leftrightarrow$$

$$E(\angle \vec{e}_1 \vec{e}_2, \uparrow m_{2j}) = \uparrow m_{2j} + a_0 \Leftrightarrow$$

$$\vec{e}_1 \times \vec{e}_2 \Uparrow (0, 0, +1), \vec{e}_1 \cdot \vec{m}_{2j} \leq 0, \quad \vec{e}_2 \cdot \vec{m}_{2j} \leq 0.$$

The equations are defined as contact condition, and $E(a_i, b_j b_{j+1})$ is defined as contact edge.

$$int(\angle b_j \cap \uparrow m_{1i}) =$$

$$int(\angle \vec{h}_1 \vec{h}_2 \cap \uparrow m_{1i}) = \emptyset \Leftrightarrow$$

$$E(\angle \vec{h}_1 \vec{h}_2, \uparrow m_{1i}) = \uparrow m_{1i} + a_0 \Leftrightarrow$$

$$\vec{h}_1 \times \vec{h}_2 \Uparrow (0, 0, +1), \vec{h}_1 \cdot \vec{m}_{1i} \leq 0, \quad \vec{h}_2 \cdot \vec{m}_{1i} \leq 0.$$

The equations are defined as contact condition, and $E(a_i a_{i+1}, b_j)$ is defined as contact edge.

## 8.7.4 Finite covers of 2D parallel edge to edge entrance

When blocks $A$ and $B$ contact along a pair of parallel edges, there are infinite contact point pairs. Among these infinite contact point pairs, only a few point pairs in vertex-to-edge contact will control the movement of blocks, especially the rotation and deformation. Each overlapped contact edge contains two vertex-to-edge contact point pairs. These vertex-to-edge contact point pairs are called "contact points" corresponding to overlapped contact edges. These contact pairs may have two states: open or closed, which will be determined by an open-close iteration. After small rotation or deformation, the contacts of blocks $A$ and $B$ happen only along these contact points.

All closed contact points together define the movements, rotations and deformations of all blocks just as in real cases.

Assuming $A$ and $B$ are the same 2D blocks as in Section 8.7.2, with the only difference that

$$(a_i a_{i+1}) \parallel (b_j b_{j+1}).$$

The inner normal vectors

$$\vec{m}_{1i} = (0, 0, 1) \times (a_i a_{i+1}),$$

$$\vec{m}_{2j} = (0, 0, 1) \times (b_j b_{j+1})$$

point the inside of blocks $A$ and $B$.

### Theorem of finite covers of 1D entrance blocks

Assuming $A$ and $B$ are edges $a_i a_{i+1}$ and $b_j b_{j+1}$,

$$(a_i a_{i+1}) \parallel (b_j b_{j+1})$$
$$E(A, B) = \cup_{m=0}^1 E(A(m), B(1 - m)).$$
$$E(a_i a_{i+1}, b_j b_{j+1}) =$$
$$E(a_i a_{i+1}, b_j) \cup E(a_i a_{i+1}, b_{j+1}) \cup$$
$$E(a_i, b_j b_{j+1}) \cup E(a_{i+1}, b_j b_{j+1}).$$

$E(a_i a_{i+1}, b_j b_{j+1})$ is a line segment. All

$$E(a_i a_{i+1}, b_j),$$
$$E(a_i a_{i+1}, b_{j+1}),$$
$$E(a_i, b_j b_{j+1}),$$
$$E(a_{i+1}, b_j b_{j+1})$$

are overlapped covers of this line segment (Figure 8.64). Each cover refers to a possible vertex-to-edge contact. Therefore the entrance of parallel edges can be transferred to the entrances of a vertex and an edge.

### Theorem of parallel edge contact

$$(a_i a_{i+1}) \parallel (b_j b_{j+1}),$$
$$\exists a \in int(a_i a_{i+1}), \quad \exists b \in int(b_j b_{j+1}),$$
$$E(a, b) \in \partial E(A, B) \Rightarrow \vec{m}_{1i} \upuparrows -\vec{m}_{2j}.$$

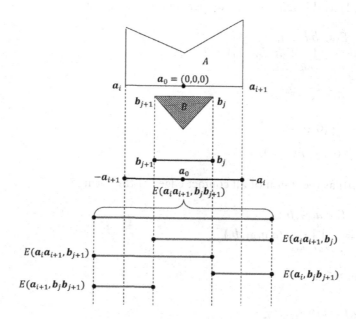

*Figure 8.64* Contact of two parallel edges.

**Proof:**

$$\vec{m}_{1i} \uparrow\uparrow \vec{m}_{2j} \Rightarrow$$
$$\exists \delta,$$
$$C = \{|x| \leq \delta\},$$
$$(C \cap \uparrow m_{1i}) + a \subset A,$$
$$(C \cap \uparrow m_{2j}) + b \subset B \Rightarrow$$
$$E(a, b) \in$$
$$int((C \cap \uparrow m_{2j}) - (C \cap \uparrow m_{1i}) + b - a + a_0) =$$
$$int(E((C \cap \uparrow m_{1i}) + a, (C \cap \uparrow m_{2j}) + b)) \subset$$
$$int(E(A, B)).$$
$$\Rightarrow E(a, b) \in int(E(A, B)), \quad E(a, b) \notin \partial E(A, B).$$

### 8.7.5  Contact edges of 2D entrance block

Assume $A$ and $B$ are the same 2D blocks as in Sections 8.7.2 and 8.7.3. From Section 8.7.2,

$$\partial E(A, B) \subset E(A(0), B(1)) \bigcup E(A(1), B(0)).$$
$$= \left( \bigcup_{i,j} E(a_i, b_j b_{j+1}) \right) \bigcup \left( \bigcup_{i,j} E(a_i a_{i+1}, b_j) \right)$$

Denote $C(0, 1)$ as the union of all contact edges of the form

$$E(a_i, b_j b_{j+1}),$$
$$C(0, 1) = \bigcup_{\text{contact edge}} E(a_i, b_j b_{j+1})$$

From Section 8.7.3,

$$\vec{e}_1 \times \vec{e}_2 \uparrow\uparrow (0, 0, +1),$$
$$\vec{e}_1 \cdot \vec{m}_{2j} \leq 0, \vec{e}_2 \cdot \vec{m}_{2j} \leq 0.$$

Denote $C(1, 0)$ as the union of all contact edges of the form

$$E(a_i a_{i+1}, b_j),$$
$$C(1, 0) = \bigcup_{\text{contact edge}} E(a_i a_{i+1}, b_j)$$

From Section 8.7.3,

$$\vec{h}_1 \times \vec{h}_2 \uparrow\uparrow (0, 0, +1),$$
$$\vec{h}_1 \cdot \vec{m}_{1i} \leq 0, \quad \vec{h}_2 \cdot \vec{m}_{1i} \leq 0.$$

The following theorem is the conclusion of Section 8.7.3.

**Theorem of 2D contact edges**

$$\partial E(A, B) \subset C(0, 1) \cup C(1, 0) \subset E(A, B).$$

## 8.8 CONTACT EDGES OF 2D BLOCKS

The contact surface of 2D blocks is a cover system formed by contact edges. Each contact edge corresponds to a contact point, resulting in a possible vertex-to-edge contact.

In the process of computing, time steps are used. In each time step, the displacement should be smaller than the set maximum displacement $\rho$. At the beginning of each time step, the contact edges are computed. From the relative position of reference point $a_0$ and the contact edges, the closed contact edges are found. The closed contact points control the movements, rotations and deformations throughout this time step.

### 8.8.1 Contact edges of 2D convex blocks

$A$ and $B$ are 2D convex blocks on $x$–$y$ plane with $z = 0$,

$$A = a_1 a_2 \cdots a_{p-1} a_p, \quad a_{p+1} = a_1,$$
$$B = b_1 b_2 \cdots b_{q-1} b_q, \quad b_{q+1} = b_1,$$

which rotates in the right hand rule. As convex, $A$ and $B$ can be represented by simultaneous inequality equations. For block $A$, denote:

$$\vec{e}_1 = (a_i a_{i+1}), \quad \vec{e}_2 = (a_i a_{i-1}),$$
$$\vec{m}_{1i} = (0, 0, 1) \times \vec{e}_1 \Rightarrow$$
$$A = \cap_{i=1,\ldots,p}\{(x - a_i) \cdot \vec{m}_{1i} \geq 0\}.$$

For block $B$, denote:

$$\vec{b}_1 = (b_j b_{j+1}), \quad \vec{b}_2 = (b_j b_{j-1}),$$
$$\vec{m}_{2j} = (0, 0, 1) \times \vec{b}_1 \Rightarrow$$
$$B = \cap_{i=1,\ldots,q}\{(x - b_j) \cdot \vec{m}_{2j} \geq 0\}.$$

Normal vectors $\vec{m}_{1j}$ and $\vec{m}_{2j}$ point the inside of block.

### Theorem of vertex-edge contact of 2D convex blocks

$$int(\angle a_i \cap \uparrow m_{2j}) = \emptyset \Rightarrow E(a_i, b_j b_{j+1}) \subset \partial E(A, B)$$

### Proof:

Based on Section 8.7.3,

$$int(\angle a_i \cap \uparrow m_{2j}) = \emptyset \Leftrightarrow$$
$$E(\angle a_i, \uparrow m_{2j}) = \uparrow m_{2j} + a_0.$$

As $A$ and $B$ are convex blocks,

$$\angle a_i + a_i \supset A, \quad \uparrow m_{2j} + b_j \supset B, \quad \Rightarrow$$
$$E(\angle a_i + a_i, \uparrow m_{2j} + b_j) \supset E(A, B).$$

On the other side,

$$E(a_i, b_j b_{j+1}) = b_j b_{j+1} - a_i + a_0,$$
$$\partial E(\angle a_i + a_i, \uparrow m_{2j} + b_j) = \perp m_{2j} + b_j - a_i + a_0$$
$$\Rightarrow E(a_i, b_j b_{j+1}) \subset \partial E(\angle a_i + a_i, \uparrow m_{2j} + b_j),$$
$$E(a_i, b_j b_{j+1}) \subset E(A, B) \Rightarrow$$
$$E(a_i, b_j b_{j+1}) \subset \partial E(A, B).$$

$E(a_i, b_j b_{j+1})$ is not only a contact but also a boundary edge of $E(A, B)$.

### Theorem of edge-vertex contact of 2D convex blocks

$$int(\angle b_j \cap \uparrow m_{1i}) = \emptyset \Rightarrow E(a_i a_{i+1}, b_j) \subset \partial E(A, B)$$

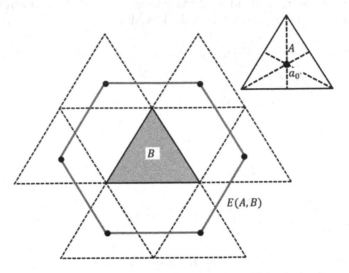

*Figure 8.65* Entrance block of two identical equal lateral triangles.

**Proof:**

Based on Section 8.7.3,

$$int(\angle b_j \cap \uparrow m_{1i}) = \emptyset \Leftrightarrow$$
$$E(\angle b_j, \uparrow m_{1i}) = \uparrow m_{1i} + a_0.$$

As $A$ and $B$ are convex blocks,

$$\uparrow m_{1i} + a_i \supset A, \quad \angle b_j + b_j \supset B \Rightarrow$$
$$E(\uparrow m_{1i} + a_i, \angle b_j + b_j) \supset E(A, B).$$

On the other side

$$E(a_i a_{i+1}, b_j) = b_j - a_i a_{i+1} + a_0.$$
$$\partial E(\uparrow m_{1i} + a_i, \angle b_j + b_j)$$
$$= b_j - \perp \vec{m}_{1i} - a_i + a_0 \Rightarrow$$
$$E(a_i a_{i+1}, b_j) \subset \partial E(\uparrow m_{1i} + a_i, \angle b_j + b_j),$$
$$E(a_i a_{i+1}, b_j) \subset E(A, B) \Rightarrow$$
$$E(a_i a_{i+1}, b_j) \subset \partial E(A, B).$$

$E(a_i a_{i+1}, b_j)$ is both a contact edge and a boundary edge of $E(A, B)$ (Figures 8.65 to 8.68).

**Theorem of contact edges of 2D convex blocks**

$$\partial E(A, B) = C(0, 1) \cup C(1, 0).$$

*Figure 8.66* Entrance block of two equal lateral blocks which are in different directions.

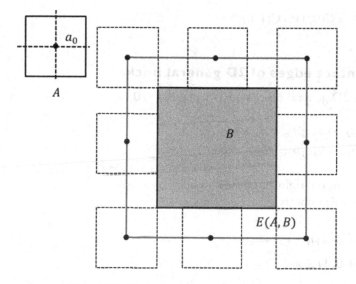

*Figure 8.67* Entrance squares of two squares.

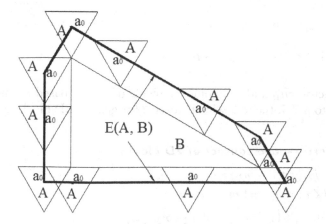

*Figure 8.68* Entrance block of two triangles.

**Proof:**

From Section 8.7.5,

$$\partial E(A, B) \subset C(0, 1) \cup C(1, 0).$$

From previous theorems in this section,

$$\partial E(A, B) \supset C(0, 1) \cup C(1, 0).$$

Therefore,

$$\partial E(A, B) = C(0, 1) \cup C(1, 0).$$

## 8.8.2  Contact edges of 2D general bocks

$A$ and $B$ are 2D general blocks on $x$–$y$ plane, $z = 0$.

$$A = a_1 a_2 \cdots a_{p-1} a_p, \quad a_{p+1} = a_1,$$
$$B = b_1 b_2 \cdots b_{q-1} b_q, \quad b_{q+1} = b_1,$$

which rotates in the right-hand rule.
For block $A$, denote:

$$\vec{e}_1 = (a_i a_{i+1}), \quad \vec{e}_2 = (a_i a_{i-1}),$$
$$\vec{m}_{1i} = (0, 0, 1) \times \vec{e}_1, \quad i = 1, \ldots, p.$$

For block $B$, denote:

$$\vec{h}_1 = (b_j b_{j+1}), \quad \vec{h}_2 = (b_j b_{j-1}),$$
$$\vec{m}_{2j} = (0, 0, 1) \times \vec{h}_1, \quad j = 1, \ldots, q.$$

Normal vectors $\vec{m}_{1i}$ and $\vec{m}_{2j}$ point the inside of block. Based on the theorems of Sections 8.7.3 to 8.7.5, the following theorems for general 2D blocks exist.

**Theorem of vertex-edge contact of 2D blocks**

$$int(\angle a_i \cap \uparrow m_{2j}) = \emptyset \Leftrightarrow$$
$$E(\angle a_i, \uparrow m_{2j}) = \uparrow m_{2j} + a_0 \Leftrightarrow$$
$$\vec{e}_1 \times \vec{e}_2 \uparrow\uparrow (0, 0, +1), \vec{e}_1 \cdot \vec{m}_{2j} \leq 0, \vec{e}_2 \cdot \vec{m}_{2j} \leq 0,$$
$$\Leftrightarrow E(a_i, b_j b_{j+1}) \subset C(0, 1).$$

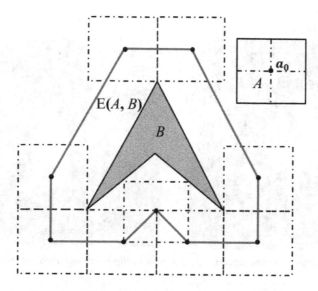

*Figure 8.69* Entrance block of generally shaped blocks.

### Theorem of edge-vertex contact of 2D blocks

$$int(\angle b_j \cap \uparrow m_{1i}) = \emptyset \Leftrightarrow$$
$$E(\angle b_j, \uparrow m_{1j}) = \uparrow m_{1i} + a_0 \Leftrightarrow$$
$$\vec{b}_1 \times \vec{b}_2 \upuparrows (0, 0, +1), \vec{b}_1 \cdot \vec{m}_{1i} \leq 0, \vec{b}_2 \cdot \vec{m}_{1i} \leq 0,$$
$$\Leftrightarrow E(a_i a_{i+1} b_j) \subset C(1, 0).$$

### Theorem of contact edges of 2D blocks

$$\partial E(A, B) \subset C(0, 1) \cup C(1, 0) \subset E(A, B).$$

All contact edges are connected or intersected with each other to form the boundaries of the entrance block $\partial E(A, B)$ (Figure 8.69).

## 8.8.3 Applications of the theory of contact in 2D DDA

The current version of 2D DDA is consistent with the new theory of contact while the following two conditions are fulfilled: the time step is small enough and the stiffness of the contact spring is large enough. Under these two conditions, DDA can simulate any possible complex movements of simply deformable block systems. Below are some examples. Figure 8.70 illustrates stability analysis of underground powerhouses using 2D DDA. Figure 8.71 illustrates 2D DDA computations of parallel tunnels under shock waves. Figures 8.72 and 8.73 illustrate stability results of arch dams and gravity dams using 2D DDA. Figure 8.74 illustrates slope stability results using 2D DDA. Figure 8.75 illustrates 2D DDA computation results of borehole blasting.

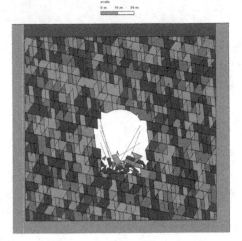
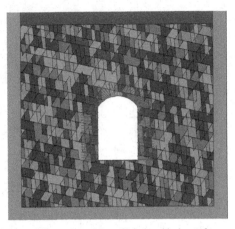

(a) Computed collapse with no support     (b) Final configuration with modeled rockbolt reinforcement

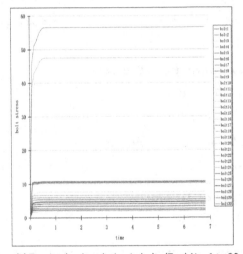

(c) Tension load evolution in bolts (Ton) No. 1 to 30,
numbered from top center down and right to left in each
elevation

*Figure 8.70* Deformation and rockbolt reinforcement modeling with 2D-DDA for a 30 m span underground powerhouse.

DDA works on deformable block systems. Each block has linear displacements, a constant stress and a constant strain. In DDA, multi-time steps are used. Both static and dynamic processes are performed in the dynamic computation. Static computation is the stabilized dynamic computation. Therefore, discontinuous and large displacement computation can be realized for both static and dynamic processes in DDA.

For each time step, there are usually several open-close iterations in DDA. Before going to the next time step, open or close modes are adjusted within iterations until

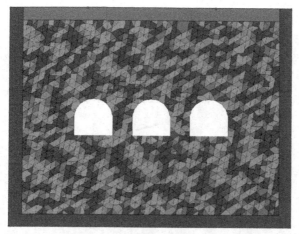

(a) Initial stable state

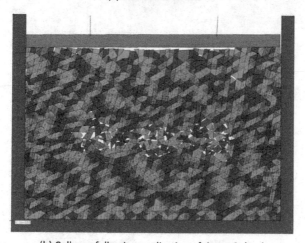

(b) Collapse following application of dynamic load

Figure 8.71   2D DDA block mesh of parallel tunnels and collapse under shock wave.

every contact position has the same contact mode before and after the equations are solved. Here, for each open-close iteration in each time step, DDA solves the global equilibrium equations. As the principal law of stability analysis, a friction law is ensured in DDA computation. The friction law is mathematically translated to inequality equations.

Every single block of 2D DDA can be a generally shaped convex or concave polygon. Based on simplex integration (Shi 1996), the stiffness matrices, the inertia matrices and all other matrices of DDA are analytical solutions.

DDA is a discontinuous version of the FEM and a visible version of the limit equilibrium method. DDA also serves as an implicit version of the DEM method.

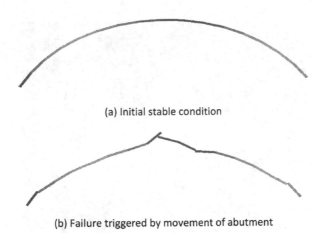

(a) Initial stable condition

(b) Failure triggered by movement of abutment

*Figure 8.72*  2D DDA computed results of a thin arch dam under five times normal water pressure.

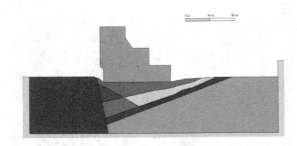

(a) Initial stable condition under normal water pressure

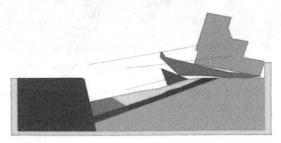

(b) Failure under three times normal water pressure

*Figure 8.73*  Stability analysis of a RCC gravity dam as computed with 2D DDA where the friction angle on the joints is 17°.

DDA has all advantages of dynamic relaxation yet the convergence is strict and the result is accurate.

More importantly, DDA is a very well-tested method by analytical solutions, physical model tests and large engineering projects.

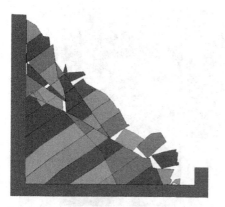 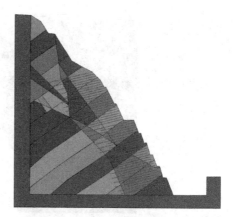

(a) Possible failure process of rock slope without support    (b) Stabilized slope with rockbolt reinforcement

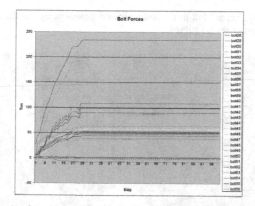

(c) Tension load evolution in rockbolts 28 to 56 counting from left to right

*Figure 8.74* 2D DDA computation of the critical section of a dam abutment slope with the assumed sliding line and using a friction angle of 18° for the joints.

*Figure 8.75* 2D DDA computation of borehole blasting.

## 8.8.4 Applications of the new theory of contact in 2D NMM

Aimed at global analysis, the well-known mathematical manifold is perhaps the most important subject of modern mathematics. Based on the concept of mathematical manifold, the numerical manifold method is a developing numerical method

*Figure 8.76* Rock and shotcrete fall in an underground powerhouse without horizontal initial stress as computed with 2D NMM.

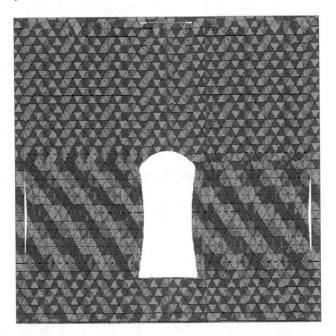

*Figure 8.77* 2D NMM simulation of the deformation of an underground powerhouse where the side-walls are parallel to the bedding planes direction with one tenth of the real elastic modulus used as input.

(Shi, 1991, 1996). In this method, the movements and deformations of continuous and discontinuous structures or materials are computed in a unified form. The meshes of the numerical manifold method consist of finite covers. The finite covers overlap each other and cover the entire material volume. On each cover, the manifold method

*Figure 8.78* Circular sliding as computed with 2D NMM.

*Figure 8.79* Coalmine deformation as computed with 2D NMM.

defines an independent displacement function, called cover function. The global displacement functions are the weighted averages of local independent cover functions on the common part of several covers. For more details, see Shi (1991).

Using the finite cover systems, continuous, jointed or blocky materials can be computed in a mathematically consistent manner. For manifold computation, the mathematical covers and physical mesh are independent. Therefore, the mathematical covers are free to define and are easily changed. The mathematical covers can be easily moved, split, removed, or added. Moving the covers, the large deformations and moving boundaries can be computed. Because the joints can divide a cover into two or more independent covers, which have independent displacement functions, the general discontinuity can be modeled.

Both the finite element method (FEM) for continua and the discontinuous deformation analysis (DDA) for block systems are special cases of NMM.

In the current development stage of NMM, by using the finite cover approach, more flexible deformations and movements of joint and block systems can be computed.

As for the applications of the new theory of contact, several engineering cases of 2D NMM are shown in Figures 8.76 to 8.80. The current version of 2D NMM adopts the same contact algorithm used in 2D DDA, which is consistent with the new theory of contact while the following two conditions are fulfilled: the time step is

(a) Failure with an assumed 30° joint friction angle
without lining

(b) Stabilization with concrete lining assuming only
5° joint friction angle

*Figure 8.80* 2D NMM results of a multi-lane highway tunnel in jointed rocks with and without concrete lining.

small enough and the stiffness of the contact spring is large enough. Figure 8.76 shows NMM computation results of rock and shotcrete fall of an underground powerhouse. Figure 8.77 shows the computed deformation of an underground powerhouse by 2D NMM. Figure 8.78 shows circular sliding computed by 2D NMM. Figure 8.79 shows the deformation of a coalmine computed by 2D NMM. Figure 8.80 is the 2D NMM computation results of a highway tunnel with and without concrete lining.

## 8.9 BOUNDARIES OF ENTRANCE SOLID ANGLE OF 3D SOLID ANGLES

Contacts between two 3D solid angles are much more complex and difficult than these between two 2D solid angles or blocks. However, under the entrance theory, the 3D entrance solid angles can be computed in a similar way as that in the 2D cases. Similarly, a 3D entrance solid angle can be defined and represented by its boundaries. The following theorems are used to find the boundaries of a 3D entrance solid angle.

## 8.9.1 Local entrance angle of 3D angles

Discontinuous computation adopts time steps. If the step displacements are small enough, all contacts will be independent and local. It will be proved that these local contacts are equivalent to 3D solid angle to solid angle contacts, 3D solid angle to vector edge contacts, 3D solid angle to plane contacts and 3D vector edge to vector edge contacts, respectively.

Based on the theorem of distance of Section 8.3.9,

$$|A, B| = \varepsilon \Leftrightarrow |a_0, \quad E(A, B)| = \varepsilon.$$

Assume the maximum step movement is $\rho$. The contact distance is $\varepsilon = 2\rho$. Denote $D$ as a ball

$$D = \{|x - a_0| \leq \varepsilon\}.$$

Therefore, for a given time step, only a small part of entrance block $E(A, B)$

$$E(A, B) \cap D \neq \emptyset$$

is needed to be considered.

## 8.9.2 Existence of entrance solid angle boundary of 3D solid angles

Assuming

$$A = \angle \vec{e}_1 \vec{e}_2 \cdots \vec{e}_{u-1} \vec{e}_u + e,$$
$$B = \angle \vec{h}_1 \vec{h}_2 \cdots \vec{h}_{v-1} \vec{h}_v + h.$$

### Theorem of empty boundary of 3D entrance solid angle

If

$$\vec{v}_0 \in int(\angle \vec{e}_1 \vec{e}_2 \cdots \vec{e}_{u-1} \vec{e}_u \cap \angle \vec{h}_1 \vec{h}_2 \cdots \vec{h}_{v-1} \vec{h}_v),$$

then $E(A, B)$ is the whole space (Figure 8.81)

$$\partial E(A, B) = \emptyset.$$

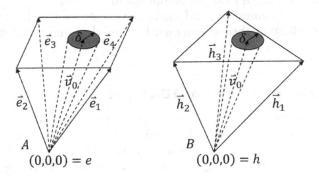

*Figure 8.81* 3D solid angles $A$ and $B$ have common vectors.

**Proof:**

$$\exists\, \varepsilon > 0,$$

$$C = \{\vec{v}_0 \cdot x / |x| \le \varepsilon\},$$

$$C \subset (\angle \vec{e}_1 \vec{e}_2 \cdots \vec{e}_{u-1} \vec{e}_u \cap \angle \vec{h}_1 \vec{h}_2 \cdots \vec{h}_{v-1} \vec{h}_v).$$

$$C - \angle \vec{v}_0$$

is the whole space and

$$\partial(C - \angle \vec{v}_0) = \emptyset$$

$$E(A, B) = B - A + a_0$$

$$= \angle \vec{h}_1 \vec{h}_2 \cdots \vec{h}_{v-1} \vec{h}_v - \angle \vec{e}_1 \vec{e}_2 \cdots \vec{e}_{u-1} \vec{e}_u + h - e + a_0$$

$$\supset C - \angle \vec{v}_0 + h - e + a_0.$$

$$\partial(C - \angle \vec{v}_0 + h - e + a_0) = \emptyset$$

$$\Rightarrow \partial E(A, B) = \emptyset.$$

Therefore, $E(A, B)$ is the whole space.
The necessary condition of

$$\partial E(A, B) \ne \emptyset$$

is the contact condition

$$\angle \vec{e}_1 \vec{e}_2 \cdots \vec{e}_{u-1} \vec{e}_u \cap \angle \vec{h}_1 \vec{h}_2 \cdots \vec{h}_{v-1} \vec{h}_v \subset$$
$$(\partial \angle \vec{e}_1 \vec{e}_2 \cdots \vec{e}_{u-1} \vec{e}_u \cap \partial \angle \vec{h}_1 \vec{h}_2 \cdots \vec{h}_{v-1} \vec{h}_v).$$

### 8.9.3   Entrance of 3D solid angle inner points

*Theorem of entrance of 3D solid angle inner points*

Assuming $A$ and $B$ are 3D solid angles, then

$$\partial E(A, B) \subset E(\partial A, \partial B).$$

**Proof:**

If $a \in int(A), b \in B$,

$$E(a, b) = b - a + a_0 \in int(b - A + a_0)$$
$$= int(E(A, b)) \subset int(E(A, B)).$$

If $a \in A, b \in int(B)$,

$$E(a, b) = b - a + a_0 \in int(B - a + a_0)$$
$$= int(E(a, B)) \subset int(E(A, B)).$$

If $a \in int(A)$ or $b \in int(B)$,

$$E(a, b) = b - a + a_0$$

is an inner point of $E(A, B)$.
   Therefore,

$$\partial E(A, B) \subset E(\partial A, \partial B).$$

### 8.9.4   Entrance of boundary solid angle to angle of 3D solid angles

Assume $A$ and $B$ are 3D angles as defined in Section 8.9.2.

*Theorem of entrance of boundary angle-angle of 3D solid angles*

$$\partial E(A, B) \subset E(A(1), B(2)) \cup E(A(2), B(1)).$$

**Proof:**

From Section 8.9.3,

$$\partial E(A, B) \subset E(\partial A, \partial B) = E(A(2), B(2)).$$
$$\angle \vec{e}_r \vec{e}_{r+1} + e, \quad r = 1, \ldots, u,$$
$$\angle \vec{h}_s \vec{h}_{s+1} + h, \quad s = 1, \ldots, v.$$

are boundary angles of 3D angles $A$ and $B$ respectively.

If solid angle $\angle \vec{e}_r \vec{e}_{r+1}$ and $\angle h_s h_{s+1}$ are not parallel, for

$$a \in int(\angle \vec{e}_r \vec{e}_{r+1} + e), \quad b \in int(\angle \vec{h}_s \vec{h}_{s+1} + h),$$

$$\exists \angle \vec{v}_0 + e, a \in int(\angle \vec{v}_0 + e) \subset int(\angle \vec{e}_r \vec{e}_{r+1} + e),$$

solid angle $\angle \vec{v}_0$ and $\angle h_s h_{s+1}$ are not parallel.

$$E(a, b) = b - a + a_0$$

$$\in int(\angle \vec{h}_s \vec{h}_{s+1} + h - \angle \vec{v}_0 - e + a_0)$$

$$= int(E(\angle \vec{v}_0 + e, \angle \vec{h}_s \vec{h}_{s+1} + h))$$

$$\subset int(E(\angle \vec{e}_r \vec{e}_{r+1} + e, \angle \vec{h}_s \vec{h}_{s+1} + h))$$

$$\subset int(E(A, B)).$$

$$E(a, b) \in int(E(A, B)).$$

$$E(\angle \vec{e}_r \vec{e}_{r+1} + e, \angle \vec{h}_s \vec{h}_{s+1} + h) \cap \partial E(A, B) \subset$$

$$(E(\partial \angle \vec{e}_r \vec{e}_{r+1} + e, \angle \vec{h}_s \vec{h}_{s+1} + h) \cup E(\angle \vec{e}_r \vec{e}_{r+1} + e, \partial \angle \vec{h}_s \vec{h}_{s+1} + h)).$$

If solid angle $\angle \vec{e}_r \vec{e}_{r+1}$ and $\angle h_s h_{s+1}$ are parallel (Figure 8.82), from theorem in Section 8.4.2

$$E(\angle \vec{e}_r \vec{e}_{r+1} + e, \angle \vec{h}_s \vec{h}_{s+1} + h) =$$

$$E(\partial \angle \vec{e}_r \vec{e}_{r+1} + e, \angle \vec{h}_s \vec{h}_{s+1} + h)$$

$$\cup E(\angle \vec{e}_r \vec{e}_{r+1} + e, \partial \angle \vec{h}_s \vec{h}_{s+1} + h).$$

Therefore

$$\partial E(A, B) \subset E(A(1), B(2)) \cup E(A(2), B(1)).$$

### 8.9.5 Entrance of boundary angle to vector of 3D solid angles

Assume $A$ and $B$ are the same angles as in Sections 8.9.2 and 8.9.4.

**_Theorem of entrance of boundary vector-angle of 3D solid angles_**

$$\partial E(A, B) \subset E(A(0), B(2)) \cup E(A(2), B(1)).$$

**_Proof:_**

$$\angle \vec{e}_r + e, \quad r = 1, \ldots, u,$$

$$\angle \vec{h}_s \vec{h}_{s+1} + h, \quad s = 1, \ldots, v,$$

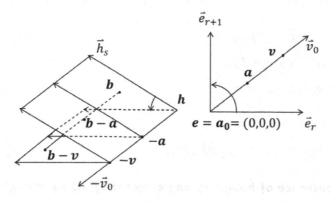

Figure 8.82   Entrance point of face angle inner points.

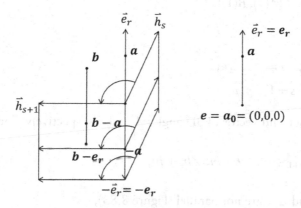

Figure 8.83   Entrance point of a face angle inner point and an edge vector inner point.

are boundary edge vector and angle of 3D angles $A$ and $B$, respectively. For any points

$$a \in int(\angle \vec{e}_r + e), \quad b \in int(\angle \vec{h}_s \vec{h}_{s+1} + h),$$

If $\angle e_r$ and $\angle \vec{h}_s \vec{h}_{s+1}$ are not parallel (Figure 8.83),

$$E(a, b) = b - a + a_0$$

$$\in int(\angle \vec{h}_s \vec{h}_{s+1} + h - \angle \vec{e}_r - e + a_0)$$

$$= int(E(\angle \vec{e}_r + e, \angle \vec{h}_s \vec{h}_{s+1} + h)$$
$$\subset int(E(A, B)).$$
$$E(a, b) \in int(E(A, B)).$$

$$E(\angle \vec{e}_r + e, \angle \vec{h}_s \vec{h}_{s+1} + h) \cap \partial E(A, B) \subset$$

$$(E(\partial \angle \vec{e}_r + e, \angle \vec{h}_s \vec{h}_{s+1} + h) \cup E(\angle \vec{e}_r + e, \partial \angle \vec{h}_s \vec{h}_{s+1} + h)).$$

If solid angle $\angle \vec{e}_r$ and $\angle h_s h_{s+1}$ are parallel, from theorem of Section 8.4.2

$$E(\angle \vec{e}_r + e, \angle \vec{h}_s \vec{h}_{s+1} + h) =$$
$$E(\partial \angle \vec{e}_r + e, \angle \vec{h}_s \vec{h}_{s+1} + h)$$
$$\cup E(\angle \vec{e}_r + e, \partial \angle \vec{h}_s \vec{h}_{s+1} + h).$$

Therefore

$$\partial E(A, B) \subset E(A(0), B(2)) \cup E(A(2), B(1)).$$

### Theorem of entrance of boundary angle-vector of 3D solid angles

$$\partial E(A, B) \subset E(A(0), B(2)) \cup E(A(2), B(0)) \cup$$
$$E(A(1), B(1)).$$

### Proof:

$$\angle \vec{e}_r \vec{e}_{r+1} + e, \quad r = 1, \ldots, u,$$
$$\angle \vec{h}_s + h, \quad s = 1, \ldots, v,$$

are boundary angle and vector of 3D angles $A$ and $B$ respectively. For any points

$$a \in int(\angle \vec{e}_r \vec{e}_{r+1} + e), \quad b \in int(\angle \vec{h}_s + h),$$

If $\angle e_r \vec{e}_{r+1}$ and $\angle \vec{h}_s$ are not parallel (Figure 8.83),

$$E(a, b) = b - a + a_0$$
$$\in int(\angle \vec{h}_s + h - \angle \vec{e}_r \vec{e}_{r+1} - e + a_0)$$
$$= int(E(\angle \vec{e}_r \vec{e}_{r+1} + e, \angle \vec{h}_s + h)$$
$$\subset int(E(A, B)).$$
$$E(a, b) \in int(E(A, B)).$$
$$E(\angle \vec{e}_r \vec{e}_{r+1} + e, \angle \vec{h}_s + h) \cap \partial E(A, B) \subset$$
$$(E(\partial \angle \vec{e}_r \vec{e}_{r+1} + e, \angle \vec{h}_s + h) \cup E(\angle \vec{e}_r \vec{e}_{r+1} + e, \partial \angle \vec{h}_s + h)).$$

If solid angles $\angle \vec{e}_r \vec{e}_{r+1}$ and $\angle h_s$ are parallel, from theorem of section 8.3.2

$$E(\angle \vec{e}_r \vec{e}_{r+1} + e, \angle \vec{h}_s + h) =$$
$$E(\partial \angle \vec{e}_r \vec{e}_{r+1} + e, \angle \vec{h}_s + h)$$
$$\cup E(\angle \vec{e}_r \vec{e}_{r+1} + e, \partial \angle \vec{h}_s + h).$$

Therefore

$$\partial E(A, B) \subset E(A(0), B(2)) \cup E(A(2), B(0)) \cup$$
$$E(A(1), B(1)).$$

## 8.9.6   Contact solid angle of 3D vertex to boundary angle contact

Assume $A$ and $B$ are the same angles as in Sections 8.9.2 and 8.9.4.

$$A = \angle \vec{e}_1 \vec{e}_2 \cdots \vec{e}_{u-1} \vec{e}_u + e,$$
$$B = \angle \vec{h}_1 \vec{h}_2 \cdots \vec{h}_{v-1} \vec{h}_v + h.$$

The inner normal vectors of $\angle \vec{e}_r \vec{e}_{r+1}$ and $\angle \vec{h}_s \vec{h}_{s+1}$ are $\vec{n}_{1r}$ and $\vec{n}_{2s}$ respectively. The following

$$\angle \vec{e}_1 \vec{e}_2 \cdots \vec{e}_{u-1} \vec{e}_u + e,$$
$$\angle \vec{h}_s \vec{h}_{s+1} + h, \quad s = 1, \ldots, v.$$

are 3D angle $A$ and a boundary angle of $B$ respectively.

### Theorem of 3D vertex-angle contact

$$\exists b \in int(\angle \vec{h}_s \vec{h}_{s+1} + h), \quad E(e, b) \in \partial E(A, B)$$
$$\Rightarrow int(\angle \vec{e}_1 \vec{e}_2 \cdots \vec{e}_{u-1} \vec{e}_u \cap \uparrow n_{2s}) = \emptyset.$$

### Proof

If

$$int(\angle \vec{e}_1 \vec{e}_2 \cdots \vec{e}_{u-1} \vec{e}_u \cap \uparrow n_{2s}) \neq \emptyset,$$
$$\exists \vec{v}_0 \in \angle \vec{e}_1 \vec{e}_2 \cdots \vec{e}_{u-1} \vec{e}_u, \quad \vec{v}_0 \cdot \vec{n}_{2s} > 0.$$
$$E(e, b) = b - e + a_0 \in int(E(e, \angle \vec{h}_s \vec{h}_{s+1} + h)) \subset$$
$$int((\angle \vec{h}_s \vec{h}_{s+1} - \angle \vec{v}_0 + h - e + a_0) \cup (B - e + a_0))$$
$$= int(E(\angle \vec{v}_0 + e, \angle \vec{h}_s \vec{h}_{s+1} + h) \cup E(e, B))$$
$$\subset int(E(A, B)).$$
$$\Rightarrow E(e, b) \in int(E(A, B)), \quad E(e, b) \notin \partial E(A, B).$$

Similarly, the following

$$\angle \vec{h}_1 \vec{h}_2 \cdots \vec{h}_{v-1} \vec{h}_v + h,$$
$$\angle \vec{e}_r \vec{e}_{r+1} + e, \quad r = 1, \ldots, u.$$

are 3D angle $B$ and a boundary angle of $A$ respectively. $\vec{n}_{1r}$ is the inner normal vector of $\angle \vec{e}_r \vec{e}_{r+1}$.

### Theorem of 3D angle-vertex contact

$$\exists a \in int(\angle \vec{e}_r \vec{e}_{r+1} + e), \quad E(a,h) \in \partial E(A,B)$$
$$\Rightarrow int(\angle \vec{h}_1 \vec{h}_2 \cdots \vec{h}_{v-1} \vec{h}_v \cap \uparrow n_{1r}) = \emptyset.$$

### Proof:

If

$$int(\angle \vec{h}_1 \vec{h}_2 \cdots \vec{h}_{v-1} \vec{h}_v \cap \uparrow n_{1r}) \neq \emptyset,$$
$$\exists \vec{v}_0 \in \angle \vec{h}_1 \vec{h}_2 \cdots \vec{h}_{v-1} \vec{h}_v, \quad \vec{v}_0 \cdot \vec{n}_{1r} > 0.$$

$$E(a,h) = h - a + a_0 \in int(E(\angle \vec{e}_r \vec{e}_{r+1} + e, h)) \subset$$
$$int((h - A + a_0) \cup (\angle \vec{v}_0 + h - e - \angle \vec{e}_r \vec{e}_{r+1} + a_0))$$
$$= int(E(A,h) \cup E(\angle \vec{e}_r \vec{e}_{r+1} + e, \angle \vec{v}_0 + h))$$
$$\subset int(E(A,B)).$$
$$\Rightarrow E(a,h) \in int(E(A,B)), \quad E(a,h) \notin \partial E(A,B).$$

### Theorem of entrance of solid angle and half space

$$int(\angle \vec{e}_1 \vec{e}_2 \cdots \vec{e}_{u-1} \vec{e}_u \cap \uparrow n_{2s}) = \emptyset \Leftrightarrow$$
$$\angle \vec{e}_1 \vec{e}_2 \cdots \vec{e}_{u-1} \vec{e}_u \cap \uparrow n_{2s}$$
$$\subset (\partial \angle \vec{e}_1 \vec{e}_2 \cdots \vec{e}_{u-1} \vec{e}_u \cap \partial \uparrow n_{2s}) \Leftrightarrow$$
$$E(\angle \vec{e}_1 \vec{e}_2 \cdots \vec{e}_{u-1} \vec{e}_u, \uparrow n_{2s}) = \uparrow n_{2s} + a_0 \Leftrightarrow$$
$$\vec{e}_r \vec{n}_{2s} \leq 0, \quad r = 1, \ldots, u,$$
$$\angle \vec{n}_{2s} \cap \angle \vec{e}_1 \vec{e}_2 \cdots \vec{e}_{u-1} \vec{e}_u = (0,0,0).$$

The equations are defined as contact condition, and $E(e, \angle \vec{h}_j + h)$ is defined as contact vector.

$$int(\angle \vec{h}_1 \vec{h}_2 \cdots \vec{h}_{v-1} \vec{h}_v \cap \uparrow n_{1r}) = \emptyset \Leftrightarrow$$
$$\angle \vec{h}_1 \vec{h}_2 \cdots \vec{h}_{v-1} \vec{h}_v \cap \uparrow n_{1r}$$
$$\subset (\partial \vec{h}_1 \vec{h}_2 \cdots \vec{h}_{v-1} \vec{h}_v \cap \partial \uparrow n_{1r}) \Leftrightarrow$$
$$E(\angle \vec{h}_1 \vec{h}_2 \cdots \vec{h}_{v-1} \vec{h}_v, \uparrow n_{1r}) = \uparrow n_{1r} + a_0 \Leftrightarrow$$
$$\vec{h}_s \cdot \vec{n}_{1r} \leq 0, \quad s = 1, \ldots, v,$$
$$\angle \vec{n}_{1r} \cap \angle \vec{h}_1 \vec{h}_2 \cdots \vec{h}_{v-1} \vec{h}_v = (0,0,0).$$

The equations are defined as contact condition, and $E(\angle \vec{e}_1 \vec{e}_2 + e, h)$ is a contact vector (Figures 8.84 & 8.85).

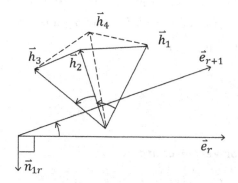

*Figure 8.84* Contact solid angle of a 3D solid angle and a face solid angle.

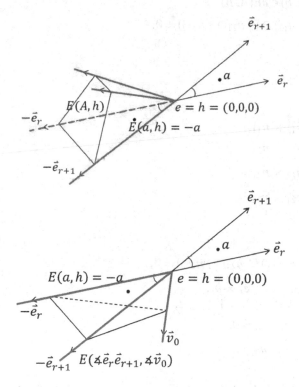

*Figure 8.85* Contact condition of a 3D solid angle and a face solid angle.

### 8.9.7   Contact solid angle of 3D boundary vector to vector contact

Assume $A$ and $B$ are the same angles as in Sections 8.9.2 and 8.9.6.

$$A = \angle \vec{e}_1 \vec{e}_2 \cdots \vec{e}_{u-1} \vec{e}_u + e,$$
$$B = \angle \vec{h}_1 \vec{h}_2 \cdots \vec{h}_{v-1} \vec{h}_v + h.$$

The inner normal vectors of $\angle \vec{e}_{r-1}\vec{e}_r$ and $\angle \vec{e}_r\vec{e}_{r+1}$ are $\vec{n}_{11}$ and $\vec{n}_{12}$ respectively. The inner normal vectors of $\angle \vec{h}_{s-1}\vec{h}_s$ and $\angle \vec{h}_s\vec{h}_{s+1}$ are $\vec{n}_{21}$ and $\vec{n}_{22}$ respectively.

Assuming edges of $\vec{e}_r$ and $\vec{h}_s$ are convex,

$$\vec{e}_r \upuparrows \vec{n}_{12} \times \vec{n}_{11},$$
$$\vec{h}_s \upuparrows \vec{n}_{22} \times \vec{n}_{21}.$$

## Theorem of 3D vector-vector contact

$$\exists a \in int(\angle \vec{e}_r + e), b \in int(\angle \vec{h}_s + h),$$
$$E(a,b) \in \partial E(A,B) \Rightarrow$$
$$int((\uparrow n_{11} \cap \uparrow n_{12}) \cap (\uparrow n_{21} \cap \uparrow n_{22})) = \emptyset.$$

## Proof:

Denote

$$\vec{e}_{r1} = \vec{n}_{11} \times \vec{n}_{12} \times \vec{n}_{11},$$
$$\vec{e}_{r2} = \vec{n}_{12} \times \vec{n}_{11} \times \vec{n}_{12},$$
$$\vec{h}_{s1} = \vec{n}_{21} \times \vec{n}_{22} \times \vec{n}_{21},$$
$$\vec{h}_{s2} = \vec{n}_{22} \times \vec{n}_{21} \times \vec{n}_{22}.$$

If

$$int((\uparrow n_{11} \cap \uparrow n_{12}) \cap (\uparrow n_{21} \cap \uparrow n_{22})) \neq \emptyset,$$
$$\exists \vec{n}_0, \quad \vec{n}_0 \parallel \vec{e}_r \times \vec{h}_s,$$
$$\exists i, \quad j, \quad 1 \leq i \leq 2, \quad 1 \leq j \leq 2,$$
$$\vec{e}_{ri} \cdot \vec{n}_0 > 0,$$
$$\vec{h}_{sj} \cdot \vec{n}_0 > 0.$$

If $i = 1, P_1 = \angle \vec{e}_{r-1}\vec{e}_r$, if $i = 2, P_1 = \angle \vec{e}_r\vec{e}_{r+1}$.
If $j = 1, P_2 = \angle \vec{h}_{s-1}\vec{h}_s$, if $j = 2, P_2 = \angle \vec{h}_s\vec{h}_{s+1}$.
Assume $\vec{e}_r$ and $\vec{h}_s$ are not parallel,

$$E(\angle \vec{e}_r, \angle \vec{h}_s) \subset E(P_1, \angle \vec{h}_s),$$
$$E(\angle \vec{e}_r, \angle \vec{h}_s) \subset E(\angle \vec{e}_r, P_2).$$
$$E(a,b) \in int(E(\angle \vec{e}_r, \angle \vec{h}_s))$$

$$\subset int(E(P_1, \angle \vec{h}_s) \cup E(\angle \vec{e}_r, P_2))$$
$$\subset int(E(P_1, P_2) \cup E(P_1, P_2))$$
$$= int(E(P_1, P_2))$$
$$\subset int(E(A, B)) \Rightarrow$$
$$E(\boldsymbol{a}, \boldsymbol{b}) \notin \partial E(A, B).$$

Denote

$$\vec{e}_{r1} = \vec{n}_{11} \times \vec{n}_{12} \times \vec{n}_{11},$$
$$\vec{e}_{r2} = \vec{n}_{12} \times \vec{n}_{11} \times \vec{n}_{12},$$
$$\vec{h}_{s1} = \vec{n}_{21} \times \vec{n}_{22} \times \vec{n}_{21},$$
$$\vec{h}_{s2} = \vec{n}_{22} \times \vec{n}_{21} \times \vec{n}_{22},$$

then

$$\uparrow n_{11} \cap \uparrow n_{12} = \angle \vec{e}_{r1} \vec{e}_{r2} + \vec{e}_r,$$
$$\uparrow n_{21} \cap \uparrow n_{22} = \angle \vec{h}_{s1} \vec{h}_{s2} + \vec{h}_s,$$
$$\partial(\uparrow n_{11} \cap \uparrow n_{12}) = (\angle \vec{e}_{r1} + \vec{e}_r) \cup (\angle \vec{e}_{r2} + \vec{e}_r),$$
$$\partial(\uparrow n_{21} \cap \uparrow n_{22}) = (\angle \vec{h}_{s1} + \vec{h}_s) \cup (\angle \vec{h}_{s2} + \vec{h}_s).$$

**Theorem of entrance of solid convex edge to solid convex edge**

$$int((\angle \vec{e}_{r1} \vec{e}_{r2} + \vec{e}_r) \cap (\angle \vec{h}_{s1} \vec{h}_{s2} + \vec{h}_s)) = \emptyset \Leftrightarrow$$
$$(\angle \vec{e}_{r1} \vec{e}_{r2} + \vec{e}_r) \cap (\angle \vec{h}_{s1} \vec{h}_{s2} + \vec{h}_s) \subset$$
$$\partial(\angle \vec{e}_{r1} \vec{e}_{r2} + \vec{e}_r) \cap \partial(\angle \vec{h}_{s1} \vec{h}_{s2} + \vec{h}_s) \Leftrightarrow$$
$$\exists \vec{n}_0, \quad \vec{n}_0 \parallel \vec{e}_r \times \vec{h}_s,$$
$$E(\angle \vec{e}_{r1} \vec{e}_{r2} + \vec{e}_r, \angle \vec{h}_{s1} \vec{h}_{s2} + \vec{h}_s) = \uparrow n_0 + a_0 \Leftrightarrow$$
$$\exists \vec{n}_0, \quad \vec{n}_0 \parallel \vec{e}_r \times \vec{h}_s,$$
$$\vec{e}_{r1} \cdot \vec{n}_0 \leq 0, \quad \vec{e}_{r2} \cdot \vec{n}_0 \leq 0,$$
$$\vec{h}_{s1} \cdot \vec{n}_0 \geq 0, \quad \vec{h}_{s2} \cdot \vec{n}_0 \geq 0.$$

The equations are defined as contact condition, and $E(\angle \vec{e}_r + e, \angle \vec{h}_s + h)$ is defined as the contact solid angle. (Figures 8.86 & 8.87)

## 8.9.8 Finite covers of parallel boundary angle to vector entrance

Assume $A$ and $B$ are the same angles as in Section 8.9.7.

$$\angle \vec{e}_r + e, \quad r = 1, \ldots, u,$$
$$\angle \vec{h}_s \vec{h}_{s+1} + h, \quad s = 1, \ldots, v.$$

are parallel edge vector and boundary angle of 3D angles $A$ and $B$ respectively.

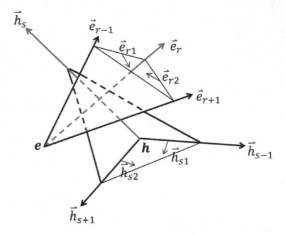

Figure 8.86   Contact of two 3D vector edges.

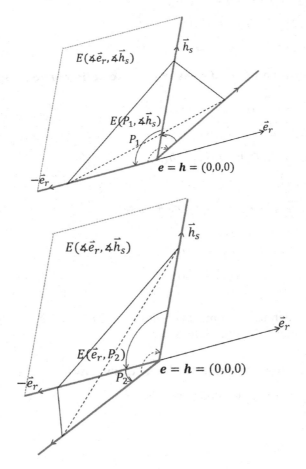

Figure 8.87   Contact condition of two 3D vector edges.

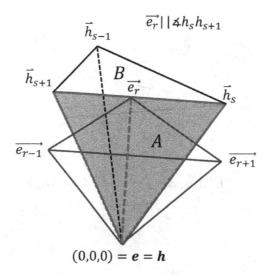

$$\vec{e_r} \,||\, \measuredangle h_s h_{s+1}$$

$$(0,0,0) = e = h$$

*Figure 8.88* Contact of an edge vector with a parallel face solid angle $int(\measuredangle \vec{e_r} \cap \measuredangle h_s \vec{h}_{s+1}) \neq \emptyset$.

### *Theorem of finite covers of entrance blocks of a vector and a 2D solid angle*

Assuming

$$A_0 = \measuredangle \vec{e_r} + e, \quad B_0 = \measuredangle \vec{h}_s \vec{h}_{s+1} + h,$$

$$\measuredangle \vec{e_r} \,||\, \measuredangle \vec{h}_s \vec{h}_{s+1}.$$

Similar to Section 8.4.2,

$$E(A_0, B_0) = \cup_{m=0}^{1} E(A_0(m), B_0(2 - m)).$$

$$E(\measuredangle \vec{e_r} + e, \measuredangle \vec{h}_s \vec{h}_{s+1} + h) =$$

$$E(e, \measuredangle \vec{h}_s \vec{h}_{s+1} + h) \cup$$

$$E(\measuredangle \vec{e_r} + e, \measuredangle \vec{h}_s + h) \cup$$

$$E(\measuredangle \vec{e_r} + e, \measuredangle \vec{h}_{s+1} + h).$$

Figures 8.88–8.91 illustrate the contacts of several cases where an edge vector is parallel to the face of a solid angle.

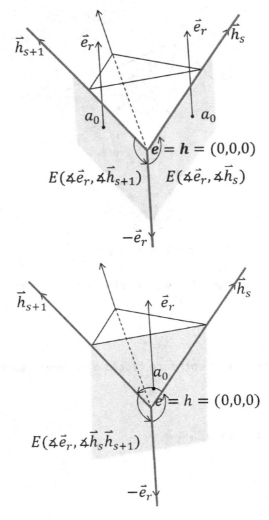

Figure 8.89 Contact solid angles of an edge vector with a parallel face solid angle $int(\angle \vec{e}_r \cap \angle \vec{h}_s \vec{h}_{s+1}) \neq \emptyset$.

Assuming

$$A = \angle \vec{e}_1 \vec{e}_2 \cdots \vec{e}_{u-1} \vec{e}_u + e,$$

$$B = \angle \vec{h}_1 \vec{h}_2 \cdots \vec{h}_{v-1} \vec{h}_v + h,$$

$$\angle \vec{e}_r \parallel \angle \vec{h}_s \vec{h}_{s+1},$$

The inner normal vectors of $\angle \vec{e}_{r-1} \vec{e}_r$ and $\angle \vec{e}_r \vec{e}_{r+1}$ are $\vec{n}_{11}$ and $\vec{n}_{12}$, respectively (Figure 8.92).

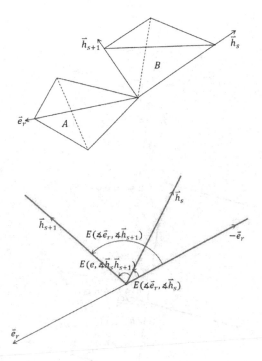

Figure 8.90  Contact of an edge vector with a parallel face solid angle $int(\angle \vec{e}_r \cap \angle \vec{h}_s \vec{h}_{s+1}) = \emptyset$.

## Theorem of parallel vector and 2D solid angle contact

$\exists a \in int(\angle \vec{e}_r + e), b \in int(\angle \vec{h}_s \vec{h}_{s+1} + h),$
$\quad E(a, b) \in \partial E(A, B) \Rightarrow$
$\quad int((\uparrow n_{11} \cap \uparrow n_{12}) \cap \uparrow n_{21}) = \emptyset.$

## Proof:

If

$$int((\uparrow n_{11} \cap \uparrow n_{12}) \cap \uparrow n_{21}) \neq \emptyset.$$
$$\exists \vec{v}_0, \angle \vec{v}_0 + \angle \vec{e}_r \subset \angle \vec{e}_1 \vec{e}_2 \cdots \vec{e}_{u-1} \vec{e}_u, \vec{v}_0 \cdot \vec{n}_{21} > 0.$$
$$E(a, b) \in int(E(\angle \vec{e}_r + e, \angle \vec{h}_s \vec{h}_{s+1} + h)) \subset$$
$$int(E(\angle \vec{e}_r + \angle \vec{v}_0 + e, \angle \vec{h}_s \vec{h}_{s+1} + h) \cup E(\angle \vec{e}_r + e, B)) \subset$$
$$\subset int(E(A, B) \cup E(A, B)) = int(E(A, B)).$$
$$\Rightarrow E(a, b) \in int(E(A, B)), \quad E(a, b) \notin \partial E(A, B).$$

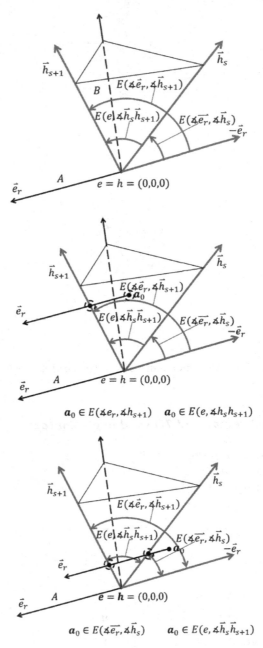

Figure 8.91 Contact solid angles of an edge vector with a parallel face solid angle $int(\measuredangle \vec{e}_r \cap \measuredangle \vec{h}_s \vec{h}_{s+1}) = \emptyset$.

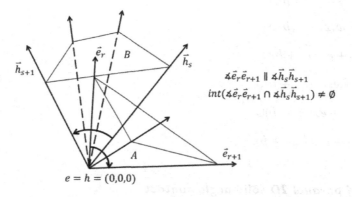

$$\sphericalangle\vec{e}_r\vec{e}_{r+1} \parallel \sphericalangle\vec{h}_s\vec{h}_{s+1}$$
$$int(\sphericalangle\vec{e}_r\vec{e}_{r+1} \cap \sphericalangle\vec{h}_s\vec{h}_{s+1}) \neq \emptyset$$

$e = h = (0,0,0)$

*Figure 8.92*  Contact of parallel face solid angles.

## 8.9.9  Finite covers of parallel boundary angle entrance

Assume $A$ and $B$ are the same 3D solid angles as in Sections 8.9.2 and 8.9.6.

$$\angle\vec{e}_r\vec{e}_{r+1} + e, \quad r=1,\ldots,u,$$
$$\angle\vec{h}_s\vec{h}_{s+1} + h, \quad s=1,\ldots,v.$$

are boundary angles.

The inner normal vectors of angles

$$\angle\vec{e}_r\vec{e}_{r+1}, \quad \angle\vec{h}_s\vec{h}_{s+1}$$

are

$$\vec{n}_{11}, \quad \vec{n}_{21}.$$

Assuming

$$\vec{n}_{11} \parallel \vec{n}_{21}.$$

### Theorem of finite covers of entrance blocks of 2D solid angles

Assuming

$$A_0 = \angle\vec{e}_r\vec{e}_{r+1} + e, \quad B_0 = \angle\vec{h}_s\vec{h}_{s+1} + h,$$
$$\angle\vec{e}_r\vec{e}_{r+1} \parallel \angle\vec{h}_s\vec{h}_{s+1}.$$

From Section 8.4.2,

$$E(A_0, B_0) = \cup_{m=0}^{2} E(A_0(m), B_0(2-m)).$$
$$E(\angle\vec{e}_r\vec{e}_{r+1} + e, \angle\vec{h}_s\vec{h}_{s+1} + h) =$$

$$E(e, \angle \vec{h_s}\vec{h}_{s+1} + h) \cup$$

$$E(\angle \vec{e_r}\vec{e}_{r+1} + e, h) \cup$$

$$E(\angle \vec{e_r} + e, \angle \vec{h_s} + h) \cup$$

$$E(\angle \vec{e_r} + e, \angle \vec{h}_{s+1} + h) \cup$$

$$E(\angle \vec{e}_{r+1} + e, \angle \vec{h_s} + h) \cup$$

$$E(\angle \vec{e}_{r+1} + e, \angle \vec{h}_{s+1} + h).$$

### Theorem of parallel 2D solid angle contact

$$\exists a \in int(\angle \vec{e_r}\vec{e}_{r+1} + e), b \in int(\angle \vec{h_s}\vec{h}_{s+1} + h),$$
$$E(a, b) \in \partial E(A, B) \Rightarrow \vec{n}_{11} \upuparrows -\vec{n}_{21}.$$

### Proof:

If

$$\vec{n}_{11} \upuparrows \vec{n}_{21} \Rightarrow$$

$$\exists \vec{v}_0, \angle \vec{e_r}\vec{e}_{r+1} + \angle \vec{v}_0 \subset \angle \vec{e_1}\vec{e_2} \cdots \vec{e}_{u-1}\vec{e_u},$$

$$\vec{v}_0 \cdot \vec{n}_{2s} > 0.$$

$$E(a, b) \in int(E(\angle \vec{e_r}\vec{e}_{r+1} + e, \angle \vec{h_s}\vec{h}_{s+1} + h)) \subset$$

$$int(E(\angle \vec{e_r}\vec{e}_{r+1} + \angle \vec{v}_0 + e, \angle \vec{h_s}\vec{h}_{s+1} + h) \cup E(\angle \vec{e_r}\vec{e}_{r+1} + e, B)) \subset$$

$$\subset int(E(A, B) \cup E(A, B)) = int(E(A, B))$$

$$\Rightarrow E(a, b) \in int(E(A, B)), \quad E(a, b) \notin \partial E(A, B) \Rightarrow$$

$$\Rightarrow \vec{n}_{11} \upuparrows -\vec{n}_{21}.$$

Figures 9.93 to 9.96 illustrate the contacts of several cases where the faces of two solid angles are parallel.

### 8.9.10 Contact angles of 3D entrance solid angle

Assume $A$ and $B$ are the same angles as in Sections 8.9.2 and 8.9.6.

$$A = \angle \vec{e_1}\vec{e_2} \cdots \vec{e}_{u-1}\vec{e_u} + e,$$
$$B = \angle \vec{h_1}\vec{h_2} \cdots \vec{h}_{v-1}\vec{h_v} + h.$$

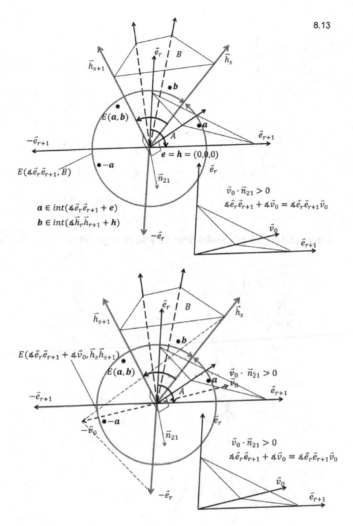

Figure 8.93 Contact of parallel boundary solid angles $int(\angle\vec{e}_r\vec{e}_{r+1} \cap \angle\vec{h}_s\vec{h}_{s+1}) \neq \emptyset$.

From Section 8.9.5,

$$\partial E(A, B) \subset E(A(0), B(2)) \cup E(A(2), B(0)) \cup$$
$$E(A(1), B(1)).$$

$$= (\cup_{s=1,\dots,v} E(e, \angle\vec{h}_s\vec{h}_{s+1} + h))$$

$$\cup(\cup_{r=1,\dots,u} E(\angle\vec{e}_r\vec{e}_{r+1} + e, \quad h))$$

$$\cup(\cup_{r=1,\dots,u,s=1,\dots,v} E(\angle\vec{e}_r + e, \angle\vec{h}_s + h))$$

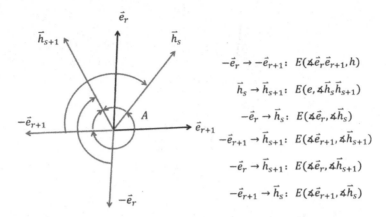

$$-\vec{e}_r \rightarrow -\vec{e}_{r+1}: \quad E(\angle\vec{e}_r\vec{e}_{r+1}, h)$$
$$\vec{h}_s \rightarrow \vec{h}_{s+1}: \quad E(e, \angle\vec{h}_s\vec{h}_{s+1})$$
$$-\vec{e}_r \rightarrow \vec{h}_s: \quad E(\angle\vec{e}_r, \angle\vec{h}_s)$$
$$-\vec{e}_{r+1} \rightarrow \vec{h}_{s+1}: \quad E(\angle\vec{e}_{r+1}, \angle\vec{h}_{s+1})$$
$$-\vec{e}_r \rightarrow \vec{h}_{s+1}: \quad E(\angle\vec{e}_r, \angle\vec{h}_{s+1})$$
$$-\vec{e}_{r+1} \rightarrow \vec{h}_s: \quad E(\angle\vec{e}_{r+1}, \angle\vec{h}_s)$$

*Figure 8.94* Contact solid angles of parallel boundary solid angles $int(\angle\vec{e}_r\vec{e}_{r+1} \cap \angle\vec{h}_s\vec{h}_{s+1}) \neq \emptyset$.

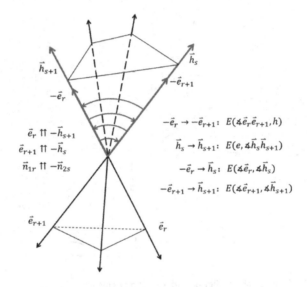

$$-\vec{e}_r \rightarrow -\vec{e}_{r+1}: \quad E(\angle\vec{e}_r\vec{e}_{r+1}, h)$$
$$\vec{h}_s \rightarrow \vec{h}_{s+1}: \quad E(e, \angle\vec{h}_s\vec{h}_{s+1})$$
$$-\vec{e}_r \rightarrow \vec{h}_s: \quad E(\angle\vec{e}_r, \angle\vec{h}_s)$$
$$-\vec{e}_{r+1} \rightarrow \vec{h}_{s+1}: \quad E(\angle\vec{e}_{r+1}, \angle\vec{h}_{s+1})$$

*Figure 8.95* Contact solid angles of parallel boundary solid angles $int(\angle\vec{e}_r\vec{e}_{r+1} \cap \angle\vec{h}_s\vec{h}_{s+1}) = \emptyset$.

Denote $C(0, 2)$ as the union of all contact vectors of the form

$$E(e, \angle\vec{h}_s\vec{h}_{s+1} + h),$$
$$C(0, 2) = \cup_{contact\ angle} E(e, \angle\vec{h}_s\vec{h}_{s+1} + h)$$

From Section 8.9.6, the contact condition is

$$int(\angle\vec{e}_1\vec{e}_2 \cdots \vec{e}_{u-1}\vec{e}_u \cap \uparrow n_{2s}) = \emptyset.$$

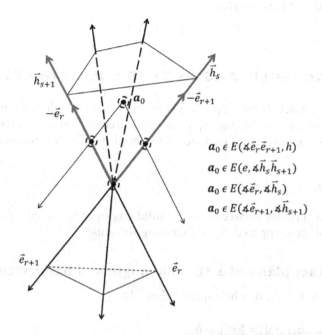

$a_0 \in E(\sphericalangle\vec{e}_r\vec{e}_{r+1}, h)$

$a_0 \in E(e, \sphericalangle\vec{h}_s\vec{h}_{s+1})$

$a_0 \in E(\sphericalangle\vec{e}_r, \sphericalangle\vec{h}_s)$

$a_0 \in E(\sphericalangle\vec{e}_{r+1}, \sphericalangle\vec{h}_{s+1})$

*Figure 8.96* Contact condition of parallel boundary solid angles.

Denote $C(2, 0)$ as the union of all contact vectors of the form

$$E(\sphericalangle\vec{e}_r\vec{e}_{r+1} + e, \quad h),$$
$$C(2, 0) = \cup_{contact\ angle} E(\sphericalangle\vec{e}_r\vec{e}_{r+1} + e, \quad h)$$

From Section 8.9.6, the contact condition is

$$int(\sphericalangle\vec{h}_1\vec{h}_2 \cdots \vec{h}_{v-1}\vec{h}_v \cap \uparrow n_{1r}) = \emptyset$$

Denote $C(1, 1)$ as the union of all contact vectors of the form

$$E(\sphericalangle\vec{e}_r + e, \sphericalangle\vec{h}_s + h),$$
$$C(1, 1) = \cup_{contact\ angle} E(\sphericalangle\vec{e}_r + e, \sphericalangle\vec{h}_s + h)$$

From Section 8.9.7, the contact condition is

$$int((\uparrow n_{11} \cap \uparrow n_{12}) \cap (\uparrow n_{21} \cap \uparrow n_{22})) = \emptyset.$$

The following theorem is the conclusion of Sections 8.9.6 and 8.9.7.

### Theorem of 3D contact angles

$$\partial E(A, B) \subset C(0, 2) \cup C(2, 0) \cup C(1, 1) \subset E(A, B).$$

## 8.10  CONTACT SOLID ANGLES OF 3D SOLID ANGLES

Based on Section 8.9.1, if the step movement $\rho$ is small enough, the entrances will be simplified as a series of solid angles. There are different entrance cases between two 3D solid angles. Only the first entrance is essential for discontinuous computations. This case is

$$a_0 \in \partial E(A, B).$$

The contact surfaces between two 3D solid angles are covered by contacts of 2D solid angles and are computed in the following sections.

### 8.10.1  Contact plane of a 3D solid angle and half space

Let $s = 1$ (Section 8.9.7), $B$ is half space defined by,

$$B = \{(x - h) \cdot \vec{n}_{21} \geq 0\} = \uparrow n_{21} + h.$$

$A$ is a 3D angle defined by

$$A = \angle \vec{e}_1 \vec{e}_2 \cdots \vec{e}_{u-1} \vec{e}_u + e.$$

The inner normal vectors of $\angle \vec{e}_r \vec{e}_{r+1}$ is $\vec{n}_{1r}$. From the theorem of entrance of solid angle and half space of Section 8.8.6, the following theorem takes place.

### Theorem of contact plane of a 3D solid angle and half space

$$int(\angle \vec{e}_1 \vec{e}_2 \cdots \vec{e}_{u-1} \vec{e}_u \cap \uparrow n_{21}) = \emptyset \Leftrightarrow$$
$$\angle \vec{e}_1 \vec{e}_2 \cdots \vec{e}_{u-1} \vec{e}_u \cap \uparrow n_{21}$$
$$\subset (\partial \angle \vec{e}_1 \vec{e}_2 \cdots \vec{e}_{u-1} \vec{e}_u \cap \partial \uparrow n_{21}) \Leftrightarrow$$
$$E(\angle \vec{e}_1 \vec{e}_2 \cdots \vec{e}_{u-1} \vec{e}_u, \uparrow n_{21}) = \uparrow n_{21} + a_0 \Leftrightarrow$$
$$\vec{e}_r \cdot \vec{n}_{21} \leq 0, \quad r = 1, \dots, u,$$
$$\uparrow \vec{n}_{21} \cap \angle \vec{e}_1 \vec{e}_2 \cdots \vec{e}_{u-1} \vec{e}_u = (0, 0, 0).$$

### 8.10.2  Contact plane of two 3D convex solid edges

The equations of 3D convex edges $A$ and $B$ are
   A:
$$(x - e) \cdot \vec{n}_{11} \geq 0,$$
$$(x - e) \cdot \vec{n}_{12} \geq 0.$$

B:

$$(x - h) \cdot \vec{n}_{21} \geq 0,$$

$$(x - h) \cdot \vec{n}_{22} \geq 0.$$

$$A = (\uparrow n_{11} \cap \uparrow n_{12}) + e,$$

$$B = (\uparrow n_{21} \cap \uparrow n_{22}) + h.$$

$\vec{e}_1$ and $\vec{h}_1$ are edge vectors

$$\vec{e}_1 = \vec{n}_{12} \times \vec{n}_{11},$$

$$\vec{h}_1 = \vec{n}_{22} \times \vec{n}_{21}.$$

$$E(A, B) = E(\uparrow n_{11} \cap \uparrow n_{12}, \uparrow n_{21} \cap \uparrow n_{22}) + h - e,$$

Let $r = 1, s = 1$, the theorem of intersection of half space from Section 8.9.7 can be simplified as

$$\vec{e}_{11} = \vec{n}_{11} \times \vec{n}_{12} \times \vec{n}_{11},$$

$$\vec{e}_{12} = \vec{n}_{12} \times \vec{n}_{11} \times \vec{n}_{12},$$

$$\vec{h}_{11} = \vec{n}_{21} \times \vec{n}_{22} \times \vec{n}_{21},$$

$$\vec{h}_{12} = \vec{n}_{22} \times \vec{n}_{21} \times \vec{n}_{22},$$

then

$$\uparrow n_{11} \cap \uparrow n_{12} = \angle \vec{e}_{11} \vec{e}_{12} + \vec{e}_1,$$

$$\uparrow n_{21} \cap \uparrow n_{22} = \angle \vec{h}_{11} \vec{h}_{12} + \vec{h}_1,$$

$$\partial(\uparrow n_{11} \cap \uparrow n_{12}) = (\angle \vec{e}_{11} + \vec{e}_1) \cup (\angle \vec{e}_{12} + \vec{e}_1),$$

$$\partial(\uparrow n_{21} \cap \uparrow n_{22}) = (\angle \vec{h}_{11} + \vec{h}_1) \cup (\angle \vec{h}_{12} + \vec{h}_1).$$

Let $r = 1, s = 1$, the theorem of entrance of solid edge to solid edge from Section 8.9.7 can be simplified as follows.

### Theorem of the contact plane of 3D solid convex edges

$$\exists \vec{n}_0, \quad \vec{n}_0 \parallel \vec{e}_1 \times \vec{h}_1,$$

$$E(\angle \vec{e}_{11} \vec{e}_{12} + \vec{e}_1, \angle \vec{h}_{11} \vec{h}_{12} + \vec{h}_1) = \uparrow n_0 + a_0 \Leftrightarrow$$

$$\exists \vec{n}_0, \quad \vec{n}_0 \parallel \vec{e}_1 \times \vec{h}_1,$$

$$\vec{e}_{11} \cdot \vec{n}_0 \leq 0, \quad \vec{e}_{12} \cdot \vec{n}_0 \leq 0,$$

$$\vec{h}_{11} \cdot \vec{n}_0 \geq 0, \quad \vec{h}_{12} \cdot \vec{n}_0 \geq 0.$$

It was assumed here, $\vec{e}_1$ and $\vec{h}_1$ are not parallel.

*Proof:*

Using the same $\vec{n}_0$, the sufficient condition can be proved. If the necessary condition is not true,

$$\exists \vec{n}_0, i, j, \quad 1 \le i \le 2, \quad 1 \le j \le 2,$$

$$\vec{n}_0 \parallel \vec{e}_1 \times \vec{h}_1, \quad \vec{e}_{1i} \vec{n}_0 > 0, \quad \vec{h}_{1j} \vec{n}_0 > 0 \Rightarrow$$

$$\vec{\vec{e}}_1 + \vec{e}_{1i} \subset \uparrow n_{11} \cap \uparrow n_{12},$$

$$\vec{\vec{h}}_1 + \vec{h}_{1j} \subset \uparrow n_{21} \cap \uparrow n_{22},$$

$$\vec{\vec{e}}_1 + \vec{e}_{1i} + \vec{\vec{h}}_1 = \uparrow n_0,$$

$$\vec{\vec{h}}_1 + \vec{h}_{1j} + \vec{\vec{e}}_1 = \uparrow n_0 \Rightarrow$$

$$E(\uparrow n \cap \uparrow n_{12}, \uparrow n_{21} \cap \uparrow n_{22}) \supset$$

$$E(\vec{\vec{e}}_1 + \vec{e}_{1i}, \vec{\vec{h}}_1 + \vec{h}_{1j})$$

$$= E(\vec{\vec{e}}_1 + \vec{e}_{1i} - \vec{\vec{h}}_1, \vec{\vec{h}}_1 + \vec{h}_{1j} - \vec{\vec{e}}_1)$$

$$= E(\vec{\vec{e}}_1 + \vec{e}_{1i} + \vec{\vec{h}}_1, \vec{\vec{h}}_1 + \vec{h}_{1j} + \vec{\vec{e}}_1)$$

$$= E(\perp n_0 + \vec{e}_{1i}, \perp n_0 + \vec{h}_{1j}) = E(\uparrow n_0, \uparrow n_0)$$

$$= \uparrow n_0 - \uparrow n_0 + a_0.$$

Here

$$\uparrow n_0 - \uparrow n_0$$

is the whole space. Then

$$\partial E(A, B) = \emptyset.$$

Therefore

$$E(A, B) = \uparrow n_0 + h - e + a_0,$$
$$\partial E(A, B) = \perp n_0 + h - e + a_0.$$

$\perp n_0 + h - e + a_0$ is the contact plane and boundary plane.

### 8.10.3 Contact half-planes of a 3D concave solid edge and a 3D solid angle

The equations of 3D concave edges $B$ and 3D angle $A$ are (Figure 8.97)

$$B = B_1 \cup B_2,$$
$$B_1 = \uparrow n_{21} + h,$$
$$B_2 = \uparrow n_{22} + h.$$

$$A = \angle \vec{e}_1 \vec{e}_2 \cdots \vec{e}_{u-1} \vec{e}_u + e.$$

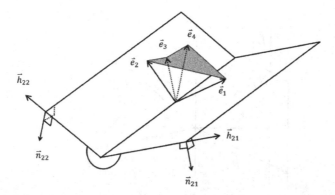

*Figure 8.97*  Contact of a 3D solid angle and a concave solid edge.

### Theorem of contact half planes of 3D solid angle to concave solid edge

$$int(A \cap B) = \emptyset \Leftrightarrow$$
$$int(A \cap B_1) = \emptyset, int(A \cap B_2) = \emptyset \Leftrightarrow$$
$$E(A, B) = B + a_0 - e.$$

### Proof:

$$E(A, B) = E(A, B_1 \cup B_2) = E(A, B_1) \cup E(A, B_2).$$
$$int(A \cap B) = \emptyset \Leftrightarrow$$
$$int(A \cap B_1) = \emptyset, int(A \cap B_2) = \emptyset.$$
$$int(A \cap B_1) = \emptyset, \Leftrightarrow E(A, B_1) = B_1 + a_0 - e \Leftrightarrow$$
$$E(A, B_1) = \uparrow n_{21} + h + a_0 - e.$$
$$int(A \cap B_2) = \emptyset \Leftrightarrow E(A, B_2) = B_2 + a_0 - e \Leftrightarrow$$
$$E(A, B_2) = \uparrow n_{22} + h + a_0 - e.$$

Therefore

$$int(A \cap B) = \emptyset \Leftrightarrow$$
$$E(A, B) = \uparrow n_{21} \cup \uparrow n_{22} + h + a_0 - e = B + a_0 - e.$$
$$\partial E(A, B) = \partial B - e + a_0.$$

Denote

$$\vec{h}_1 = \vec{n}_{22} \times \vec{n}_{21},$$

$$\vec{h}_{21} = \vec{n}_{22} \times \vec{n}_{21} \times \vec{n}_{21},$$

$$\vec{h}_{22} = \vec{n}_{21} \times \vec{n}_{22} \times \vec{n}_{22}.$$

$$\partial E(A, B) = (\vec{\vec{h}}_1 + \angle \vec{h}_{21} + h - e + a_0) \cup$$

$$(\vec{\vec{h}}_1 + \angle \vec{h}_{22} + h - e + a_0).$$

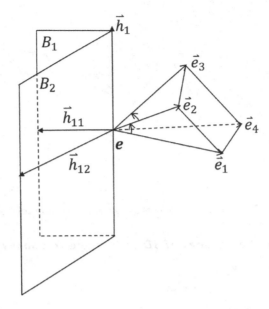

Figure 8.98 Contact of a 3D solid angle and a convex vector edge.

### 8.10.4 Contact half-planes of a 3D convex solid edge and a 3D solid angle

The equations of 3D convex edges $B$ and 3D angle $A$ are

$$B = B_1 \cap B_2,$$
$$B_1 = \uparrow n_{21} + h,$$
$$B_2 = \uparrow n_{22} + h.$$
$$A = \angle \vec{e}_1 \vec{e}_2 \cdots \vec{e}_{u-1} \vec{e}_u + e.$$
$$\vec{h}_1 = \vec{n}_{22} \times \vec{n}_{21}.$$

The inner normal vector of $\angle e_r e_{r+1}$ is $\angle \vec{n}_{1r}$ (Figures 8.98 & 8.99).

**Theorem of contact half planes of 3D solid angle to convex solid edge**

$$int(A \cap B) = \emptyset \Leftrightarrow \exists \angle \vec{d}_1, \angle \vec{d}_2,$$
$$E(A, B) = \angle \vec{d}_1 \vec{d}_2 + \vec{h}_1 + h - e + a_0$$

**Proof:**

Denote

$$\vec{h}_{11} = \vec{n}_{21} \times \vec{n}_{22} \times \vec{n}_{21},$$

$$\vec{h}_{12} = \vec{n}_{22} \times \vec{n}_{21} \times \vec{n}_{22}.$$

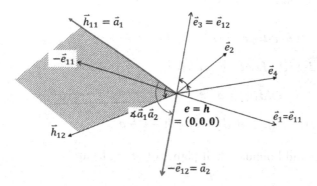

Figure 8.99 Projection along edge vector of the contact of a 3D solid angle and a convex vector edge.

$$B = \vec{\vec{h}}_1 + \angle \vec{h}_{11} \vec{h}_{12} + h,$$
$$E(A, B) =$$

$$E(\angle \vec{e}_1 \vec{e}_2 \cdots \vec{e}_{u-1} \vec{e}_u + e, \vec{\vec{h}}_1 + \angle \vec{h}_{11} \vec{h}_{12} + h) =$$

$$E(\angle \vec{e}_1 \vec{e}_2 \cdots \vec{e}_{u-1} \vec{e}_u, \vec{\vec{h}}_1 + \angle \vec{h}_{11} \vec{h}_{12}) + h - e =$$

$$E(\angle \vec{e}_1 \vec{e}_2 \cdots \vec{e}_{u-1} \vec{e}_u - \vec{\vec{h}}_1, \vec{\vec{h}}_1 + \angle \vec{h}_{11} \vec{h}_{12}) + h - e =$$

$$E(\angle \vec{e}_1 \vec{e}_2 \cdots \vec{e}_{u-1} \vec{e}_u + \vec{\vec{h}}_1, \vec{\vec{h}}_1 + \angle \vec{h}_{11} \vec{h}_{12}) + h - e$$

$$\exists \vec{e}_i, \quad \exists \vec{e}_j,$$

$$\angle \vec{e}_1 \vec{e}_2 \cdots \vec{e}_{u-1} \vec{e}_u + \vec{\vec{h}}_1 = \angle \vec{e}_i \vec{e}_j + \vec{\vec{h}}_1.$$

$$\vec{e}_{11} = \vec{h}_1 \times \vec{e}_i \times \vec{h}_1,$$

$$\vec{e}_{12} = \vec{h}_1 \times \vec{e}_j \times \vec{h}_1.$$

It can be assumed

$$\vec{e}_{11} \times \vec{e}_{12} = \vec{h}_1.$$

$$\angle \vec{e}_1 \vec{e}_2 \cdots \vec{e}_{u-1} \vec{e}_u + \vec{\vec{h}}_1 = \angle \vec{e}_{11} \vec{e}_{12} + \vec{\vec{h}}_1.$$

The computation of $E(A, B)$ is reduced to 2D computation of entrance block of an angle and an angle.

$$E(A, B) =$$

$$E(\angle \vec{e}_1 \vec{e}_2 \cdots \vec{e}_{u-1} \vec{e}_u + \vec{\vec{h}}_1, \vec{\vec{h}}_1 + \angle \vec{h}_{11} \vec{h}_{12}) + h - e$$

$$= E(\angle \vec{e}_{11} \vec{e}_{12} + \vec{\vec{h}}_1, \angle \vec{h}_{11} \vec{h}_{12} + \vec{\vec{h}}_1) + h - e$$

$$= E(\angle \vec{e}_{11} \vec{e}_{12}, \angle \vec{h}_{11} \vec{h}_{12}) + \vec{\vec{h}}_1 + h - e.$$

Based on Section 8.6.2, under the condition

$$int(A \cap B) = \emptyset \Leftrightarrow int(\angle \vec{e}_{11}\vec{e}_{12} \cap \angle \vec{h}_{11}\vec{h}_{12}) \neq \emptyset \Leftrightarrow$$

$$\exists \angle \vec{d}_1, \angle \vec{d}_2, E(\angle \vec{e}_{11}\vec{e}_{12}, \angle \vec{h}_{11}\vec{h}_{12})$$

$$= \angle \vec{d}_1\vec{d}_2 + a_0 \Leftrightarrow$$

$$\exists \angle \vec{d}_1, \angle \vec{d}_2, E(A, B) = \angle \vec{d}_1\vec{d}_2 + \vec{h}_1 + h - e + a_0.$$

The contact and boundary half-planes of $\partial E(A, B)$ are

$$\vec{\vec{h}}_1 + \angle \vec{d}_1 + h - e + a_0,$$

$$\vec{\vec{h}}_1 + \angle \vec{d}_2 + h - e + a_0.$$

### 8.10.5 Contact solid angles of two 3D convex solid angles

Convex angles $A$ and $B$ are defined by

$$A = \angle \vec{e}_1\vec{e}_2 \cdots \vec{e}_{u-1}\vec{e}_u + e,$$

$$B = \angle \vec{h}_1\vec{h}_2 \cdots \vec{h}_{v-1}\vec{h}_v + h.$$

The inner normal vector of $\angle e_r e_{r+1}$ is $\vec{n}_{1r}$,

$$\vec{n}_{1r} = +\vec{e}_{r+1} \times \vec{e}_r, \quad r = 1, \ldots, u.$$

The inner normal vector of $\angle h_s h_{s+1}$ is $\vec{n}_{2s}$,

$$\vec{n}_{2s} = +\vec{h}_{s+1} \times \vec{h}_s, \quad s = 1, \ldots, v.$$

As convex, $A$ and $B$ can be represented by simultaneous inequality equations.

$$A = \cap_{r=1,\ldots,u}\{(x - e) \cdot \vec{n}_{1r} \geq 0\}.$$

$$B = \cap_{s=1,\ldots,v}\{(x - h) \cdot \vec{n}_{2s} \geq 0\}.$$

Based on Section 8.9.10, all contact angles are searched and examined as follows (Figure 8.100).

### *Theorem of vertex-angle contact of 3D convex solid angles*

$$int(\angle \vec{e}_1\vec{e}_2 \cdots \vec{e}_{u-1}\vec{e}_u \cap \uparrow n_{2s}) = \emptyset \Rightarrow$$

$$E(e, \angle \vec{h}_s\vec{h}_{s+1} + h) \subset \partial E(A, B)$$

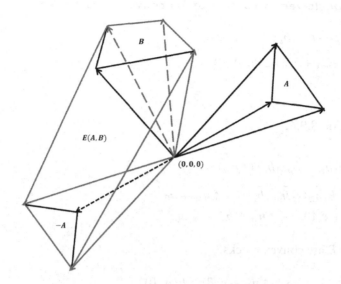

*Figure 8.100*  Entrance 3D solid angle of two 3D convex solid angles.

**Proof:**

Based on Section 8.9.6,

$$int(\angle \vec{e_1}\vec{e_2}\cdots\vec{e}_{u-1}\vec{e}_u \cap \uparrow n_{2s}) = \emptyset \Rightarrow$$
$$E(\angle \vec{e_1}\vec{e_2}\cdots\vec{e}_{u-1}\vec{e}_u, \uparrow n_{2s}) = \uparrow n_{2s} + a_0.$$

$E(A, \uparrow n_{2s} + h) = \uparrow n_{2s} + h - e + a_0$
   As $A$ and $B$ are convex blocks,

$$\uparrow n_{2s} + h \supset B, \quad \Rightarrow E(A, \uparrow n_{2s} + h) \supset E(A, B).$$

On the other side,

$$E(e, \angle \vec{h_s}\vec{h}_{s+1} + h) = \angle \vec{h_s}\vec{h}_{s+1} + h - e + a_0,$$
$$\partial E(A, \uparrow n_{2s} + h) = \perp n_{2s} + h - e + a_0 \Rightarrow$$

$$E(e, \angle \vec{h_s}\vec{h}_{s+1} + h) \subset \partial E(A, \uparrow n_{2s} + h),$$

$$E(e, \angle \vec{h_s}\vec{h}_{s+1} + h) \subset E(A, B) \Rightarrow$$

$$E(e, \angle \vec{h_s}\vec{h}_{s+1} + h) \subset \partial E(A, B).$$

The following theorem uses the results of Section 8.9.6.

### Theorem of angle-vertex contact of 3D convex solid angles

$$int(\angle \vec{h}_1 \vec{h}_2 \cdots \vec{h}_{v-1} \vec{h}_v \cap \uparrow n_{1r}) = \emptyset \Rightarrow$$
$$E(\angle \vec{e}_r \vec{e}_{r+1} + e, h) \subset \partial E(A, B)$$

**Proof:**

Based on Section 8.9.6,

$$int(\angle \vec{h}_1 \vec{h}_2 \cdots \vec{h}_{v-1} \vec{h}_v \cap \uparrow n_{1r}) = \emptyset \Rightarrow$$
$$E(\uparrow n_{1r}, \angle \vec{h}_1 \vec{h}_2 \cdots \vec{h}_{v-1} \vec{h}_v) = - \uparrow n_{1r} + a_0$$
$$E(\uparrow n_{1r} + e, B) = - \uparrow n_{1r} + h - e + a_0.$$

As $A$ and $B$ are convex blocks,

$$\uparrow n_{1r} + e \supset A, \quad \Rightarrow E(\uparrow n_{1r} + e, B) \supset E(A, B).$$

On the other side,

$$E(\angle \vec{e}_r \vec{e}_{r+1} + e, h) = -\angle \vec{e}_{r+1} \vec{e}_r + h - e + a_0,$$
$$\partial E(\uparrow n_{1r} + e, B) = -\perp n_{1r} + h - e + a_0 \Rightarrow$$
$$E(\angle \vec{e}_r \vec{e}_{r+1} + e, h) \subset \partial E(\uparrow n_{1r} + e, B),$$
$$E(\angle \vec{e}_r \vec{e}_{r+1} + e, h) \subset E(A, B) \Rightarrow$$
$$E(\angle \vec{e}_r \vec{e}_{r+1} + e, h) \subset \partial E(A, B).$$

The following theorem uses the results of Section 8.9.7.

### Theorem of vector-vector contact of 3D convex solid angles

$$int(\angle \vec{e}_{r1} \vec{e}_{r2} + \vec{\vec{e}}_r \cap \angle \vec{h}_{s1} \vec{h}_{s2} + \vec{\vec{h}}_s) = \emptyset \Rightarrow$$
$$E(\angle \vec{e}_r + e, \angle \vec{h}_s + h) \subset \partial E(A, B)$$

**Proof:**

From Section 8.9.7,

$$int(\angle \vec{e}_{r1} \vec{e}_{r2} + \vec{\vec{e}}_r \cap \angle \vec{h}_{s1} \vec{h}_{s2} + \vec{\vec{h}}_s) = \emptyset \Rightarrow$$
$$\exists \vec{n}_0, \quad \vec{n}_0 \parallel \vec{e}_r \times \vec{h}_s,$$
$$E(\angle \vec{e}_{r1} \vec{e}_{r2} + \vec{\vec{e}}_r, \angle \vec{h}_{s1} \vec{h}_{s2} + \vec{\vec{h}}_s) = \uparrow n_0 + a_0.$$

As $A$ and $B$ are convex blocks,

$$\angle \vec{e}_{r1} \vec{e}_{r2} + \vec{e}_r + e \supset A, \quad \angle \vec{h}_{s1} \vec{h}_{s2} + \vec{h}_s + h \supset B \Rightarrow$$

$$E(\angle \vec{e}_{r1} \vec{e}_{r2} + \vec{e}_r + e, \quad \angle \vec{h}_{s1} \vec{h}_{s2} + \vec{h}_s + h \supset E(A, B).$$

$$\Rightarrow \uparrow n_0 + h - e + a_0 \supset E(A, B).$$

On the other side,

$$E(\angle \vec{e}_r + e, \angle \vec{h}_s + h) = -\angle \vec{e}_r + \angle \vec{h}_s + h - e + a_0,$$

$$= \angle(-\vec{e}_r)\vec{h}_s + h - e + a_0 \subset \perp n_0 + h - e + a_0 \Rightarrow$$

$$E(\angle \vec{e}_r + e, \angle \vec{h}_s + h)$$

$$\subset \partial E(\angle \vec{e}_{r1} \vec{e}_{r2} + \vec{e}_r + e, \angle \vec{h}_{s1} \vec{h}_{s2} + \vec{h}_s + h).$$

$$E(\angle \vec{e}_r + e, \angle \vec{h}_s + h) \subset E(A, B) \Rightarrow$$

$$E(\angle \vec{e}_r + e, \angle \vec{h}_s + h) \subset \partial E(A, B).$$

### Theorem of contact edges of 3D convex solid angles

$$\partial E(A, B) = C(0, 2) \cup C(2, 0) \cup C(1, 1).$$

### Proof:

From Section 8.9.10,

$$\partial E(A, B) \subset C(0, 2) \cup C(2, 0) \cup C(1, 1).$$

From previous theorems of this section,

$$\partial E(A, B) \supset C(0, 2) \cup C(2, 0) \cup C(1, 1).$$

Therefore,

$$\partial E(A, B) = C(0, 2) \cup C(2, 0) \cup C(1, 1).$$

## 8.10.6   Contact solid angles of two 3D general solid angles

General 3D angles A and B (Figures 8.101 to 8.103) are defined by

$$A = \angle \vec{e}_1 \vec{e}_2 \cdots \vec{e}_{u-1} \vec{e}_u + e,$$

$$B = \angle \vec{h}_1 \vec{h}_2 \cdots \vec{h}_{v-1} \vec{h}_v + h.$$

The inner normal vector of $\angle \vec{e}_r \vec{e}_{r+1}$ is $\vec{n}_{1r}$.

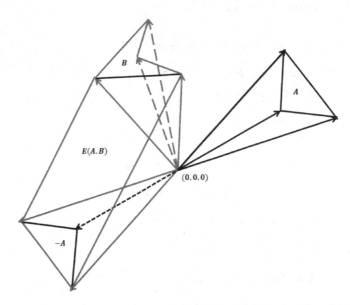

*Figure 8.101* Entrance 3D solid angle of two 3D general solid angles.

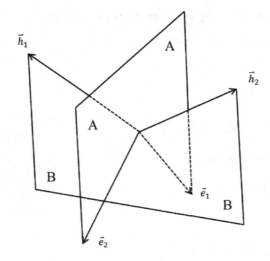

*Figure 8.102* 3D contact of two 2D concave solid angles.

For all face angles of $A$

$$\angle \vec{e}_r \vec{e}_{r+1}, \quad r = 1, \ldots, u,$$

if face $\angle \vec{e}_r \vec{e}_{r+1}$ is convex,

$$\vec{n}_{1r} = +\vec{e}_{r+1} \times \vec{e}_r,$$

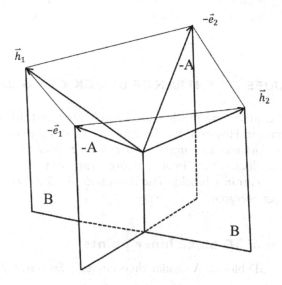

*Figure 8.103* Entrance 3D concave solid angle of two 2D concave solid angles.

if face $\angle \vec{e}_r \vec{e}_{r+1}$ is concave,

$$\vec{n}_{1r} = -\vec{e}_{r+1} \times \vec{e}_r.$$

The inner normal vector of $\angle \vec{h}_s \vec{h}_{s+1}$ is $\vec{n}_{2s}$. For all face angles of $B$

$$\angle \vec{h}_s \vec{h}_{s+1}, \quad s = 1, \ldots, v,$$

If face $\angle \vec{h}_s \vec{h}_{s+1}$ is convex,

$$\vec{n}_{2s} = +\vec{h}_{s+1} \times \vec{h}_s.$$

If face $\angle \vec{h}_s \vec{h}_{s+1}$ is concave,

$$\vec{n}_{2s} = -\vec{h}_{s+1} \times \vec{h}_s.$$

Based on Section 8.9.10, all contact angles are searched and examined as follows. The following theorem is the conclusion of Sections 8.9.6 and 8.9.7.

**Theorem of contact solid angles of 3D solid angles**

$$\partial E(A, B) \subset C(0, 2) \cup C(2, 0) \cup C(1, 1) \subset E(A, B).$$

The contact angles connected and intersected with each other to form a 3D angle which is the 3D entrance angle $E(A, B)$. If no 3D angle can be formed, $E(A, B)$ is the whole space.

## 8.11 BOUNDARIES OF ENTRANCE BLOCK OF 3D BLOCKS

Three dimensional contacts are much more complicated and difficult than the two dimensional counterpart. However, the 3D entrance blocks are computed in a similar way with 2D cases. The way to compute 3D entrance block is also to compute the boundary of entrance block. The following theorems are for understanding and finding the boundaries of 3D entrance blocks. The boundaries of 3D entrance blocks are the subsets of the contact polygons.

### 8.11.1 Entrance of 3D block inner points

Assume $A$ and $B$ are 3D blocks. A similar theorem as in Section 8.9.3 exists.

**Theorem of entrance of 3D block inner points**

$$\partial E(A, B) \subset E(\partial A, \partial B).$$

**Proof:**

If $a \in int(A), b \in B$,

$$E(a, b) = b - a + a_0 \in int(b - A + a_0)$$
$$= int(E(A, b)) \subset int(E(A, B)).$$

If $a \in A, b \in int(B)$,

$$E(a, b) = b - a + a_0 \in int(B - a + a_0)$$
$$= int(E(a, B)) \subset int(E(A, B)).$$

If $a \in int(A)$ or $b \in int(B)$,

$$E(a, b) = b - a + a_0$$

is an inner point of $E(A, B)$.
Therefore

$$\partial E(A, B) \subset E(\partial A, \partial B)$$

### 8.11.2 Entrance of boundary polygons of 3D blocks

Assume $A$ and $B$ are two 3D blocks.

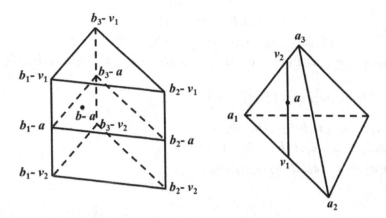

*Figure 8.104*  Entrance point of polygon faces.

### Theorem of entrance of boundary polygon-polygon of 3D blocks

$$\partial E(A, B) \subset E(A(1), B(2)) \cup E(A(2), B(1)).$$

### Proof:

From Section 8.11.1,

$$\partial E(A, B) \subset E(\partial A, \partial B) = E(A(2), B(2)).$$

Assume

$$a_1 a_2 \cdots a_{p-1} a_p, \quad a_{p+1} = a_1$$

is a polygon of $A$. The inner normal vector is $\vec{m}_{1k}$,

$$b_1 b_2 \cdots b_{q-1} b_q, \quad b_{q+1} = b_1$$

is a polygon of $B$. The inner normal vector is $\vec{m}_{2l}$, (Figures 8.104).

If $\vec{m}_{1k}$ and $\vec{m}_{2l}$ are not parallel, for

$$a \in int(a_1 a_2 \cdots a_{p-1} a_p),$$
$$b \in int(b_1 b_2 \cdots b_{q-1} b_q),$$
$$E(a, b) = b - a + a_0,$$
$$\exists v_1, \quad v_2, \quad (v_1 v_2) \cdot \vec{m}_{2l} \neq 0,$$
$$a \in int(v_1 v_2) \subset int(a_1 a_2 \cdots a_{p-1} a_p).$$
$$E(a, b) \in int(E(v_1 v_2, b_1 b_2 \cdots b_{q-1} b_q)$$
$$\subset int(E(a_1 a_2 \cdots a_{p-1} a_p, b_1 b_2 \cdots b_{q-1} b_q)$$

$$\subset int(E(A, B)).$$
$$E(a, b) \in int(E(A, B)).$$
$$E(a_1 a_2 \cdots a_{p-1} a_p, b_1 b_2 \cdots b_{q-1} b_q) \cap \partial E(A, B) \subset$$
$$(E(\partial a_1 a_2 \cdots a_{p-1} a_p, b_1 b_2 \cdots b_{q-1} b_q) \cup E(a_1 a_2 \cdots a_{p-1} a_p, \partial b_1 b_2 \cdots b_{q-1} b_q)).$$

If $\vec{m}_{1k} \parallel \vec{m}_{2l}$, from theorem of Section 8.3.2

$$E(a_1 a_2 \cdots a_{p-1} a_p, b_1 b_2 \cdots b_{q-1} b_q) =$$
$$E(\partial a_1 a_2 \cdots a_{p-1} a_p, b_1 b_2 \cdots b_{q-1} b_q)$$
$$\cup E(a_1 a_2 \cdots a_{p-1} a_p, \partial b_1 b_2 \cdots b_{q-1} b_q).$$

Therefore

$$\partial E(A, B) \subset E(A(1), B(2)) \cup E(A(2), B(1)).$$

### 8.11.3   Entrance of polygon to edge of 3D blocks

Assume $A$ and $B$ are the same 3D blocks as in Sections 8.11.1 and 8.11.2.

#### Theorem of entrance of boundary edge-polygon of 3D blocks

$$\partial E(A, B) \subset E(A(0), B(2)) \cup E(A(2), B(0)) \cup$$
$$E(A(1), B(1)).$$

#### Proof:

$$a_i a_{i+1}, \quad i = 1, \ldots, p,$$
$$b_1 b_2 \cdots b_{q-1} b_q.$$

are edge and polygon of 3D blocks $A$ and $B$ respectively, (Figure 8.105). For any points

$$a \in int(a_i a_{i+1}),$$
$$b \in int(b_1 b_2 \cdots b_{q-1} b_q),$$

If

$$(a_i a_{i+1}) \cdot \vec{m}_{2l} \neq 0,$$
$$E(a, b) \in int(E(a_i a_{i+1}, b_1 b_2 \cdots b_{q-1} b_q)$$
$$\subset int(E(A, B)).$$

Therefore, if

$$(a_i a_{i+1}) \vec{m}_{2l} \neq 0,$$
$$E(a_i a_{i+1}, b_1 b_2 \cdots b_{q-1} b_q) \cap \partial E(A, B) =$$
$$(E(\partial a_i a_{i+1}, b_1 b_2 \cdots b_{q-1} b_q) \cup E(a_i a_{i+1}, \partial b_1 b_2 \cdots b_{q-1} b_q)).$$

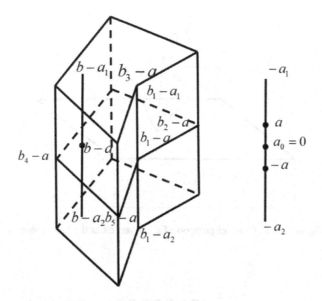

*Figure 8.105* Entrance point of a face polygon inner point and an edge inner point.

If $(a_i a_{i+1}) \cdot \vec{m}_{2l} = 0$, from the theorem in Section 8.4.2

$E(a_i a_{i+1}, b_1 b_2 \cdots b_{q-1} b_q) \subset$
$E(\partial a_i a_{i+1}, b_1 b_2 \cdots b_{q-1} b_q)$
$\cup E(a_i a_{i+1}, \partial b_1 b_2 \cdots b_{q-1} b_q).$

Therefore

$\partial E(A, B) \subset E(A(0), B(2)) \cup E(A(2), B(0)) \cup$
$\qquad E(A(1), B(1)).$

## 8.11.4  Contact polygon of 3D vertex to polygon contact

Assume $A$ and $B$ are the same blocks as in Section 8.11.2.

$b_1 b_2 \cdots b_{q-1} b_q, \quad b_{q+1} = b_1$

is a polygon of $B$. The normal vector of polygon is $\vec{m}_{2l}$. Given a vertex $e$ of $A$, all vertices connected with $e$ by edges are (Figures 8.106 & 8.107)

$e_1 e_2 \cdots e_{u-1} e_u, \quad e_{u+1} = e_1.$

Denote $\vec{e}_r = (ee_r)$, $\vec{n}_{1r}$ as inner normal of $\angle e_r e_{r+1}$.

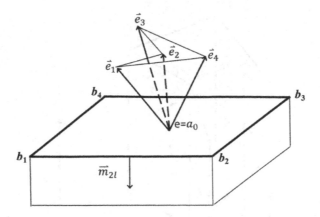

*Figure 8.106* Contact polygon of a 3D vertex and a face polygon.

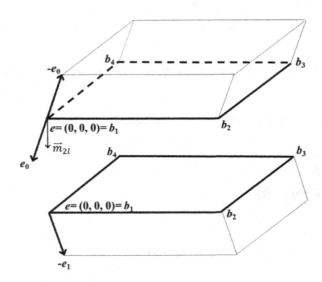

*Figure 8.107* Contact condition of a 3D vertex and a face polygon.

### Theorem of 3D vertex-polygon contact

$$\exists b \in int(b_1 b_2 \cdots b_{q-1} b_q), \quad E(e, b) \in \partial E(A, B)$$
$$\Rightarrow int(\angle \vec{e}_1 \vec{e}_2 \cdots \vec{e}_{u-1} \vec{e}_u \cap \uparrow m_{2l}) = \emptyset.$$

### Proof:

If

$$int(\angle \vec{e}_1 \vec{e}_2 \cdots \vec{e}_{u-1} \vec{e}_u \cap \uparrow m_{2l}) \neq \emptyset,$$
$$\exists ee_0 \subset A, \quad (ee_0) \cdot \vec{m}_{2l} > 0,$$
$$E(e, b) = b - e + a_0 \in$$

$$int((B + a_0 - e) \cup (b_1 b_2 \cdots b_{q-1} b_q - e e_0 + a_0))$$
$$= int(E(e, B) \cup E(e e_0, b_1 b_2 \cdots b_{q-1} b_q))$$
$$\subset int(E(A, B))$$
$$\Rightarrow E(e, b) \in int(E(A, B)), \quad E(e, b) \notin \partial E(A, B).$$

Similarly,

$$a_1 a_2 \cdots a_{p-1} a_p, \quad a_{p+1} = a_1$$

is a polygon of $A$. The normal vector of polygon is $\vec{m}_{1k}$.

Given a vertex $h$ of $B$, all vertices connected with $h$ by edges are

$$b_1 b_2 \cdots b_{v-1} b_v, \quad b_{v+1} = b_1.$$

Denote $\vec{b}_s = (h b_s)$, $\vec{n}_{2s}$ as the inner normal of $\angle h_s h_{s+1}$.

### Theorem of 3D polygon-vertex contact

$$\exists a \in int(a_1 a_2 \cdots a_{p-1} a_p), \quad E(a, h) \in \partial E(A, B)$$

$$\Rightarrow int(\angle \vec{b}_1 \vec{b}_2 \cdots \vec{b}_{v-1} \vec{b}_v \cap \uparrow m_{1k}) = \emptyset.$$

### Proof:

If

$$int(\angle \vec{b}_1 \vec{b}_2 \cdots \vec{b}_{v-1} \vec{b}_v \cap \uparrow m_{1k}) \neq \emptyset,$$
$$\exists h_0, \quad h h_0 \subset B, \quad (h h_0) \cdot \vec{m}_{1k} > 0.$$
$$E(a, h) = h - a + a_0 \in$$
$$int(h - a_1 a_2 \cdots a_{p-1} a_p + a_0) \subset$$
$$int((h - A + a_0) \cup (h h_0 - a_1 a_2 \cdots a_{p-1} a_p + a_0))$$
$$= int(E(A, h) \cup E(a_1 a_2 \cdots a_{p-1} a_p, h h_0))$$
$$\subset int(E(A, B)) \Rightarrow$$
$$E(a, h) \subset int(E(A, B)) \Rightarrow E(a, h) \notin \partial E(A, B).$$

## 8.11.5  Contact polygon of 3D edge to edge contact

Assume $A$ and $B$ are the same blocks as in Section 8.11.2. As defined in Section 8.11.4, $e e_r$ is an edge of $A$, and $e$ and $e_r$ are vertices of block $A$. $h h_s$ is an edge of $B$, and $h$ and $h_s$ are vertices of block $B$ (Figures 8.108 & 8.109).

The inner normal vectors of $\angle \vec{e}_{r-1} \vec{e}_r$ and $\angle \vec{e}_r \vec{e}_{r+1}$ are $\vec{n}_{11}$ and $\vec{n}_{12}$ respectively. The inner normal vectors of $\angle \vec{h}_{s-1} \vec{h}_s$ and $\angle \vec{h}_s \vec{h}_{s+1}$ are $\vec{n}_{21}$ and $\vec{n}_{22}$ respectively.

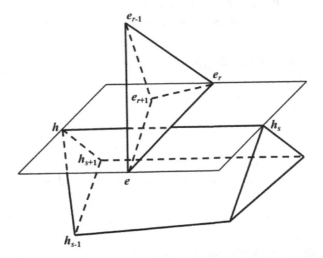

Figure 8.108 An example of 3D edge to edge contact.

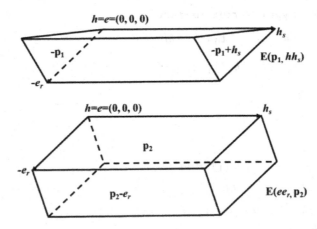

Figure 8.109 Another example of 3D edge to edge contact.

Assume the edges of $\vec{e}_r$ and $\vec{h}_s$ are convex

$$\vec{e}_r \uparrow\uparrow \vec{n}_{12} \times \vec{n}_{11},$$

$$\vec{h}_s \uparrow\uparrow \vec{n}_{22} \times \vec{n}_{21}.$$

Denote

$$\vec{e}_{r1} = \vec{n}_{11} \times \vec{n}_{12} \times \vec{n}_{11},$$

$$\vec{e}_{r2} = \vec{n}_{12} \times \vec{n}_{11} \times \vec{n}_{12},$$

$$\vec{h}_{s1} = \vec{n}_{21} \times \vec{n}_{22} \times \vec{n}_{21},$$

$$\vec{h}_{s2} = \vec{n}_{22} \times \vec{n}_{21} \times \vec{n}_{22},$$

### Theorem of 3D edge-edge contact

$$\exists a \in int(ee_r), b \in int(hh_s),$$
$$E(a,b) \in \partial E(A,B) \Rightarrow$$
$$int((\uparrow n_{11} \cap \uparrow n_{12}) \cap (\uparrow n_{21} \cap \uparrow n_{22})) = \emptyset.$$

### Proof:

If

$$int((\uparrow n_{11} \cap \uparrow n_{12}) \cap (\uparrow n_{21} \cap \uparrow n_{22})) \neq \emptyset,$$

$$\exists \vec{n}_0, \quad \vec{n}_0 \parallel \vec{e}_r \times \vec{h}_s,$$
$$\exists i, \quad j, \quad 1 \leq i \leq 2, \quad 1 \leq j \leq 2,$$
$$\vec{e}_{ri} \cdot \vec{n}_0 > 0,$$
$$\vec{h}_{sj} \cdot \vec{n}_0 > 0.$$

If $i=1, P_1 = P(e_r e)$, if $i=2, P_1 = P(ee_r)$.
If $j=1, P_2 = P(h_s h)$, if $j=2, P_2 = P(hh_s)$.

Assuming $\vec{e}_r$ and $\vec{h}_s$ are not parallel,

$$E(ee_r, hh_s) \subset E(P_1, hh_s),$$
$$E(ee_r, hh_s) \subset E(ee_r, P_2).$$
$$E(a,b) \in int(E(ee_r, hh_s))$$
$$\subset int(E(P_1, hh_s) \cup E(ee_r, P_2))$$
$$\subset int(E(P_1, P_2) \cup E(P_1, P_2))$$
$$= int(E(P_1, P_2))$$
$$\subset int(E(A,B)) \Rightarrow$$
$$E(a,b) \notin \partial E(A,B).$$

Therefore there is a vector $\vec{n}_1$ that satisfies the related equations.

## 8.11.6   Finite covers of parallel polygon to edge entrance

When blocks $A$ and $B$ contact along parallel edges and polygons or along a pair of parallel polygons, there are infinite contact point pairs. Among these infinite contact point pairs, only a few contact point pairs are vertex-to-polygon contact point pairs and edge-to-edge contact point pairs, and will control the movements, will especially control the rotations. These vertex-to-polygon contact point pairs and edge-to-edge contact point pairs are called contact points which correspond to overlapped contact polygons. After small rotation or deformation, blocks $A$ and $B$ only contact along these contact points.

All closed contact points together define the movements, rotations and deformations of all blocks just as in the real cases.

The following theorems are based on the assumption of blocks $A$ and $B$ contacting along the parallel edges and polygons. Assume $A$ and $B$ are the same blocks as in Sections 8.11.2 and 8.11.3.

$$a_i a_{i+1}, \quad i = 1, \ldots, p,$$

$$b_1 b_2 \cdots b_{q-1} b_q, \quad j = 1, \ldots, q.$$

are edge and polygon of 3D blocks $A$ and $B$ respectively.

### Theorem of finite covers of entrance blocks of an edge and a polygon

$$A_0 = a_i a_{i+1}, \quad B_0 = b_1 b_2 \cdots b_{q-1} b_q,$$
$$a_i a_{i+1} \parallel b_1 b_2 \cdots b_{q-1} b_q \Rightarrow$$
$$E(A_0, B_0) = \cup_{m=0}^{1} E(A_0(m), B_0(2 - m)).$$

### Proof:

Similar as in Section 8.4.2,

$$E(A_0, B_0) = \cup_{m=0}^{1} E(A_0(m), B_0(2 - m)).$$
$$E(a_i a_{i+1}, b_1 b_2 \cdots b_{q-1} b_q) =$$
$$E(a_i, b_1 b_2 \cdots b_{q-1} b_q) \cup$$
$$E(a_{i+1}, b_1 b_2 \cdots b_{q-1} b_q) \cup$$
$$(\cup_{j=1}^{q} E(a_i a_{i+1}, b_j b_{j+1})).$$

Here

$$\partial a_i a_{i+1} = a_i \cup a_{i+1},$$
$$\partial b_1 b_2 \cdots b_{q-1} b_q = \cup_{j=1}^{q} b_j b_{j+1}.$$

$P(a_i a_{i+1})$ is a polygon having $a_i a_{i+1}$ as its boundary, and $P(a_{i+1} a_i)$ is another polygon having $a_{i+1} a_i$ as its boundary.
The inner normal vectors of

$$P_1 = P(a_i a_{i+1}), \quad P_2 = P(a_{i+1} a_i)$$

are $\vec{n}_{11}$ and $\vec{n}_{12}$ respectively.

$$\vec{e}_{r1} = \vec{n}_{11} \times \vec{n}_{12} \times n_{11},$$
$$\vec{e}_{r2} = \vec{n}_{12} \times \vec{n}_{11} \times \vec{n}_{12},$$

## Theorem of parallel edge and 2D polygon contact

Assuming

$$a_i a_{i+1} \cdot \vec{m}_{2l} = 0,$$

$$\exists a \in int(a_i a_{i+1}), b \in int(b_1 b_2 \cdots b_{q-1} b_q),$$

$$E(a,b) \in \partial E(A,B) \Rightarrow$$

$$int((\uparrow n_{11} \cap \uparrow n_{12}) \cap \uparrow m_{2l}) = \emptyset.$$

### Proof:

If

$$int((\uparrow n_{11} \cap \uparrow n_{12}) \cap \uparrow m_{2l}) \neq \emptyset,$$

$$\exists k, \quad 1 \le k \le 2, \quad \vec{e}_{rk} \vec{m}_{2l} > 0.$$

$$E(a,b) \in int(E(a_i a_{i+1}, b_1 b_2 \cdots b_{q-1} b_q)) \subset$$

$$int(E(P_k, b_1 b_2 \cdots b_{q-1} b_q) \cup E(a_i a_{i+1}, B)) \subset$$

$$\subset int(E(A,B) \cup E(A,B)) = int(E(A,B)).$$

$$\Rightarrow E(a,b) \in int(E(A,B)), \quad E(a,b) \notin \partial E(A,B).$$

## 8.11.7 Finite covers of parallel polygon entrance

The following theorems are based on the assumption of blocks $A$ and $B$ contacting along two parallel polygons. Here $A$ and $B$ are two blocks. Assuming

$$a_1 a_2 \cdots a_{p-1} a_p, \quad a_{p+1} = a_1$$

is a polygon of $A$. The inner normal vector is $\vec{m}_{1k}$.

$$b_1 b_2 \cdots b_{q-1} b_q, \quad b_{q+1} = b_1$$

is a polygon of $B$ (Figures 8.110), where the inner normal vector is $\vec{m}_{2l}$.

## Theorem of finite covers of entrance blocks of 2D polygons

Assuming

$$A_0 = a_1 a_2 \cdots a_{p-1} a_p,$$
$$B_0 = b_1 b_2 \cdots b_{q-1} b_q,$$
$$\vec{m}_{1k} \parallel \vec{m}_{2l}.$$

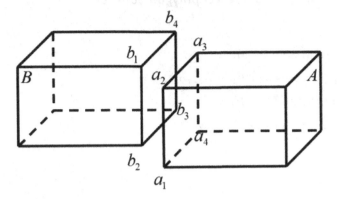

Figure 8.110  Contact of two parallel face polygons.

From Section 8.4.2,

$$E(A_0, B_0) = \cup_{m=0}^{2} E(A_0(m), B_0(2-m)).$$
$$E(a_1 a_2 \cdots a_{p-1} a_p, b_1 b_2 \cdots b_{q-1} b_q) =$$
$$\cup_{i=1}^{p} E(a_i, b_1 b_2 \cdots b_{q-1} b_q) \cup$$
$$\cup_{j=1}^{q} E(a_1 a_2 \cdots a_{p-1} a_p, b_j) \cup$$
$$(\cup_{i=1}^{p} \cup_{j=1}^{q} E(a_i a_{i+1}, b_j b_{j+1})).$$

Here

$$\partial a_1 a_2 \cdots a_{p-1} a_p = \cup_{i=1}^{p} a_i a_{i+1}.$$
$$\partial b_1 b_2 \cdots b_{q-1} b_q = \cup_{j=1}^{q} b_j b_{j+1}.$$

### Theorem of parallel polygon contact

Assuming

$$\vec{m}_{1k} \parallel \vec{m}_{2l}$$
$$\exists a \in int(a_1 a_2 \cdots a_{p-1} a_p), b \in int(b_1 b_2 \cdots b_{q-1} b_q),$$
$$E(a, b) \in \partial E(A, B) \Rightarrow \vec{m}_{1k} \uparrow\uparrow -\vec{m}_{2l}.$$

as shown in Figures 8.111 to 8.114.

### Proof:

If

$$\vec{m}_{1k} \uparrow\uparrow \vec{m}_{2l} \Rightarrow$$
$$\exists \delta,$$
$$C = \{|x| \leq \delta\},$$

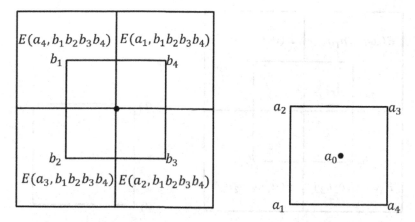

Figure 8.111 Contact polygons of the form $E(A_0(0), B_0(2))$.

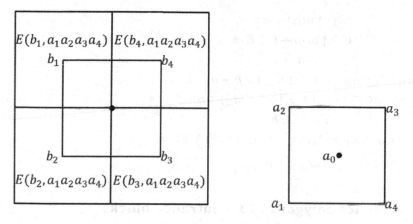

Figure 8.112 Contact polygons of the form $E(A_0(2), B_0(0))$.

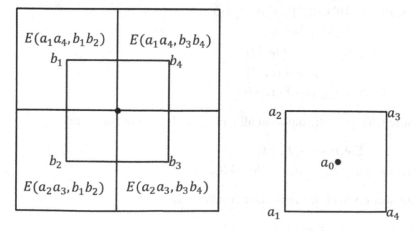

Figure 8.113 Contact polygons of the form $E(A_0(1), B_0(1))$.

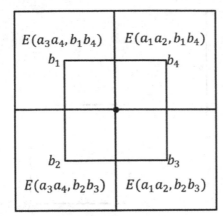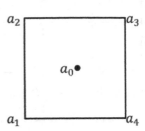

Figure 8.114  Contact polygons of the form $E(A_0(1), B_0(1))$.

$$(C \cap \uparrow m_{1k}) + a \subset A,$$
$$(C \cap \uparrow m_{2l}) + b \subset B \Rightarrow$$
$$E(a, b) \in$$
$$int((C \cap \uparrow m_{2l}) - (C \cap \uparrow m_{1k}) + b - a + a_0) =$$
$$int(E((C \cap \uparrow m_{1k}) + a, (C \cap \uparrow m_{2l}) + b)) \subset$$
$$int(E(A, B)).$$
$$\Rightarrow E(a, b) \in int(E(A, B)), \quad E(a, b) \notin \partial E(A, B) \Rightarrow$$
$$\Rightarrow \vec{m}_{1k} \upuparrows -\vec{m}_{2l}.$$

## 8.11.8  Contact polygons of 3D entrance blocks

Assume $A$ and $B$ are the same blocks as in Section 8.11.1. From Section 8.11.3,

$$\partial E(A, B) \subset E(A(0), B(2)) \cup E(A(2), B(0)) \cup$$
$$E(A(1), B(1)).$$
$$= (\cup_{e \in A(0), P \subset B(2)} E(e, P))$$
$$\cup (\cup_{P \subset A(2), e \in B(0)} E(P, h))$$
$$\cup (\cup_{ea \in A(1), hb \subset B(1)} E(ea, hb))$$

Denote $C(0, 2)$ as the union of all contact polygons of the form

$$E(e, b_1 b_2 \cdots b_{q-1} b_q)$$
$$C(0, 2) = \cup_{contact\ polygon} E(e, b_1 b_2 \cdots b_{q-1} b_q)$$

From Section 8.11.4, the contact condition is

$$int(\angle \vec{e}_1 \vec{e}_2 \cdots \vec{e}_{u-1} \vec{e}_u \cap \uparrow m_{2l}) = \emptyset.$$

Denote $C(2, 0)$ as the union of all contact polygons of the form

$$E(a_1 a_2 \cdots a_{p-1} a_p, \quad b),$$
$$C(2, 0) = \cup_{contact \ polygon} E(a_1 a_2 \cdots a_{p-1} a_p, \quad b)$$

From Section 8.11.4, the contact condition is

$$int(\angle \vec{m}_{1k} \cap \angle \vec{b}_1 \vec{b}_2 \cdots \vec{b}_{v-1} \vec{b}_v) = \emptyset$$

Denote $C(1, 1)$ as the union of all contact polygons of the form

$$E(ee_r, \quad bb_s),$$
$$C(1, 1) = \cup_{contact \ polygon} E(ee_r, bb_s)$$

From Section 8.11.5, the contact condition is

$$int((\uparrow n_{11} \cap \uparrow n_{12}) \cap (\uparrow n_{21} \cap \uparrow n_{22})) = \emptyset.$$

The following theorem is the conclusion of Sections 8.11.4 and 8.11.5.

**Theorem of 3D contact polygons**

$$\partial E(A, B) \subset C(0, 2) \cup C(2, 0) \cup C(1, 1) \subset E(A, B).$$

## 8.12 CONTACT POLYGONS OF 3D BLOCKS

The simplest 3D contact case is the ball-to-ball contact with an obvious entrance ball. Geometrically, balls are convex. Most of the blocks are convex. According to the convex block theorem (Section 8.4.9), the entrance block of two convex blocks is convex. Therefore, the computation of contact polygons of 3D convex entrance blocks is about few times of the computation of ball cases. Complex blocks are union of convex blocks.

The contact surface of a 3D entrance block is formed by contact polygons. Each contact polygon corresponds to a vertex-to-polygon contact or an edge-to-edge contact.

In the process of computing, time steps are used. At the beginning of each time step, the contact polygons are computed. From the relative position of reference point $a_0$ and the contact polygons, the closed contact point pairs are found. The closed contact point pairs are defined by contact polygons and control the movements, rotations and deformations throughout this time step.

### 8.12.1 Contact polygons of two 3D convex blocks

The 3D blocks $A$ and $B$ are the same as defined in Section 8.2.5. The faces of block $A$ are polygons which rotate outward. The vertices of face $k$ are

$$a_1 a_2 \cdots a_{p-1} a_p, \quad a_{p+1} = a_1.$$

The normal vector $\vec{m}_{1k}$ of face $k$ points inward. The faces of block $B$ are polygons which rotate the outward block. The vertices of face $l$ are

$$b_1 b_2 \cdots b_{q-1} b_q, \quad b_{q+1} = b_1.$$

The normal vector $\vec{m}_{1k}$ of face $l$ points the inside of block $B$.

$a_{1k}$ is any point on the face plane $k$ of $A$, $(a_{1k} a_i) = \vec{a}_i$.

$b_{1l}$ is any point on the face plane $l$ of $B$, $(b_{1l} b_j) = \vec{b}_j$.

$$\vec{m}_{1k} = \sum_{i=1}^{p} \vec{a}_{i+1} \times \vec{a}_i.$$
$$\vec{m}_{2l} = \sum_{j=1}^{q} \vec{b}_{j+1} \times \vec{b}_j.$$

Denote $e$ as a vertex of $A$, and all vertices connecting with $e$ by edges are

$$e_1 e_2 \cdots e_{u-1} e_u, \quad e_{u+1} = e_1, \quad (ee_r) = \vec{e}_r.$$

The inner normal vector of angle $\angle e_r e_{r+1}$ is $\vec{n}_{1r}$. Since $A$ is convex, $\angle e_r e_{r+1}$ is a convex angle,

$$\vec{n}_{1r} = +\vec{e}_{r+1} \times \vec{e}_r.$$

Denote $h$ as a vertex of $B$, and all vertices connected with $h$ by edges are:

$$h_1 h_2 \cdots h_{v-1} h_v, \quad h_{v+1} = h_1, \quad (hh_s) = \vec{h}_s.$$

The inner normal vector of angle $\angle h_s h_{s+1}$ is $\vec{n}_{2s}$, which points the inside of block $B$. $B$ is convex, and $\angle h_s h_{s+1}$ is a convex angle,

$$\vec{n}_{2s} = +\vec{h}_{s+1} \times \vec{h}_s.$$

As convex, $A$ and $B$ can be represented by simultaneous inequality equations.

$$A = \cap_{k=1,\dots,f} \{(x - a_{1k}) \cdot \vec{m}_{1k} \geq 0\}$$
$$B = \cap_{l=1,\dots,g} \{(x - b_{1l}) \cdot \vec{m}_{2l} \geq 0\}$$

### Theorem of vertex-polygon contact of 3D convex blocks

$$int(\angle \vec{e}_1 \vec{e}_2 \cdots \vec{e}_{u-1} \vec{e}_u \cap \uparrow m_{2l}) = \emptyset \Rightarrow$$
$$E(e, b_1 b_2 \cdots b_{q-1} b_q) \subset \partial E(A, B)$$

## Proof:

Based on Section 8.9.6,

$$int(\angle\vec{e}_1\vec{e}_2\cdots\vec{e}_{u-1}\vec{e}_u \cap \uparrow m_{2l}) = \emptyset \Rightarrow$$
$$E(\angle\vec{e}_1\vec{e}_2\cdots\vec{e}_{u-1}\vec{e}_u, \uparrow m_{2l}) = \uparrow m_{2l} + a_0.$$

As $A$ and $B$ are convex blocks,

$$\angle\vec{e}_1\vec{e}_2\cdots\vec{e}_{u-1}\vec{e}_u + e \supset A \Rightarrow$$
$$E(A, \uparrow m_{2l} + b_1) \subset \uparrow m_{2l} + b_1 - e + a_0.$$
$$\uparrow m_{2l} + b_1 \supset B \Rightarrow E(A, \uparrow m_{2l} + b_1) \supset E(A, B).$$

On the other side,

$$E(e, b_1 b_2 \cdots b_{q-1} b_q) = b_1 b_2 \cdots b_{q-1} b_q - e + a_0,$$
$$\partial E(A, \uparrow m_{2l} + b_1) \subset \perp m_{2l} + b_1 - e + a_0 \Rightarrow$$
$$E(e, b_1 b_2 \cdots b_{q-1} b_q) \subset \partial E(A, \uparrow m_{2l} + b_1),$$
$$E(e, b_1 b_2 \cdots b_{q-1} b_q) \subset E(A, B) \subset E(A, \uparrow m_{2l} + b_1)$$
$$\Rightarrow E(e, b_1 b_2 \cdots b_{q-1} b_q) \subset \partial E(A, B).$$

### Theorem of angle-vertex contact of 3D convex blocks

$$int(\angle\vec{h}_1\vec{h}_2\cdots\vec{h}_{v-1}\vec{h}_v \cap \uparrow m_{1k}) = \emptyset \Rightarrow$$
$$E(a_1 a_2 \cdots a_{p-1} a_p, \quad h) \subset \partial E(A, B)$$

## Proof:

Based on Section 8.9.6,

$$int(\angle\vec{h}_1\vec{h}_2\cdots\vec{h}_{v-1}\vec{h}_v \cap \uparrow m_{1k}) = \emptyset \Rightarrow$$
$$E(\uparrow m_{1k}, \angle\vec{h}_1\vec{h}_2\cdots\vec{h}_{v-1}\vec{h}_v) = -\uparrow m_{1k} + a_0.$$

As $A$ and $B$ are convex blocks,

$$\angle\vec{h}_1\vec{h}_2\cdots\vec{h}_{v-1}\vec{h}_v + h \supset B \Rightarrow$$
$$E(\uparrow m_{1k} + a_1, B) \subset -\uparrow m_{1k} + h - a_1 + a_0.$$
$$\uparrow m_{1k} + a_1 \supset A, \quad \Rightarrow E(\uparrow m_{1k} + a_1, B) \supset E(A, B).$$

On the other side,

$$E(a_1 a_2 \cdots a_{p-1} a_p, \quad h) =$$
$$-a_1 a_2 \cdots a_{p-1} a_p + h + a_0,$$
$$\partial E(\uparrow m_{1k} + a_1, B) \subset -\perp m_{1k} + h - a_1 + a_0 \Rightarrow$$
$$E(a_1 a_2 \cdots a_{p-1} a_p, \quad h) \subset \partial E(\uparrow m_{1k} + a_1, B),$$
$$E(a_1 a_2 \cdots a_{p-1} a_p, h) \subset E(A, B) \subset E(\uparrow m_{1k} + a_1, B)$$
$$\Rightarrow E(a_1 a_2 \cdots a_{p-1} a_p, \quad h) \subset \partial E(A, B).$$

As defined in Section 8.11.5, $ee_r$ is an edge of $A$; $e$ and $e_r$ are vertices of block $A$. $hh_s$ is an edge of $B$; $h$ and $h_s$ are vertices of block $B$. The inner normal vectors of $\angle \vec{e}_{r-1} \vec{e}_r$ and $\angle \vec{e}_r \vec{e}_{r+1}$ are $\vec{n}_{11}$ and $\vec{n}_{12}$ respectively. The inner normal vectors of $\angle \vec{h}_{s-1} \vec{h}_s$ and $\angle \vec{h}_s \vec{h}_{s+1}$ are $\vec{n}_{21}$ and $\vec{n}_{22}$ respectively. As blocks $A, B$ are convex, the edges of $\vec{e}_r$ and $\vec{h}_s$ are convex

$$\vec{e}_r = (ee_r) \upuparrows \vec{n}_{12} \times \vec{n}_{11},$$
$$\vec{h}_s = (hh_s) \upuparrows \vec{n}_{22} \times \vec{n}_{21}.$$

Denote

$$\vec{e}_{r1} = \vec{n}_{11} \times \vec{n}_{12} \times \vec{n}_{11},$$
$$\vec{e}_{r2} = \vec{n}_{12} \times \vec{n}_{11} \times \vec{n}_{12},$$
$$\vec{h}_{s1} = \vec{n}_{21} \times \vec{n}_{22} \times \vec{n}_{21},$$
$$\vec{h}_{s2} = \vec{n}_{22} \times \vec{n}_{21} \times \vec{n}_{22},$$

**Theorem of edge-edge contact of 3D convex blocks**

$$int \left( \angle \vec{e}_{r1} \vec{e}_{r2} + \vec{\vec{e}}_r \cap \angle \vec{h}_{s1} \vec{h}_{s2} + \vec{\vec{h}}_s \right) = \emptyset \Rightarrow$$

$$E(ee_r, hh_s) \subset \partial E(A, B).$$

**Proof:**

From Section 8.9.7,

$$int(\angle \vec{e}_{r1} \vec{e}_{r2} + \vec{\vec{e}}_r \cap \angle \vec{h}_{s1} \vec{h}_{s2} + \vec{\vec{h}}_s) = \emptyset \Rightarrow$$
$$\exists \vec{n}_0, \quad \vec{n}_0 \parallel \vec{e}_r \times \vec{h}_s,$$
$$E(\angle \vec{e}_{r1} \vec{e}_{r2} + \vec{\vec{e}}_r, \angle \vec{h}_{s1} \vec{h}_{s2} + \vec{\vec{h}}_s) = \uparrow n_0 + a_0.$$

As $A$ and $B$ are convex blocks,

$$\angle \vec{e}_{r1}\vec{e}_{r2} + \vec{e}_r + e \supset A, \quad \angle \vec{h}_{s1}\vec{h}_{s2} + \vec{h}_s + h \supset B \Rightarrow$$
$$E(\angle \vec{e}_{r1}\vec{e}_{r2} + \vec{e}_r + e, \angle \vec{h}_{s1}\vec{h}_{s2} + \vec{h}_s + h) \supset E(A, B).$$
$$\Rightarrow \uparrow n_0 + h - e + a_0 \supset E(A, B).$$

On the other side,

$$E(ee_r, hh_s) = hh_s - ee_r + a_0,$$
$$\subset \perp n_0 + h - e + a_0 \Rightarrow$$
$$E(ee_r, hh_s)$$

$$\subset \partial E(\angle \vec{e}_{r1}\vec{e}_{r2} + \vec{e}_r + e, \angle \vec{h}_{s1}\vec{h}_{s2} + \vec{h}_s + h),$$
$$E(ee_r, hh_s) \subset E(A, B)$$

$$\subset E(\angle \vec{e}_{r1}\vec{e}_{r2} + \vec{e}_r + e, \angle \vec{h}_{s1}\vec{h}_{s2} + \vec{h}_s + h)$$
$$\Rightarrow E(ee_r, hh_s) \subset \partial E(A, B).$$

### Theorem of contact edges of 3D convex blocks

Assuming $A$ and $B$ are convex blocks,

$$\partial E(A, B) = C(0, 2) \cup C(2, 0) \cup C(1, 1).$$

### Proof:

From Section 8.11.8,

$$\partial E(A, B) \subset C(0, 2) \cup C(2, 0) \cup C(1, 1).$$

From previous theorems of this section,

$$\partial E(A, B) \supset C(0, 2) \cup C(2, 0) \cup C(1, 1).$$

Therefore,

$$\partial E(A, B) = C(0, 2) \cup C(2, 0) \cup C(1, 1).$$

## 8.12.2  Contact polygons of two general 3D blocks

This section is about the computation of entrance blocks of 3D general blocks. The 3D blocks $A$ and $B$ are the same as defined in Section 8.2.5. The faces of block $A$ are polygons which rotate outward. The vertices of face $k$ are

$$a_1 a_2 \cdots a_{p-1} a_p, \quad a_{p+1} = a_1.$$

*Figure 8.115*  Failure of an arch structure computed by 3D DDA.

Normal vector $\vec{m}_{1k}$ of polygon $k$ points the inside of block $A$. The faces of block $B$ are polygons which rotate outward. The vertices of face $l$ are

$$b_1 b_2 \cdots b_{q-1} b_q, \quad b_{q+1} = b_1.$$

The normal vector $\vec{m}_{2l}$ of polygon $l$ points the inside of block $B$. $a_{1k}$ is any point on the polygon plane $k$ of $A$,

$$(a_{1k} a_i) = \vec{a}_i$$

$b_{1l}$ is any point on the polygon plane $l$ of $B$,

$$(b_{1l} b_j) = \vec{b}_j$$

$$\vec{m}_{1k} = \sum_{i=1}^{p} \vec{a}_{i+1} \times \vec{a}_i.$$

$$\vec{m}_{2l} = \sum_{j=1}^{q} \vec{b}_{j+1} \times \vec{b}_j.$$

Based on Section 8.11.8, all contact polygons are searched and examined. The following theorem is the conclusion of Sections 8.11.4 and 8.11.5.

### Theorem of contact polygons of 3D blocks

$$\partial E(A, B) \subset C(0, 2) \cup C(2, 0) \cup C(1, 1) \subset E(A, B).$$

The contact polygons of 3D entrance block $E(A, B)$ are connected and intersected with each other to form $E(A, B)$. This process, which is called cutting, is a standard computing geometric process to form 3D entrance blocks from contact polygons. In this process, all contact polygons are input polygons.

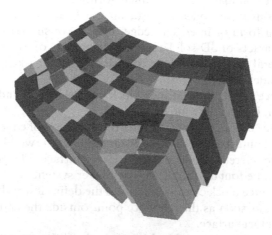

Figure 8.116 3D DDA computation of subsidence of coalmine excavation.

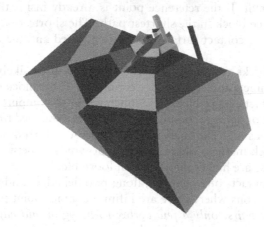

Figure 8.117 3D DDA computation of impact.

## 8.12.3 Simple examples of the new contact theory in 3D DDA

The current version of 3D DDA is mostly for dealing with convex blocks. 3D DDA has to be improved based on this newly developed theory of contact, even in the special cases of convex blocks. Figures 8.115–8.117 are the simple trial examples computed by 3D DDA.

## 8.13 CONCLUSIONS

In this study, a general contact theory for 2D and 3D discontinuous computation was proposed along with a new definition for operating the point sets named as the entrance block. The entrance block is defined and represented by its boundary.

Further, the solutions of the contact boundary are analytically derived. The contact boundary is proved to be a cover system, with each cover corresponding to a possible contact. The contact boundaries are of similar formulas in all conditions (2D or 3D, angles or blocks, convex or concave), e.g., all the local contacts which are either in vertex-to-polygon form or in edge-to-edge form, have similar or almost identical formulas for the contacts of 3D solid angles, the contacts of 3D convex blocks, and the contacts of general 3D blocks. These consistencies will make the computation of an entrance block much easier.

Given a reference point, the entrance block can be solved and thus simplify the contact computation in the following ways:

First, the entrance block concept simplifies the processes of contact searching prior to the occurrence of contacts. The shortest distance between two blocks is the shortest distance between the reference point and the contact surface. The point pairs with the shortest distance can be found from the contact cover system.

Second, the entrance block concept simplifies the definition and searching routines of the first entrance, as soon as the reference point outside the entrance block moves onto one certain contact surface.

Last but not least, the entrance block concept simplifies the searching routines of the shortest path of exit. If the reference point is already inside the entrance block, it will exit the entrance block in the shortest path. The shortest path connected with the closest point on the contact surface can be computed and the contact points are determined.

When two real blocks approach each other, there are most likely two possibilities: vertex to polygon contacts and edge to edge contacts in loose block system. Due to the small number of contacts, loose blocks are much easier to compute.

In destabilization or damage computation, solids are changed from the continuous state to discontinuous state. The analyses were based on starting with newly divided and unperturbed block meshes, where there are no voids in between. The quantities of contacts in this case are higher than those in loose blocks.

In such a case, contacts often happen along parallel edges and polygons or along a pair of parallel polygons where there are infinite contact point pairs. *Among these infinite contact point pairs, only a few vertex-to-polygon and edge-to-edge contact point pairs control the movements of blocks, especially rotations.*

With this new theory, it is now possible to compute contacts with infinite contact point pairs, such as contacts in densely compacted blocks. In this theory, these vertex-to-polygon contact point pairs and edge-to-edge contact point pairs are called *contact points*, which correspond with overlapping contact polygons. As a result, after small rotation or deformation, blocks contact only along these few contact points.

Based on this theory, all closed contact points control the movements, rotations and deformations of blocks, as in real cases.

Given a block system, the contact points should be the same as in any computation method, as well as the means of finding contact points.

As the discontinuous computation follows time steps, if one contact point in one time step was not correct, the whole computation would not be useful and large penetration would happen. However the method of "shortest exit", i.e., using the shortest path for the reference point to exit the entrance block, will recover the block system from a penetrated state "to a" contact state. This will make the discontinuous computation more reliable.

The proposed contact theory is applicable to all computational methods, in areas of civil, structural, geological, mechanical, and robotic engineering.

# References

Ahn, T.Y. & Song, J.J. (2011) New contact-definition algorithm using inscribed spheres for 3D discontinuous deformation analysis. *International Journal of Computational Methods*, 8 (2), 171–191.

Akao, S., Ohnish, Y., Nishiyama, S. & Nishimura, T. (2007) Comprehending DDA for block behaviour under dynamic condition. In: Ju, Y., Fang, X. & Bian, H. (eds.) *The Eighth International Symposium on Analysis of Discontinuous Deformation, Beijing, China.* pp. 135–140.

Amadei, B., Lin, C. & Dewyer, J. (1996) Recent extensions to DDA. In: Salami, M.R. & Banks, D. (eds.) *First International Forum on Discontinuous Deformation Analysis and Simulations of Discontinuous Media.* Berkeley, CA, TSI Press. pp. 1–30.

Ashby, J.P. (1971) *Sliding and Toppling Modes of Failure in Models and Jointed Rock Slopes.* M Sc Thesis. London, Inst. of Geol. Sciences., University of London, Imperial College.

Ashford, S.A., Sitar, N., Lysmer, J. & Deng, N. (1997) Topographic effects on the seismic response of steep slopes. *Bulletin of the Seismological Society of America*, 87, 701–709.

Bahat, D. (1991) *Tectonofractography.* Berlin, Springer-Verlag. 453 pp.

Bai, T. & Gross, M.R. (1999) Theoretical analysis of cross-joint geometries and their classification. *Journal of Geophysical Research-Solid Earth*, 104, 1163–1177.

Bakun-Mazor, D. (2011) *Modeling Dynamic Rock Mass Deformation with the Numerical DDA Method.* PhD Thesis. Beer Sheva, Dept. of Geological and Environmental Sciences, Ben-Gurion University of the Negev. 102 pp.

Bakun-Mazor, D., Hatzor, Y.H. & Dershowitz, W.S. (2009a) Modeling mechanical layering effects on stability of underground openings in jointed sedimentary rocks. *International Journal of Rock Mechanics and Mining Sciences*, 46, 262–271.

Bakun-Mazor, D., Hatzor, Y.H. & Glaser, S.D. (2009b) 3D DDA vs. analytical solutions for dynamic sliding of a tetrahedral wedge. In: *ICADD9 Nanyang Technological University, Singapore, 25–27 November 2009.*

Bakun-Mazor, D., Hatzor, Y.H. & Glaser, S.D. (2012) Dynamic sliding of tetrahedral wedge: The role of interface friction. *International Journal for Numerical and Analytical Methods in Geomechanics*, 36, 327–343.

Bakun-Mazor, D., Hatzor, Y.H., Glaser, S.D. & Santamarina, J.C. (2013) Thermally vs. seismically induced block displacements in Masada rock slopes. *International Journal of Rock Mechanics and Mining Sciences*, 61, 196–211.

Balden, V., Scheele, F. & Nurick, G. (2001) Discontinuous deformation analysis in ball milling. In: Bicanic, N. (ed.) *The Fourth International Conference on Analysis of Discontinuous Deformation, Glasgow, Scotland, UK.* pp. 337–348.

Bao, H.R. & Zhao, Z. (2010a) An alternative scheme for the corner-corner contact in the two-dimensional discontinuous deformation analysis. *Advances in Engineering Software*, 41, 206–212.

Bao, H.R. & Zhao, Z.Y. (2010b) Modelling crack propagation with nodal-based discontinuous deformation analysis. In: Ma, G. & Zhou, Y. (eds.) *Analysis of Discontinuous Deformation: New Developments and Applications*. Singapore, Nanyang Technological University. pp. 161–167.

Bao, H.R. & Zhao, Z.Y. (2013) Modelling brittle fracture with the nodal-based discontinuous deformation analysis. *International Journal of Computational Methods*, 10 (6), 1350040 (26 pp.).

Bao, H.R., Hatzor, Y.H. & Huang, X. (2012) A new viscous boundary condition in the two-dimensional discontinuous deformation analysis method for wave propagation problems. *Rock Mechanics and Rock Engineering*, 45, 919–928.

Bao, H.R., Yagoda-Biran, G. & Hatzor, Y.H. (2014a) Site response analysis with two-dimensional numerical discontinuous deformation analysis. *Earthquake Engineering and Structural Dynamics*, 43, 225–246.

Bao, H.R., Zhao, Z. & Tian, Q. (2014b) On the implementation of augmented Lagrangian method in the tow-dimensional discontinuous deformation analysis. *International Journal for Numerical and Analytical Methods in Geomechanics*, 38, 551–571.

Bard, P.Y. & Tucker, B.E. (1985) Underground and ridge and site effects: Comparison of observation and theory. *Bulletin of the Seismological Society of America*, 75, 905–922.

Barla, G. & Paronuzzi, P. (2013) The 1963 Vajont landslide: 50th anniversary. *Rock Mechanics and Rock Engineering*, 46, 1267–1270.

Barla, G., Monacis, G., Perino, A. & Hatzor, Y.H. (2010) Distinct element modelling in static and dynamic conditions with application to an underground archaeological site. *Rock Mechanics and Rock Engineering*, 43, 877–890.

Barton, N. (1976) The shear strength of rock and rock joints. *International Journal of Rock Mechanics and Mining Sciences*, 13, 255–279.

Barton, N., Lien, R. & Lunde, J. (1974) Engineering classification of jointed rock masses for the design of tunnel support. *Rock Mechanics*, 6, 189–236.

Beer, M. & Meek, J.L. (1982) Design curves for roofs and hanging-walls in bedded rock based on Voussoir beam and plate solutions. *Transactions of the Institution of Mining and Metallurgy*, 91, 18–22.

Belytschko, T. & Neal, M.O. (1991) Contact-impact by the pinball algorithm with penalty and Lagrangian methods. *International Journal for Numerical Methods in Engineering*, 31, 547–572.

Ben, Y.X., Xue, J., Miao, Q.H. & Wang, Y. (2012) Coupling fluid flow with discontinuous deformation analysis. In: Zhao, J., Ohnishi, Y., Zhao, G.F. & Sasaki, T. (eds.) *Advances in Discontinuous Numerical Methods and Applications in Geomechanics and Geoengineering*. pp. 107–112.

Ben, Y.X., Wang, Y. & Shi, G.H. (2013) Development of a model for simulating hydraulic fracturing with DDA. In: *Proceedings of the 11th International Conference on Analysis of Discontinuous Deformation (ICADD'13), 27–29 August*. Fukuoka, Taylor & Francis Group. pp. 169–175.

Benson, D.J. & Hallquist, J.O. (1990) A single surface contact algorithm for the post-buckling analysis of shell structures. *Computers Methods in Applied Mechanics and Engineering*, 78, 141–163.

Beyabanaki, S.A.R. & Jafari, A. (2005) Modified point-to-face frictionless contact constraints for three-dimensional discontinuous deformation analysis (3-D DDA). In: *Proceedings of the Seventh International Conference on Analysis of Discontinuous Deformation (ICADD-7), Hawaii*. pp. 24–36.

Beyabanaki, S.A.R., Mikola, R.G. & Hatami, K. (2008) Three-dimensional discontinuous deformation analysis (3-D DDA) using a new contact resolution algorithm. *Computers and Geotechnics*, 35 (3), 346–356.

Beyabanaki, S.A.R., Jafari, A. & Yeung, M.R. (2009a) Second-order displacement functions for three-dimensional discontinuous deformation analysis (3-D DDA). *International Journal of Science and Technology*, 16 (3), 216–225.

Beyabanaki, S.A.R., Jafari, A., Biabanaki, S.O.R. & Yeung, M.R. (2009b) A coupling model of 3-D discontinuous deformation analysis (3-D DDA) and finite element method. *Arabian Journal for Science and Engineering*, 34 (1B), 108.

Beyabanaki, S.A.R., Jafari, A., Biabanaki, S.O.R. & Yeung, M.R. (2009c) Nodal-based three-dimensional discontinuous deformation analysis (3-D DDA). *Computers and Geotechnics*, 36 (3), 359–372.

Beyabanaki, S.A.R., Jafari, A. & Yeung, M.R. (2010) High-order three-dimensional discontinuous deformation analysis (3-D DDA). *International Journal for Numerical Methods in Biomedical Engineering*, 26 (12), 1522–1547.

Bicanic, N. & Stirling, C. (2001) DDA analysis of the Couplet/Heyman minimum thickness arch problem. In: Bicanic, N. (ed.) *The Fourth International Conference on Analysis of Discontinuous Deformation, Glasgow, Scotland, UK*. pp. 165–170.

Bicanic, N., Stirling, C. & Pearce, C.J. (2003) Discontinuous modelling of masonry bridges. *Computational Mechanics*, 31, 60–68.

Bieniawski, Z.T. (1970) Time-dependent behaviour of fractured rock. *Rock Mechanics*, 2, 123–137.

Bieniawski, Z.T. (1974) Geomechanics classification of the rock masses and its application in tunneling. In: *Proceedings of the Third Congress ISRM, Denver*. pp. 7–32.

Bieniawski, Z.T. (1976) Rock mass classification in rock engineering. In: Bieniawski, Z.T. (ed.) *Exploration for Rock Engineering, Proc. of the Symp.* Vol. 1. Cape Town, Balkema. pp. 97–106.

Bin Shi, G., Bai, J., Wang, M., Sun, B., Wang, Y. & Shi, G. (2010) Study on failure characteristics and support measure of layer structure-cataclasm rock mass. In: Ma, G. & Zhou, Y. (eds.) *Analysis of Discontinuous Deformation: New Developments and Applications*. Singapore, Nanyang Technological University. pp. 135–143.

Bonnet, J. & Peraire, J. (1991) An alternating digital tree (ADT) algorithm for 3D geometric searching and intersection problem. *International Journal for Numerical Methods in Engineering*, 31, 1–17.

Boore, D.M. & Joyner, W.B. (1997) Site amplifications for generic rock sites. *Bulletin of the Seismological Society of America*, 87, 327–341.

Bouchon, M. & Barker, J.S. (1996) Seismic response of a hill: The example of Tarzana, California. *Bulletin of the Seismological Society of America*, 86, 66–72.

Brady, B.H.G. & Brown, E.T. (2004) *Rock Mechanics for Underground Mining*. 3rd edition. Dordrecht, Kluwer Academic Publisher.

Bray, J.W. & Goodman, R.E. (1981) The theory of base friction models. *International Journal of Rock Mechanics and Mining Sciences*, 18, 453–468.

Brekke, T.L. Personal Communication. U. C. Berkeley graduate level course entitled: Geological engineering of underground openings.

Cai, M. & Kaiser, P.K. (2005) Assessment of excavation damaged zone using a micromechanics model. *Tunnelling and Underground Space Technology*, 20, 301–310.

Cao, C.Y., Zhong, Y. & Shi, G.-H. (2007) Application of DDA-FEM coupled method in pavement analysis. In: Ju, Y., Fang, X. & Bian, H. (eds.) *The Eighth International Symposium on Analysis of Discontinuous Deformation, Beijing, China*. pp. 111–116.

Celebi, M. (1987) Topographical and geological amplifications determined from strong-motion and aftershock records of the 3 March 1985 Chile earthquake. *Bulletin of the Seismological Society of America*, 77, 1147–1167.

Chavez-Garcia, F.J., Sanchez, L.R. & Hatzfeld, D. (1996) Topographic site effects and HVSR – A comparison between observations and theory. *Bulletin of the Seismological Society of America*, 86, 1559–1573.

Chen, G. & Ohnish, Y. (1999) A non-linear model for discontinuities in DDA. In: Amadei, B. (ed.) *The Third International Conference on Analysis of Discontinuous Deformation, Vail, Colorado.* pp. 57–64.

Chen, G., Zheng, L. & Zhang, Y. (2013) Practical applications of DDA to disaster prevention. In: Chen, G., Ohnishi, Y., Zheng, L. & Sasaki, T. (eds.) *The 11th International Symposium on Analysis of Discontinuous Deformation.* Fukuoka, Taylor & Francis Group. pp. 15–28.

Chen, W. & Deng, J. (2008) Application of key block theory and DDA to the stability analysis of underground powerhouse of Jinping hydropower station I. In: Ju, Y., Fang, X. & Bian, H. (eds.) *The Eighth International Symposium on Analysis of Discontinuous Deformation, Beijing, China.* pp. 243–247.

Cheng, Y.M. & Zhang, Y.H. (2000) Rigid body rotation and block internal discretization in DDA analysis. *International Journal for Numerical and Analytical Methods in Geomechanics*, 24, 567–578.

Chern, J.C., Koo, C.Y. & Chen, S. (1995) Development of second order displacement function for DDA and manifold method. In: *Working Forum on the Manifold Method of Material Analysis, Vicksburg.* pp. 183–202.

Choo, L.Q., Zhao, Z., Chen, H. & Tian, Q. (2016) Hydraulic fracturing modeling using the discontinuous deformation analysis (DDA) method. *Computers and Geotechnics*, 76, 12–22.

Clapp, K.K. & MacLaughlin, M. (2003) Preliminary development of a fully grouted rock bolt element for discontinuous deformation analysis. In: Lu, M. (ed.) *The Sixth International Conference on Analysis of Discontinuous Deformation.* Trondheim, Balkema Publishers. pp. 111–118.

Clatworthy, D. & Scheele, F. (1999) A method for sub-meshing in discontinuous deformation analysis (DDA). In: Amadei, B. (ed.) *The Third International Conference on Analysis of Discontinuous Deformation, Vail, Colorado.* pp. 85–94.

Cook, N.G.W. (1966) *Rock Mechanics Applied to the Study of Rockbursts.* Johannesburg, South African Institute of Mining and Metallurgy.

Cook, N.G.W. (1976) Seismicity associated with mining. *Engineering Geology*, 10, 99–122.

Crawford, A.M. & Curran, J.H. (1981) The influence of shear velocity on the frictional resistance of rock discontinuities. *International Journal of Rock Mechanics and Mining Sciences*, 18, 505–515.

Crawford, A.M. & Curran, J.H. (1982) The influence of rate-dependent and displacement-dependent shear resistance on the response of rock slopes to seismic loads. *International Journal of Rock Mechanics and Mining Sciences*, 19, 1–8.

Cundall, P.A. (1971) A computer model for simulating progressive, large scale movements in blocky rock system. In: *Symposium of International Society of Rock Mechanics, Nancy, France.* pp. 11–18.

Cundall, P.A. (1988) Formulation of a three-dimensional distinct element model-Part I: A scheme to detect and represent contacts in a system composed of many polyhedral blocks. *International Journal of Rock Mechanics and Mining Sciences & Geomechanics*, 25, 107–116.

Davis, L.L. & West, L.R. (1973) Observed effects of topography on ground motion. *Bulletin of the Seismological Society of America*, 63, 283–298.

De Luca, A., Giordano, A. & Mele, E. (2004) A simplified procedure for assessing the seismic capacity of masonry arches. *Engineering Structures*, 26, 1915–1929.

Dershowitz, B., LaPointe, P., Eiben, T. & Wei, L.L. (2000) Integration of discrete feature network methods with conventional simulator approaches. *SPE Reservoir Evaluation & Engineering*, 3, 165–170.

Dershowitz, W.S. & Einstein, H.H. (1988) Characterizing rock joint geometry with joint system models. *Rock Mechanics and Rock Engineering*, 21, 21–51.

Desai, C.S. & Abel, J.F. (1972) *Introduction to the Finite Element Method*. New York, NY, Van Nostrand-Reinhold.

Diederichs, M.S. & Kaiser, P.K. (1999a) Authors' reply to discussion by A.L. Sofianos regarding Diederichs M.S. and Raiser P.K. Stability of large excavations in laminated hard rock masses: The voussoir analogue revisited. *International Journal of Rock Mechanics and Mining Sciences*, 36, 97–117, 995–997.

Diederichs, M.S. & Kaiser, P.K. (1999b) Stability of large excavations in laminated hard rock masses: The voussoir analogue revisited. *International Journal of Rock Mechanics and Mining Sciences*, 36, 97–117.

Diederichs, M.S. & Kaiser, P.K. (1999c) Tensile strength and abutment relaxation as failure control mechanisms in underground excavations. *International Journal of Rock Mechanics and Mining Sciences*, 36, 69–96.

Dieterich, J.H. (1972) Time-dependent friction in rocks. *Journal of Geophysical Research*, 77, 3690–3697.

Dieterich, J.H. (1978) Time-dependent friction and the mechanism of stick-slip. *Pure and Applied Geophysics*, 116, 790–806.

Dieterich, J.H. (1979) Modeling of rock friction, 1: Experimental results and constitutive equations. *Journal of Geophysical Research*, 84, 2161–2168.

Dieterich, J.H. (1981) Constitutive properties of faults with simulated gouge. In: Carter, N.L., Friedman, M., Logan, J.M. & Stearns, D.W. (eds.) *Mechanical Behavior of Crustal Rocks. The Handin Volume*. Washington, DC, American Geophysical Union. pp. 103–120.

Dieterich, J.H. & Kilgore, B.D. (1994) Direct observation of frictional contacts: New insights for state-dependent properties. *Pure and Applied Geophysics*, 143, 283–302.

Dieterich, J.H. & Kilgore, B.D. (1996) Imaging surface contacts: Power law contact distributions and contact stresses in quartz, calcite, glass and acrylic plastic. *Tectonophysics*, 256, 219–241.

Dong, P.H, & Osada, M. (2007) Effects of dynamic friction on sliding behaviour of block in DDA. In: Ju, Y., Fang, X. & Bian, H. (eds.) *The Eighth International Symposium on Analysis of Discontinuous Deformation, Beijing, China*. pp. 129–134.

Dong, X., Wu, A. & Ren, F. (1996) A preliminary application of DDA to the three gorges project on Yangtze river, China. In: Salami, M.R. & Banks, D. (eds.) *First International Forum on Discontinuous Deformation Analysis and Simulations of Discontinuous Media*. Berkeley, CA, TSI Press. pp. 310–317.

Doolin, D.M. & Sitar, N. (2002) Displacement accuracy of discontinuous deformation analysis method applied to sliding block. *Journal of Engineering mechanics (ASCE)*, 128 (11), 1158–1168.

Doolin, D.M. & Sitar, N. (2004) Time integration in discontinuous deformation analysis. *Journal of Engineering mechanics (ASCE)*, 130 (3), 249–258.

Dunlop, P., Dunne, J.M. & Seed, H.B. (1970) Finite element analysis of slopes in soils. *Journal of the Soil Mechanics and Foundations Division, ASCE*, 96, 471–493.

Einstein, H.H., Veneziano, D., Baecher, G.B. & Oreilly, K.J. (1983) The effect of discontinuity persistence on rock slope stability. *International Journal of Rock Mechanics and Mining Sciences*, 20, 227–236.

Fairhurst, C. & Cook, N.G.W. (1966) The phenomenon of rock splitting parallel to the direction of maximum compression in the neighbourhood. In: *1st Congress ISRM, Lisbon*. pp. 296–689.

Ferrero, A.M., Forlani, G., Roncella, R. & Voyat, H.I. (2009) Advanced geostructural survey methods applied to rock mass characterization. *Rock Mechanics and Rock Engineering*, 42, 631–665.

Fisher, F.R.S. (1953) Dispersion on a sphere. *Philosophical Transactions of the Royal Society of London*, 217, 295–305.

FLAC (1998) Fast Lagrangian Analysis of Continua (FLAC). In: Group, I.C. (ed.) Minneapolis, MN.

Fredrich, J.T., Evans, B. & Wong, T.F. (1990) Effect of grain-size on brittle and semibrittle strength – Implications for micromechanical modeling of failure in compression. *Journal of Geophysical Research-Solid Earth and Planets*, 95, 10907–10920.

Fu, G.Y. (2014) *Realistic Rock Mass Modelling and Its Engineering Applications*. PhD Dissertation Thesis. Crawley, WA, School of Civil, Environmental and Mining Engineering, The University of Western Australia.

Fu, G.Y., Ma, G.W. & Qu, X.L. (2015a) Dual reciprocity boundary element based block model for discontinuous deformation analysis. *Science China-Technological Sciences*, 58 (9), 1575–1586.

Fu, X., Sheng, Q., Zhang, Y., Zhou, Y. & Dai, F. (2015b) Boundary setting method for the seismic dynamic response analysis of engineering rock mass structures using the discontinuous deformation analysis method. *International Journal for Numerical and Analytical Methods in Geomechanics*. Available from: 10.1002/nag.2374.

Fu, G.Y., Ma, G.W. & Qu, X.L. (2016) Boundary element based discontinuous deformation analysis. *International Journal for Numerical and Analytical Methods in Geomechanics*. Available from: 10.1002/nag.2661.

Gilbert, E.G., Johnson, D.W. & Keerthi, S.S. (1988) A fast procedure for computing the distance between complex objects in three-dimensional space. *IEEE Journal on Robotics and Automation*, 4, 193–203.

Goodman, R.E. (1976) *Methods of Geological Engineering in Discontinuous Rocks*. San Francisco, CA, West Publishing Company.

Goodman, R.E. (1989) *Introduction to Rock Mechanics*. 2nd edition. New York, NY, John Wiley & Sons.

Goodman, R.E. (2013) Toppling – A fundamental failure mode in discontinuous materials – Description and analysis. In: *Geotechnical Special Publication No. 231 "Geo-Congress 2013: "Stability and Performance of Slopes and Embankments III"* 231.

Goodman, R.E. & Bray, J.W. (1976) Toppling of rock slopes. In: *Specialty Conference on Rock Engineering for Foundations and Slopes*. Boulder, CO, American Society of Civil Engineering. pp. 201–234.

Goodman, R.E. & Brown, C.B. (1963) Dead load stresses and the instability of slopes. *Journal of the Soil Mechanics and Foundations Division, ASCE*, 89, 103–104.

Goodman, R.E. & Kieffer, D.S. (2000) Behavior of rock in slopes. *Journal of Geotechnical and Geoenvironmental Engineering*, 126, 675–684.

Goodman, R.E. & Seed, H.B. (1966) Earthquake induced displacements in sands and embankments. *Journal of the Soil Mechanics and Foundations Division, ASCE*, 92 (SM2), 125–146.

Goodman, R.E. & Shi, G.H. (1985) *Block Theory and Its Application to Rock Engineering*. Englewood Cliffs, NJ, Prentice-Hall, Inc.

Goodman, R.E., Taylor, R.L. & Brekke, T.L. (1968) A model for the mechanics of jointed rock. *Journal of the Soil Mechanics and Foundations Division, ASCE*, 94 (3), 637–659.

Grayeli, R. & Mortazavi, A. (2005) Implementation of a quadratic triangular mesh into the DDA method. In: MacLaughlin, M. & Sitar, N. (eds.) *The Seventh International Symposium on Analysis of Discontinuous Deformation, Honolulu, Hawaii*. pp. 91–101.

Grayeli, R. & Mortazavi, A., (2007) Elasto-plastic discontinuous deformation analysis using Mohr-Coulomb model. In: Ju, Y., Fang, X. & Bian, H. (eds.) *The Eighth International Symposium on Analysis of Discontinuous Deformation, Beijing, China*. pp. 195–200.

Gross, M.R. (1993) The origin and spacing of cross joints – Examples from the Monterey Formation, Santa-Barbara Coastline, California. *Journal of Structural Geology*, 15, 737–751.

Guo, P. & Lin, S. (2007) Numerical simulation of mechanical characteristics of coarse granular materials by discontinuous deformation analysis. In: Ju, Y., Fang, X. & Bian, H. (eds.) *The Eighth International Symposium on Analysis of Discontinuous Deformation, Beijing, China.* pp. 57–60.

Hatzor, Y.H. (1993) The block failure likelihood – A contribution to rock engineering in blocky rock masses. *International Journal of Rock Mechanics and Mining Sciences*, 30, 1591–1597.

Hatzor, Y.H. (2003) Keyblock stability in seismically active rock slopes – Snake Path Cliff, Masada. *Journal of Geotechnical and Geoenvironmental Engineering*, 129, 1069.

Hatzor, Y.H. & Benary, R. (1998) The stability of a laminated Voussoir beam: Back analysis of a historic roof collapse using DDA. *International Journal of Rock Mechanics and Mining Sciences*, 35, 165–181.

Hatzor, Y.H. & Feintuch, A. (2001) The validity of dynamic block displacement prediction using DDA. *International Journal of Rock Mechanics and Mining Sciences*, 38, 599–606.

Hatzor, Y.H. & Goodman, R.E. (1997) Three-dimensional back-analysis of saturated rock slopes in discontinuous rock – A case study. *Geotechnique*, 47, 817–839.

Hatzor, Y.H. & Palchik, V. (1997) The influence of grain size and porosity on crack initiation stress and critical flaw length in dolomites. *International Journal of Rock Mechanics and Mining Sciences*, 34, 805–816.

Hatzor, Y.H., Talesnick, M. & Tsesarsky, M. (2002) Continuous and discontinuous stability analysis of the bell-shaped caverns at Bet Guvrin, Israel. *International Journal of Rock Mechanics and Mining Sciences*, 39, 867–886.

Hatzor, Y.H., Arzi, A.A., Zaslavsky, Y. & Shapira, A. (2004) Dynamic stability analysis of jointed rock slopes using the DDA method: King Herod's Palace, Masada, Israel. *International Journal of Rock Mechanics and Mining Sciences*, 41, 813–832.

Hatzor, Y.H., Wainshtein, I. & Mazor, D.B. (2010) Stability of shallow karstic caverns in blocky rock masses. *International Journal of Rock Mechanics and Mining Sciences*, 47 (8), 1289–1303.

Hatzor, Y.H., Feng, X.T., Li, S., Yagoda-Biran, G., Jiang, Q. & Hu, L. (2015) Tunnel reinforcement in columnar jointed basalt: The role of rock mass anisotropy. *Tunneling and Underground Space Technology*, 46, 1–11.

Hayashi, M., Yamada, S., Araya, M., Koyama, T., Fukuda, M. & Iwasaki, Y. (2012) Study for reinforcement planning of masonry structure with cracks at Bayon main tower, Angkor. In: Zhao, J., Ohnishi, Y., Zhao, G.F. & Sasaki, T. (eds.) *Advances in Discontinuous Numerical Methods and Applications in Geomechanics and Geoengineering.* pp. 247–252.

He, B.G., Zelig, R., Hatzor, Y.H. & Feng, X.T. (2016) Rockburst generation in discontinuous rock masses. *Rock Mechanics and Rock Engineering*, 49, 4103–4124.

He, L. (2010) *Three Dimensional Numerical Manifold Method and Rock Engineering Applications.* PhD Thesis. Singapore, Nanyang Technological University.

He, M.C., Nie, W., Zhao, Z.Y. & Guo, W. (2012) Experimental investigation of bedding plane orientation on the rockburst behavior of sandstone. *Rock Mechanics and Rock Engineering*, 45, 311–326.

He, M.C., Sousa, L.R.E., Miranda, T. & Zhu, G.L. (2015) Rockburst laboratory tests database – Application of data mining techniques. *Engineering Geology*, 185, 116–130.

Hoek, E. (2007) *Practical Rock Engineering.* Rocscience.

Hoek, E. & Bray, J.W. (1981) *Rock Slope Engineering.* 3rd edition. London, Institution of Mining and Metallurgy.

Hoek, E. & Brown, E.T. (1997) Practical estimates of rock mass strength. *International Journal of Rock Mechanics and Mining Sciences*, 34, 1165–1186.

Hoek, E. & Diederichs, M.S. (2006) Empirical estimation of rock mass modulus. *International Journal of Rock Mechanics and Mining Sciences*, 43, 203–215.

Hoek, E., Kaiser, P.K. & Bawden, W.F. (1995) *Support of Underground Excavations in Hard Rock*. Rotterdam, A.A. Balkema. p. 215.

Huang, T., Zhang, G.X. & Peng, X.C. (2010) The analysis of structure deformation using DDA with third order displacement function. In: Ma, G. & Zhou, Y. (eds.) *Analysis of Discontinuous Deformation: New Developments and Applications*. Singapore, Nanyang Technological University. pp. 177–183.

Hudson, J.A. & Priest, S.D. (1979) Discontinuities and rock mass geometry. *International Journal of Rock Mechanics and Mining Sciences*, 16, 339–362.

Hudson, J.A. & Priest, S.D. (1983) Discontinuity frequency in rock masses. *International Journal of Rock Mechanics and Mining Sciences*, 20, 73–89.

Hudson, J.A., Backstrom, A., Rutqvist, J., Jing, L., Backers, T., Chijimatsu, M., Christiansson, R., Feng, X.T., Kobayashi, A., Koyama, T., Lee, H.S., Neretnieks, I., Pan, P.Z., Rinne, M. & Shen, B.T. (2009) Characterising and modelling the excavation damaged zone in crystalline rock in the context of radioactive waste disposal. *Environmental Geology*, 57, 1275–1297.

Hughes, T.J.R. (1983) Analysis of transient algorithms with particular reference to stability behavior. In: Belytschko, T. & Hughes, T.J.R. (eds.) *Computational Methods for Transient Analysis*. Amsterdam, Elsevier Science.

ICADD1 (1995) In: Li, J.C., Wang, C.-Y. & Sheng, J. (eds.) *The First International Conference on Analysis of Discontinuous Deformation, Changli, Taiwan*.

ICADD2 (1997) In: Ohnishi, Y. (ed.) *The Second International Conference on Analysis of Discontinuous Deformation, Kyoto, Japan*.

ICADD3 (1999) In: Amadei, B. (ed.) *The Third International Conference on Analysis of Discontinuous Deformation, Vail, Colorado*.

ICADD4 (2001) In: Bicanic, N. (ed.) *The Fourth International Conference on Analysis of Discontinuous Deformation, Glasgow, Scotland, UK*.

ICADD5 (2002) In: Hatzor, Y.H. (ed.) *The Fifth International Conference on Analysis of Discontinuous Deformation*. Wuhan, Balkema.

ICADD6 (2003) In: Lu, M. (ed.) *The Sixth International Conference on Analysis of Discontinuous Deformation*. Trondheim, Balkema Publishers.

ICADD7 (2005) In: MacLaughlin, M. & Sitar, N. (eds.) *The Seventh International Symposium on Analysis of Discontinuous Deformation, Honolulu, Hawaii*.

ICADD8 (2007) In: Ju, Y., Fang, X. & Bian, H. (eds.) *The Eighth International Symposium on Analysis of Discontinuous Deformation, Beijing, China*.

ICADD9 (2009) In: Ma, G. & Zhou, Y. (eds.) *The Ninth International Symposium on Analysis of Discontinuous Deformation*. Research Publishing, Singapore.

ICADD10 (2011) In: Zhao, J., Ohnishi, Y., Zhao, G.F. & Sasaki, K. (eds.) *The Tenth International Symposium on Analysis of Discontinuous Deformation*. Honolulu, HI, CRC Press, Taylor &Francis Group.

ICADD11 (2013) In: Chen, G., Ohnishi, Y., Zheng, L. & Sasaki, T. (eds.) *The 11th International Symposium on Analysis of Discontinuous Deformation*. Fukuoka, Taylor & Francis Group.

Ishikawa, T. & Ohnish, Y. (2001) A study on the time dependency of granular materials with DDA. In: Bicanic, N. (ed.) *The Fourth International Conference on Analysis of Discontinuous Deformation, Glasgow, Scotland, UK*. pp. 271–279.

Ishikawa, T., Ohnishi, Y. & Namura, A. (1997) DDA applied to deformation analysis of course granular material (Ballast). In: Ohnishi, Y. (ed.) *The Second International Conference on Analysis of Discontinuous Deformation, Kyoto, Japan*. pp. 252–262.

ITASCA (1993) *FLAC – Fast Lagrangian Analysis of Continua*. Minneapolis, MN, ITASCA Consulting Group Inc.

Jaeger, J., Cook, N.G. & Zimmerman, R. (2007) *Fundamentals of Rock Mechanics*. 4th edition. Wiley-Blackwell.

Jager, C. (1979) The vaiont rock slide. In: *Rock Mechanics and Rock Engineering*. Cambridge, Cambridge University Press.

Jelenić, G. & Crisfield, M.A. (1996) Non-linear master-slave relationships for joints in 3D beams with large rotations. *Computer Methods in Applied Mechanics and Engineering*, 135, 211–228.

Jiang, Q., Feng, X.T., Hatzor, Y.H., Hao, X.J. & Li, S.J. (2014) Mechanical anisotropy of columnar jointed basalts: An example from the Baihetan hydropower station, China. *Engineering Geology*, 175, 35–45.

Jiang, Q.A., Feng, X.T., Xiang, T.B. & Su, G.S. (2010) Rockburst characteristics and numerical simulation based on a new energy index: A case study of a tunnel at 2,500 m depth. *Bulletin of Engineering Geology and the Environment*, 69, 381–388.

Jiang, Q.H. & Yeung, M.R. (2004) A model of point-to-face contact for three-dimensional discontinuous deformation analysis. *Rock Mechanics and Rock Engineering*, 37 (2), 95–116.

Jiao, Y., Huang, G., Zhao, Z., Zheng, F. & Wang, L. (2015) An improved three-dimensional spherical DDA model for simulating rock failure. *Science China-Technological Sciences*, 58 (9), 1533–1541.

Jiao, Y.Y. & Zhang, X.L. (2012) DDARF-A simple solution for simulating rock fragmentation. In: Zhao, J., Ohnishi, Y., Zhao, G.F. & Sasaki, T. (eds.) *Advances in Discontinuous Numerical Methods and Applications in Geomechanics and Geoengineering*. pp. 85–98.

Jiao, Y.Y., Zhang, X.L., Zhao, J. & Liu, Q.S. (2007) Viscous boundary of DDA for modeling stress wave propagation in jointed rock. *International Journal of Rock Mechanics and Mining Sciences*, 44 (7), 1070–1076.

Jiao, Y.Y., Zhang, X.L. & Zhao, J. (2011) Two-dimensional DDA contact constitutive model for simulating rock fragmentation. *Journal of Engineering Mechanics*, 138 (2), 199–209.

Jiao, Y.Y., Zhang, X.L., Zhang, H.Q. & Huang, G.H. (2014) A discontinuous numerical model to simulate rock failure process. *Geomechanics and Geoengineering: An International Journal*, 9, 133–141.

Jiao, Y.Y., Zhang, X.L., Zhang, H.Q., Li, H.B., Yang, S.Q. & Li, J.C. (2015a) A coupled thermo-mechanical discontinuum model for simulating rock cracking induced by temperature stresses. *Computers and Geotechnics*, 67, 142–149.

Jiao, Y.Y., Zhang, H.Q., Zhang, X.L., Li, H.B. & Jiang, Q.H. (2015b) A two-dimensional coupled hydromechanical discontinuum model for simulating rock hydraulic fracturing. *International Journal for Numerical and Analytical Methods in Geomechanics*, 39 (5), 457–481.

Jibson, R.W. (2007) Regression models for estimating coseismic landslide displacement. *Engineering Geology*, 91, 209–218.

Jing, L., Ma, Y. & Fang, Z. (2001) Modeling of fluid flow and solid deformation for fractured rocks with discontinuous deformation analysis (DDA) method. *International Journal of Rock Mechanics and Mining Sciences*, 38, 343–355.

Jun, L., Xianjing, K. & Gao, L. (2004) Formulations of the three-dimensional discontinuous deformation analysis method. *Acta Mechanica Sinica*, 20 (3), 270–282.

Kaiser, P.K. & Cai, M. (2012) Design of rock support system under rockburst condition. *Journal of Rock Mechanics and Geotechnical Engineering*, 4, 215–227.

Kamai, R. (2006) *Estimation of Historical Seismic Ground-Motions Using Back Analysis of Structural Failures in Archaeological Sites*. M Sc Thesis. Beer-Sheva, The Department of Geological and Environmental Sciences, Ben-Gurion University of the Negev. 127 pp.

Kamai, R. & Hatzor, Y.H. (2005) Dynamic back analysis of structural failures in archeological sites to obtain paleo-seismic parameters using DDA. In: MacLaughlin, M. & Sitar, N. (eds.) *The Seventh International Symposium on Analysis of Discontinuous Deformation, Honolulu, Hawaii*. pp. 121–136.

Kamai, R. & Hatzor, Y.H. (2008) Numerical analysis of block stone displacements in ancient masonry structures: A new method to estimate historic ground motions. *International Journal for Numerical and Analytical Methods*, 32, 1321–1340.

Ke, T.C. (1995) DDA combined with the artificial joint concept. In: Li, J.C., Wang, C.-Y. & Sheng, J. (eds.) *The First International Conference on Analysis of Discontinuous Deformation, Changli, Taiwan*. pp. 124–139.

Ke, T.C. (1996) The issue of rigid-body rotation in DDA. In: Salami, M.R. & Banks, D. (eds.) *Proceedings of the First International Forum on Discontinuous Deformation Analysis (DDA) and Simulations of Discontinuous Media, June 12–14*. Berkeley, CA, TSI Press. pp. 318–325.

Ke, T.C. (1997) Application of DDA to simulate fracture propagation in solid. In: Ohnishi, Y. (ed.) *Proceedings of the Second International Conference on Analysis of Discontinuous Deformation*. Kyoto, Japan Institute of Systems Research. pp. 155–185.

Ke, T.C. & Bray, J. (1995) Modeling of particulate media using discontinuous deformation analysis. *Journal of Engineering Mechanics ASCE*, 121, 1234–1243.

Kemeny, J. (2005) Time-dependent drift degradation due to the progressive failure of rock bridges along discontinuities. *International Journal of Rock Mechanics and Mining Sciences*, 42, 35–46.

Keneti, A.R., Jafari, A. & Wu, J.H. (2008) A new algorithm to identify contact patterns between convex blocks for three-dimensional discontinuous deformation analysis. *Computers and Geotechnics*, 35 (5), 746–759.

Kieffer, S.D. (1998) *Rock Slumping – A Compound Failure Mode of Jointed Hard Rock Slopes*. PhD Dissertation Thesis. Berkeley, Dept. of Civil and Environmental Engineering, U. C. Berkeley.

Kilgore, B.D., Blanpied, M.L. & Dieterich, J.H. (1993) Velocity dependent friction of granite over a wide range of conditions. *Geophysical Research Letters*, 20, 903–906.

Kirsch, G. (1898) Die theorie der elastizitat und die bedurfnisse der festigkeitslehre. *Zeitschrift des Vereines deutscher Ingenieure*, 42, 797–807.

Kolsky, H. (1964) Stress waves in solids. *Journal of Sound and Vibration*, 1, 88–110.

Konyukhov, A. & Schweizerhof, K. (2013) *Computational Contact Mechanics: Geometrically Exact Theory for Arbitrary Shaped Bodies*. New York, NY, Springer.

Koo, C.Y. & Chern, J.C. (1996) The development of DDA with third order displacement function. In: Salami, M.R. & Banks, D. (eds.) *Proceedings of the First International Forum on Discontinuous Deformation Analysis (DDA) and Simulation of Discontinuous Media*. Albuquerque, TSI Press. pp. 342–349.

Koo, C.Y. & Chern, J.C. (1997) Modelling of progressive fracture in jointed rock by DDA method. In: Ohnishi, Y. (ed.) *The Second International Conference on Analysis of Discontinuous Deformation, Kyoto, Japan*. pp. 186–200.

Koo, C.Y., Chern, J.C. & Chen, S. (1995) Development of second order displacement function for DDA. In: Li, J.C., Wang, C.-Y. & Sheng, J. (eds.) *The First International Conference on Analysis of Discontinuous Deformation, Changli, Taiwan*. pp. 91–108.

Koyama, T., Irie, K., Nagano, K., Nishiyama, S., Sakai, N. & Ohnishi, Y. (2012) DDA simulations for slope failure/collapse experiment caused by torrential rainfall. In: Zhao, J., Ohnishi, Y., Zhao, G.F. & Sasaki, T. (eds.) *Advances in Discontinuous Numerical Methods and Applications in Geomechanics and Geoengineering*. pp. 119–125.

Koyama, T., Hashimoto, R., Kikumoto, M., Yamada, S., Araya, M., Iwasaki, Y. & Ohnishi, Y. (2013) Application of coupled elasto-plastic NMM–DDA procedure for the stability analysis of Prasat Suor Prat N1 Tower, Angkor, Cambodia. *Geosystem Engineering*, 16, 62–74.

Landers, J.A. & Taylor, R.L. (1986) *An Lagrangian Formulation for the Finite Element Solution of Contact Problems*. Report No. CR 86.008 to Naval Civil Eng. Lab., Cal.

Lang, T.A. (1961) Theory and practice of rock bolting. *Transactions of the American Institute of Mining Engineering*, 220, 333–348.

Lang, T.A. (1972) Rock reinforcement. *Bulletin of AEG*, 9, 215–239.

Lang, T.A. & Bischoff, J.A. (1982) Stabilization of rock excavations using rock reinforcement. In: *23rd U.S. Rock Mechanics Symposium*. pp. 935–943.

Lauffer, H. (1958) Gerbirgsklassifizierung fur den stollenbau. *Geologie und Bauwesen*, 24, 46–51.

Li, S.H., Zhao, M.H., Wang, Y.N. & Rao, Y. (2004) A new numerical method for DEM-block and particle model. *International Journal of Rock Mechanics and Mining Sciences*, 41 (3), 414–418.

Li, S.J., Feng, X.T., Li, Z.H., Chen, B.R., Jiang, Q., Wu, S., Hu, B. & Xu, J.S. (2011) In situ experiments on width and evolution characteristics of excavation damaged zone in deeply buried tunnels. *Science China-Technological Sciences*, 54, 167–174.

Li, S.J., Feng, X.T., Li, Z.H., Zhang, C.Q. & Chen, B.R. (2012) Evolution of fractures in the excavation damaged zone of a deeply buried tunnel during TBM construction. *International Journal of Rock Mechanics and Mining Sciences*, 55, 125–138.

Lin, C.T., Anadei, B., Ouyang, S. & Huang, C. (1995) Development of fracturing algorithms for jointed rock masses with the discontinuous deformation analysis. In: Li, J.C., Wang, C.-Y. & Sheng, J. (eds.) *The First International Conference on Analysis of Discontinuous Deformation, Changli, Taiwan*. pp. 64–90.

Lin, J.S. (2013) A personal perspective on the discontinuous deformation analysis. In: Chen, G., Ohnishi, Y., Zheng, L. & Sasaki, T. (eds.) *Frontiers of Discontinuous Numerical Methods and Practical Simulations in Engineering and Disaster Prevention. Proceedings of ICADD-11*. Fukuoka, Taylor and Francis. pp. 61–66.

Lin, J.S. & Al-Zahrani, R.M. (2001) A coupled DDA and boundary element analysis. In: *Proceedings of the 4th International Conference on Discontinuous Deformation Analysis, University of Glasgow, 6–8 June*.

Lin, J.-S. & Lee, D.-H. (1996) Manifold method using polynomial basis function of any order. In: Salami, M.R. & Banks, D. (eds.) *First International Forum on DDA and Simulation of Discontinuous Media*. Berkeley, CA, TSI Press.

Lin, S. & Qiu, K. (2010) Stability analysis of expansive soil slope using DDA. In: Ma, G. & Zhou, Y. (eds.) *Analysis of Discontinuous Deformation: New Developments and Applications*. Singapore, Nanyang Technological University. pp. 145–151.

Londe, P.F., Vigier, G. & Vormeringer, R. (1969) Stability of rock slopes, a three dimensional study. *Journal of the Soil Mechanics and Foundations Division, ASCE*, 95, 235–262.

Londe, P.F., Vigier, G. & Vormeringer, R. (1970) Stability of rock slopes – Graphical methods. *Journal of the Soil Mechanics and Foundations Division, ASCE*, 96, 1411–1434.

Lysmer, J. & Kuhlemeyer, R.L. (1969) Finite dynamic model for infinite media. *Journal of the Engineering Mechanics Division, ASCE*, 95 (EM4), 859–877.

Ma, G.W. & Fu, G.Y. (2011) Toward a realistic rock mass numerical model. In: Zhao, J. (ed.) *The Tenth International Symposium on Analysis of Discontinuous Deformation, Honolulu, Hawaii*.

Ma, Y.M., Zaman, M. & Zhu, J.H. (1996) Discontinuous deformation analysis using the third order displacement function. In: Salami, M.R. & Banks, D. (eds.) *First International Forum on Discontinuous Deformation Analysis and Simulations of Discontinuous Media*. Berkeley, CA, TSI Press. pp. 383–394.

Ma, Y.Z., Jiang, W., Huang, Z.C. & Zheng, H. (2007) A new meshfree displacement approximation mode for DDA method and its application. In: Ju, Y., Fang, X. & Bian, H. (eds.) *The Eighth International Symposium on Analysis of Discontinuous Deformation, Beijing, China*. pp. 81–88.

MacLaughlin, M.M. (1997) *Discontinuous Deformation Analysis of the Kinematics of Landslides*. PhD Thesis. Berkeley, CA, Dept. of Civil and Environmental Engineering, University of California.

MacLaughlin, M.M. & Doolin, D.M. (2006) Review of validation of the discontinuous deformation analysis (DDA) method. *International Journal for Numerical and Analytical Methods*, 30, 271–305.

MacLaughlin, M.M. & Hayes, M.A., (2005) Validation of DDA block motions and failure modes using laboratory models. In: MacLaughlin, M.M. & Sitar, N. (eds.) *Proceedings of ICADD-7, Honolulu, Hawaii.* pp. 71–78.

MacLaughlin, M.M. & Sitar, N. (1996) Rigid body rotations in DDA. In: *Proceedings of the First International Forum on Discontinuous Deformation Analysis (DDA) and Simulations of Discontinuous Media, June 12–14.* Berkeley, CA, TSI Press; pp. 620–635.

MacLaughlin, M.M. & Sitar, N. (1999) A gravity turn-on routine for DDA. In: Amadei, B. (ed.) *3rd International Conference on Analysis of Discontinuous Deformation (ICADD-3) ARMA, Vail, CO.* pp. 65–74.

Makris, N. & Roussos, Y.S. (2000) Rocking response of rigid blocks under near-source ground motions. *Geotechnique*, 50, 243–262.

Marone, C. (1998) The effect of loading rate on static friction and the rate of fault healing during the earthquake cycle. *Nature*, 391, 69–72.

MATLAB, version 7. *The MathWorks, Inc.*, Natick, MA.

Mauldon, M. (1998) Estimating mean fracture trace length and density from observations in convex windows. *Rock Mechanics and Rock Engineering*, 31, 201–216.

Mauldon, M. & Mauldon, J.G. (1997) Fracture sampling on a cylinder: From scanlines to boreholes and tunnels. *Rock Mechanics and Rock Engineering*, 30, 129–144.

Mauldon, M., Dunne, W.M. & Rohrbaugh, M.B. (2001) Circular scanlines and circular windows: New tools for characterizing the geometry of fracture traces. *Journal of Structural Geology*, 23, 247–258.

Mencl, V. (1966) Mechanics of landslides with non-circular slip surfaces with special reference to vaiont slide. *Geotechnique*, 16, 329–337.

Miki, S., Sasaki, T., Koyama, T., Nishiyama, S. & Ohnishi, Y. (2010) Development of coupled discontinuous deformation analysis and numerical manifold method (NMM-DDA). *International Journal of Computational Methods*, 7 (1), 131–150.

Mikola, R.G. & Sitar, N. (2013) Next generation discontinuous rock mass models: 3-D and rock-fluid interaction. In: Chen, G., Ohnishi, Y., Zheng, L. & Sasaki, T. (eds.) *The 11th International Symposium on Analysis of Discontinuous Deformation.* Fukuoka, Taylor & Francis Group. pp. 81–90.

Milev, A.M., Spottiswoode, S.M., Rorke, A.J. & Finnie, G.J. (2001) Seismic monitoring of a simulated rockburst on a wall of an underground tunnel. *Journal of the South African Institute of Mining and Metallurgy*, 101 (5), 253–260.

Minkowski, H. & Geometrie der Zahlen. Leipzig and Berlin, B.G. Teubner, 1910, JFM 41.0239.03, MR 0249269.

Moosavi, M. & Grayeli, R. (2006) A model for cable bolt-rock mass interaction: Integration with discontinuous deformation analysis (DDA) algorithm. *International Journal of Rock Mechanics and Mining Sciences*, 43, 661–670.

Moosavi, M., Jafari, A. & Beyabanaki, S.A.R. (December 2005) Dynamic three-dimensional discontinuous deformation analysis (3-D DDA) validation using analytical solution. In: *Proceedings of the Seventh International Conference on Analysis of Discontinuous Deformation (ICADD-7), Hawaii, USA.* pp. 37–48.

Morgan, W.E. & Aral, M.M. (2015) An implicitly coupled hydro-geomechanical model for hydraulic fracture simulation with the discontinuous deformation analysis. *International Journal of Rock Mechanics and Mining Sciences*, 73, 82–94.

Munjiza, A. & Andrews, K.R.F. (1998) NBS contact detection algorithm for bodies of similar size. *International Journal for Numerical Methods in Engineering*, 43 (1), 131–149.

Munjiza, A. & Latham, J.P. (2002) Grand challenge of discontinuous deformation analysis. In: Hatzor, Y.H. (ed.) *The Fifth International Conference on Analysis of Discontinuous Deformation*. Wuhan, Balkema. pp. 69–74.

Narr, W. & Suppe, J. (1991) Joint spacing in sedimentary-rocks. *Journal of Structural Geology*, 13, 1037–1048.

Newmark, N. (1965) Effects of earthquakes on dams and embankments. *Geotechnique*, 15, 139–160.

Nezami, E.G., Hashash, Y.M.A., Zhao, D. & Ghaboussi, J. (2006) Shortest link method for contact detection in discrete element method. *International Journal for Numerical and Analytical Methods in Geomechanics*, 30 (8), 783–801.

Nie, W., Zhao, Z.Y., Ning, Y.J. & Sun, J.P. (2014) Development of rock bolt elements in two-dimensional discontinuous deformation analysis. *Rock Mechanics and Rock Engineering*, 47, 2157–2170.

Ning, Y.J. & Zhao, Z. (2012a) A detailed investigation of block dynamic sliding by the discontinuous deformation analysis. *International Journal for Numerical and Analytical Methods in Geomechanics*, 37 (15), 2373–2393.

Ning, Y.J. & Zhao, Z.Y. (2012b) Nonreflecting boundaries for the discontinuous deformation analysis. In: Zhao, J., Ohnishi, Y., Zhao, G.F. & Sasaki, T. (eds.) *Advances in Discontinuous Numerical Methods and Applications in Geomechanics and Geoengineering*. pp. 147–153.

Ning, Y.J., Yang, J. & Chen, P.-W. (2007) Application study of DDA method in blasting numerical simulation. In: Ju, Y., Fang, X. & Bian, H. (eds.) *The Eighth International Symposium on Analysis of Discontinuous Deformation, Beijing, China*. pp. 117–122.

Ning, Y.J., Yang, J., Ma, G.W. & Chen, P.W. (2010) Contact algorithm modification of DDA and its verification. In: Ma, G. & Zhou, Y. (eds.) *Analysis of Discontinuous Deformation: New Developments and Applications*. Singapore, Nanyang Technological University. pp. 73–81.

Obert, L. & Duvall, W.I. (1967) *Rock Mechanics and the Design of Structures in Rock*. New York, NY, John Wiley & Sons.

Oh, Y.N., Jeng, D.S., Chen, S. & Chien, L.K. (2002) A parametric study using discontinuous deformation analysis to model wave-induced seabed response. In: Hatzor, Y.H. (ed.) *The Fifth International Conference on Analysis of Discontinuous Deformation*. Wuhan, Balkema. pp. 113–120.

Ohnishi, Y. & Miki, S. (1996) Development of circular and elliptical disc elements for DDA. In: Salami, M.R. & Banks, D. (eds.) *First International Forum on Discontinuous Deformation Analysis and Simulations of Discontinuous Media*. Berkeley, CA, TSI Press. pp. 44–51.

Ohnishi, Y., Chen, G. & Miki, S. (1995) Recent development of DDA in rock mechanics practice. In: Li, J.C., Wang, C.-Y. & Sheng, J. (eds.) *The First International Conference on Analysis of Discontinuous Deformation, Changli, Taiwan*. pp. 26–47.

Ohnishi, Y., Nishiyama, S., Akao, S., Yang, M. & Miki, S. (2005) DDA for elastic elliptic element. In: MacLaughlin, M. & Sitar, N. (eds.) *Proceedings of ICADD-7, Honolulu, Hawaii*. pp. 103–112.

Ohnishi, Y., Koyama, T., Sasaki, T., Hagiwara, I., Miki, S. & Shimauchi, T. (2012) Application of DDA and NMM to practical problems in recent new insight. In: Zhao, J., Ohnishi, Y., Zhao, G.F. & Sasaki, T. (eds.) *Advances in Discontinuous Numerical Methods and Applications in Geomechanics and Geoengineering*. pp. 31–42.

Ortlepp, W.D. & Stacey, T.R. (1994) Rockburst mechanisms in tunnels and shafts. *Tunnelling and Underground Space Technology*, 9, 59–65.

Osada, M. & Tanityama, H. (2005) Preliminary consideration for analyzing ground deformation due to fault movement. In: MacLaughlin, M. & Sitar, N. (eds.) *Proceedings of ICADD-7, Honolulu, Hawaii*. pp. 113–120.

O'Sullivan, C. & Bray, J.D. (2001) A comparative evaluation of two approaches to discrete element modeling of particulate media. In: Bicanic, N. (ed.) *The Fourth International Conference on Analysis of Discontinuous Deformation, Glasgow, Scotland, UK*. pp. 97–110.

Parker, R.N., Densmore, A.L., Rosser, N.J., de Michele, M., Li, Y., Huang, R.Q., Whadcoat, S. & Petley, D.N. (2011) Mass wasting triggered by the 2008 Wenchuan earthquake is greater than orogenic growth. *Nature Geoscience*, 4, 449–452.

Perras, M.A. & Diederichs, M.S. (2016) Predicting excavation damage zone depths in brittle rocks. *Journal of Rock Mechanics and Geotechnical Engineering*, 8, 60–74.

Pietrzak, G. & Curnier, A. (1999) Large deformation frictional contact mechanics: Continuum formulation and augmented Lagrangian treatment. *Computer Methods in Applied Mechanics and Engineering*, 177, 351–381.

Poetsch, M. (2011) *The Analysis of Rotational and Sliding Modes of Failure for Slopes, Foundations, and Underground Structures in Blocky Rock*. Institute fur Felsmechanik und Tunnel bau Tech, Univ. GRAZ, Heft.

Priest, S.D. (1985) Hemispherical Projectiom Methods in Rock Mechanics. George Allen & Unwin Publishers. Note: this publisher has been purchased by Wiley and they gave us the permision for Figure 4.20.

Priest, S.D. (1993) *Discontinuity Analysis for Rock Engineering*. London, Chapman & Hall.

Priest, S.D. & Hudson, J.A. (1976) Discontinuity spacings in rock. *International Journal of Rock Mechanics and Mining Sciences*, 13, 135–148.

Priest, S.D. & Hudson, J.A. (1981) Estimation of discontinuity spacing and trace length using scanline surveys. *International Journal of Rock Mechanics and Mining Sciences*, 18, 183–197.

Reches, Z. (1987) Determination of the tectonic stress tensor from slip along faults that obey the coulomb yield condition. *Tectonics*, 6, 849–861.

Rohrbaugh, M.B., Dunne, W.M. & Mauldon, M. (2002) Estimating fracture trace intensity, density, and mean length using circular scan lines and windows. *AAPG Bulletin*, 86, 2089–2104.

Rouainia, M., Pearce, C. & Bicanic, N. (2001) Hydro-DDA modeling of fractured mudrock seals. In: Icanic, N. (ed.) *The Fourth International Conference on Analysis of Discontinuous Deformation, Glasgow, Scotland, UK*. pp. 413–423.

Ruf, J.C., Rust, K.A. & Engelder, T. (1998) Investigating the effect of mechanical discontinuities on joint spacing. *Tectonophysics*, 295, 245–257.

Ruina, A. (1983) Slip instability and state variable friction laws. *Journal of Geophysical Research*, 88, 10359–10370.

Rutqvist, J. & Stephansson, O. (2003) The role of hydro-mechanical coupling in fractured rock engineering. *Hydrogeology Journal*, 11, 7–40.

Sagaseta, C. (1986) On the modes of instability of a rigid block on an inclined plane. *Rock Mechanics and Rock Engineering*, 19, 261–266.

Sagy, A. & Brodsky, E.E. (2009) Geometric and rheological asperities in an exposed fault zone. *Journal of Geophysical Research-Solid Earth*, 114, B02301.

Salami, M.R. & Banks, D. (1996) In: Salami, M.R. & Banks, D. (eds.) *First International Forum on Discontinuous Deformation Analysis and Simulations of Discontinuous Media*. Berkeley, CA, TSI Press.

Sammis, C.G. & Ashby, M.F. (1986) The failure of brittle porous solids under compressive stress states. *Acta Metallurgica*, 34, 511–526.

Sanchez-Sesma, F.J. & Campillo, M. (1991) Diffraction of P, SV and Rayleigh waves by topographic features: A boundary integral formulation. *Bulletin of the Seismological Society of America*, 81, 2234–2253.

Sasaki, T., Hagiwara, I., Sasaki, K., Horikawa, S., Ohnishi, Y. & Nishiyama, S. (2005) Earthquake response analysis of a rock-fall by discontinuous deformation analysis.

In: MacLaughlin, M. & Sitar, N. (eds.) *Proceedings of ICADD-7, Honolulu, Hawaii.* pp. 137–146.

Sasaki, T., Hagiwara, I., Sasaki, K., Ohnish, Y. & Ito, T. (2007) Fundamental studies for dynamic response of simple block structures by DDA. In: Ju, Y., Fang, X. & Bian, H. (eds.) *The Eighth International Symposium on Analysis of Discontinuous Deformation, Beijing, China.* pp. 140–146.

Sasaki, T., Hagiwara, I., Sasaki, K., Yoshinaka, R., Ohnishi, Y., Nishiyama, S. & Koyama, T. (2011) Stability analyses for Ancient masonry structures using discontinuous deformation analysis and numerical manifold method. *International Journal of Computational Methods,* 8, 247–275.

Scheele, F. & Bates, B. (2005) DDA benchmark testing at UCT – A summary. In: MacLaughlin, M. & Sitar, N. (eds.) *Proceedings of ICADD-7, Honolulu, Hawaii.* pp. 59–70.

Schnabel, P.B., Lysmer, J. & Seed, H.B. (1972) *SHAKE: A Computer Program for Earthquake Response Analysis of Horizontally Layered Sites.* Berkeley, CA, Earthquake Engineering Research Center, University of California. p. 102.

Scholz, C.H. (2002) *The Mechanics of Earthquakes and Faulting.* Cambridge, Cambridge University Press.

Shi, G.H. (1988) *Discontinuous Deformation Analysis – A New Numerical Model for the Static and Dynamics of Block Systems.* PhD Dissertation. Berkeley, CA, Department of Civil Engineering, U.C. Berkeley. 378 pp.

Shi, G.H. (1991) Manifold method of material analysis. In: *Transactions of the Ninth Army Conference on Applied Mathematics and Computing, Minneapolis, MN.* pp. 57–76.

Shi, G.H. (1993) *Block System Modeling by Discontinuous Deformation Analysis.* Southampton, and Boston, MA, Computational Mechanics Publications.

Shi, G.H. (1995) Numerical manifold method. In: Li, J.C., Wang, C.-Y. & Sheng, J. (eds.) *The First International Conference on Analysis of Discontinuous Deformation, Changli, Taiwan.* pp. 187–222.

Shi, G.H. (1996a) Simplex integration for manifold method, FEM, DDA and analytical analysis. In: *Proceedings of the First International Forum on Discontinuous Deformation Analysis (DDA) and Simulations of Discontinuous Media, Berkeley, CA.* Albuquerque, New Mexico, TSI Press. pp. 206–263.

Shi, G.H. (1996b) *Discontinuous Deformation Analysis Programs Version 96 User's Manual.* Revised by Man-Chu Ronald Young, 1996. Unpublished Report.

Shi, G.H. (1997) *The Numerical Manifold Method and Simplex Integration.* Vicksburg, MS, US Army Corps of Engineers, Waterways Experiment Station. p. 180.

Shi, G.H. (1999) Applications of discontinuous deformation analysis and manifold method. In: Amadei, B. (ed.) *Third International Conference on Analysis of Discontinuous Deformation, Vail, Colorado.* pp. 3–16.

Shi, G.H. (January 2001a) Three dimensional discontinuous deformation analyses. In: *DC Rocks 2001, The 38th US Symposium on Rock Mechanics (USRMS).* American Rock Mechanics Association.

Shi, G.H. (2001b) Theory and examples of three dimensional discontinuous deformation analyses. In: *Proceedings of the 2nd Asian Rock Mechanics Symposium, Beijing.* pp. 27–32.

Shi, G.H. (2005) Producing joint polygons, cutting rock blocks and finding removable blocks for general free surfaces using 3-D DDA. In: MacLaughlin, M. & Sitar, N. (eds.) *Proceedings of ICADD-7, Honolulu, Hawaii.* pp. 1–24.

Shi, G.H. (2012) Rock block stability analysis of slopes and underground power houses. In: Zhao, J., Ohnishi, Y., Zhao, G.F. & Sasaki, T. (eds.) *Advances in Discontinuous Numerical Methods and Applications in Geomechanics and Geoengineering.* pp. 3–16.

Shi, G.H. (2013a) Basic equations of two and three dimensional contacts. In: *Proceeding of the 47th US Rock Mechanics/Geomechanics Symposium, San Francisco, ARMA.* pp. 253–269.

Shi, G.H. (2013b) Basic theory of two dimensional and three dimensional contacts. In: Chen, G.Q., Ohnishi, Y., Zheng, L. & Sasaki, T. (eds.) *ICADD-11: The 11th International Conference on Analysis of Discontinuous Deformation*. London and Fukuoka, Taylor and Francis. pp. 3–14.

Shi, G.H. & Goodman, R.E. (1984) Discontinuous deformation analysis. In: *Proc. 25th U. S. Symposium on Rock Mechanics*. pp. 269–277.

Shi, G.H. & Goodman, R.E. (1985) Two dimensional discontinuous deformation analysis. *International Journal for Numerical and Analytical Methods in Geomechanics*, 9, 541–556.

Shinji, M., Ohno, H., Otsuka, Y. & Ma, G. (1997) The viscosity coefficient of the rock-fall simulation. In: Ohnishi, Y. (ed.) *The Second International Conference on Analysis of Discontinuous Deformation, Kyoto, Japan*. pp. 201–210.

Shyu, K. (1993) *Nodal-Based Discontinuous Deformation Analysis*. PhD Thesis. Berkeley, CA, University of California.

Shyu, K., Wang, X. & Chang, C.T. (1999) Dynamic behaviors in discontinuous elastic media using DDA. In: *Proceedings of the Third International Conference on Analysis of Discontinuous Deformation from Theory to Practice*. pp. 243–252.

Sitar, N., MacLaughlin, M.M. & Doolin, D.M. (2005) Influence of kinematics on landslide mobility and failure mode. *Journal of Geotechnical and Geoenvironmental Engineering*, 131, 716–728.

Skempton, A.W. (1966) Bedding-plane slip residual strength and vaiont landslide. *Geotechnique*, 16, 82–84.

Smith, I.M. & Griffiths, D.V. (1998) *Programming the Finite Element Method*. Chichester, John Wiley & Son Press.

Sofianos, A.I. (1999) Discussion of the paper by M.S. Diederichs and P.K. Kaiser "Stability of large excavations in laminated hard rock masses: The Voussoir analogue revisited". *International Journal of Rock Mechanics and Mining Sciences*, 36, 97–117, 991–993.

Sofianos, I. (1996) Analysis and design of an underground hard rock Voussoir beam roof. *International Journal of Rock Mechanics and Mining* Sciences, 33, 153–166.

Solberg, P. & Byerlee, J.D. (1984) A note on the rate sensitivity of frictional sliding of Westerly granite. *Journal of Geophysical Research*, 89, 4203–4205.

Stead, D., Eberhardt, E. & Coggan, J.S. (2006) Developments in the characterization of complex rock slope deformation and failure using numerical modelling techniques. *Engineering Geology*, 83, 217–235.

Stefanou, I., Sulem, J. & Vardoulakis, I. (2006) Continuum modeling of masonry structures under static and dynamic loading. In: Kourkoulis, S.K. (ed.) *Fracture and Failure of Natural Building Stones – Applications in the Restoration of Ancient Monuments*. Dordrecht, Springer. pp. 123–136.

Tal, Y., Hatzor, Y.H. & Feng, X.T. (2014) An improved numerical manifold method for simulation of sequential excavation in fractured rocks. *International Journal of Rock Mechanics and Mining Sciences*, 64, 116–128.

Tang, C., Zhu, J., Qi, X. & Ding, J. (2011) Landslides induced by the Wenchuan earthquake and the subsequent strong rainfall event: A case study in the Beichuan area of China. *Engineering Geology*, 122, 22–33.

Taylor, R.L. (2004) Finite element solution of contact problems: From 1974 to 2004. In: *Advances in Computational Mechanics, Celebrating the 60th Birthday of Tom Hughes, 7 April 2004*.

Terzaghi, K. (1946) Load on tunnel supports. In: Proctor, R.V. & White, T.L. (eds.) *Rock Tunneling with Steel Supports*. Ohio, Commercial Shearing Inc. pp. 47–86.

Terzaghi, R.D. (1965) Sources of error in joint surveys. *Geotechnique*, 15, 287–304.

Thomas, P.A., Bray, J.D. & Ke, T.C. (1996) Discontinuous deformation analysis for soil mechanics. In: Salami, M.R. & Banks, D. (eds.) *First International Forum on Discontinuous*

*Deformation Analysis and Simulations of Discontinuous Media*. Berkeley, CA, TSI Press. pp. 454–461.

Tian, J.Q., Nishiyama, S., Koyama, T. & Ohnishi, Y. (2012) Masonry retaining wall under static load using discontinuous deformation analysis. In: Zhao, J., Ohnishi, Y., Zhao, G.F. & Sasaki, T. (eds.) *Advances in Discontinuous Numerical Methods and Applications in Geomechanics and Geoengineering*. pp. 169–174.

Trollope, D.H. (1980) The vaiont slope failure. *Rock Mechanics*, 13, 71–88.

Tsai, J.S. & Wang, W.C. (1995) Dynamic response of sliding structure subjected to seismic excitation. In: Li, J.C., Wang, C.-Y. & Sheng, J. (eds.) *The First International Conference on Analysis of Discontinuous Deformation, Changli, Taiwan*. pp. 420–432.

Tsesarsky, M. (2005) *Stability of Underground Openings in Stratified and Jointed Rock*. PhD Thesis. Beer Sheva, Dept. of Geological and Environmental Sciences, Ben-Gurion University of the Negev. 187 pp.

Tsesarsky, M. & Hatzor, Y.H. (2006) Tunnel roof deflection in blocky rock masses as a function of joint spacing and friction – A parametric study using discontinuous deformation analysis (DDA). *Tunnelling and Underground Space Technology*, 21, 29–45.

Tsesarsky, M. & Hatzor, Y.H. (2009) Kinematics of overhanging slopes in discontinuous rock. *Journal of Geotechnical and Geoenvironmental Engineering, ASCE*, 135, 1122–1129.

Veveakis, E., Vardoulakis, I. & Di Toro, G. (2007) Thermoporomechanics of creeping landslides: The 1963 Vaiont slide, northern Italy. *Journal of Geophysical Research-Earth*, 112, F3.

Wang, C.Y., Chuang, C.C. & Sheng, J. (1996) Time integration theories for the DDA method with finite element meshes. In: Salami, M.R. & Banks, D. (eds.) *1st International Forum on Discontinuous Deformation Analysis (DDA) and Simulation of Discontinuous Media*. Berkeley, CA, TSI Press. pp. 263–287.

Wang, C.Y., Sheng, J. & Chen, M.H. (1995) Dynamic contact analysis scheme applied in the DDA method. In: Li, J.C., Wang, C.-Y. & Sheng, J. (eds.) *The First International Conference on Analysis of Discontinuous Deformation, Changli, Taiwan*. pp. 433–459.

Wang, L.-Z., Jiang, H.-Y., Yang, Z.-X., Xu, Y.-C. & Zhu, X.-B. (2013) Development of discontinuous deformation analysis with displacement-dependent interface shear strength. *Computers and Geotechnics*, 47, 91–101.

Wang, W., Zhang, H., Zheng, L., Zhang, Y.B., Wu, Y.Q. & Liu, S.G. (2017) A new approach for modeling landslide movement over 3D topography using 3D discontinuous deformation analysis. *Computers and Geotechnics*, 81, 87–97.

Wang, X.B., Ding, X.L., Lu, B. & Wu, A. (2007) DDA with higher order polynomial displacement functions for large elastic deformation problems. In: Ju, Y., Fang, X. & Bian, H. (eds.) *The Eighth International Symposium on Analysis of Discontinuous Deformation, Beijing, China*. pp. 89–94.

Wittke, W. (1965) Methods to analyze the stability of rock slopes with and without additional loading (in German). *Rock Mechanics and Engineering Geology*, Suppl. II, 52.

Wu, A., Ding, X., Lu, B. & Zhang, Q. (2007) Validation for rock block stability kinematics and its application to rock slope stability evaluation using discontinuous deformation analysis. In: Ju, Y., Fang, X. & Bian, H. (eds.) *Proceedings of ICADD-8, Beijing, China*. pp. 27–32.

Wu, A., Yang, Q., Ma, G., Ju, B. & Li, X.S. (2009) Study on the formation mechanism of Tanjiashan Landslide triggered by Wenchuan earthquake using DDA simulation. In: Ma, G. & Zhou, Y. (eds.) *The Ninth International Symposium on Analysis of Discontinuous Deformation*. Singapore, Research Publishing.

Wu, A.Q., Zhang, Y. & Lin, S.Z. (2012) Complete and high order polynomial displacement approximation and its application to elastic mechanics analysis based on DDA. In: Zhao, J.,

Ohnishi, Y., Zhao, G.F. & Sasaki, T. (eds.) *Advances in Discontinuous Numerical Methods and Applications in Geomechanics and Geoengineering.* pp. 43–53.

Wu, J.H. (2007) Applying discontinuous deformation analysis to assess the constrained area of the unstable Chiu-fen-erh-shan landslide slope. *International Journal for Numerical and Analytical Methods,* 31, 649–666.

Wu, J.H. (2008) New edge-to-edge contact calculating algorithm in three-dimensional discrete numerical analysis. *Advances in Engineering Software,* 39 (1), 15–24.

Wu, J.H. (2010) Seismic landslide simulations in discontinuous deformation analysis. *Computers and Geotechnics,* 37, 594–601.

Wu, J.H., Juang, C.H. & Lin, H.M. (2005) Vertex-to-face contact searching algorithm for three-dimensional frictionless contact problems. *International Journal for Numerical Methods in Engineering,* 63 (6), 876–897.

Wu, J.H. & Lin, H.M. (2013) Improvement of open-close iteration in DDA. In: *The Eleventh International Conference on Analysis of Discontinuous Deformation, Fukuoka, Japan.* pp. 185–191.

Wu, J.H., Ohnishi, Y. & Nishiyama, S. (2005a) A development of the discontinuous deformation analysis for rock fall analysis. *International Journal for Numerical and Analytical Methods,* 29, 971–988.

Wu, J.H., Ohnishi, Y., Shi, G.H. & Nishiyama, S. (2005b) Theory of three-dimensional discontinuous deformation analysis and its application to a slope toppling at Amatoribashi, Japan. *International Journal of Geomechanics,* 5 (3), 179–195.

Wu, W., Zhu, H., Zhuang, X., Ma, G. & Cai, Y. (2014) A multi-shell cover algorithm for contact detection in the three dimensional discontinuous deformation analysis. *Theoretical and Applied Fracture Mechanics,* 72, 136–149.

Yagoda-Biran, G. (2013) *Seismic Hazard Analysis Using the Numerical DDA Method.* PhD Thesis. Beer Sheva, Dept. of Geological and Environmental Sciences, Ben-Gurion University of the Negev. 153 pp.

Yagoda-Biran, G. & Hatzor, Y.H. (2010) Constraining paleo PGA values by numerical analysis of overturned columns. *Earthquake Engineering & Structural Dynamics,* 39, 463–472.

Yagoda Biran, G. & Hatzor, Y.H. (2013) A new failure mode chart for toppling and sliding with consideration of earthquake inertia force. *International Journal of Rock Mechanics and Mining Sciences,* 64, 122–131.

Yagoda Biran, G. & Hatzor, Y.H. (2016) Benchmarking the numerical discontinuous deformation analysis method. *Computers and Geotechnics,* 71, 30–46.

Yang, J. & Ning, Y.J. (2005) Numerical simulation of bench blasting by 2D-DDA method. In: MacLaughlin, M. & Sitar, N. (eds.) *Proceedings of ICADD-7, Honolulu, Hawaii.* pp. 159–165.

Yang, M., Fukawa, T., Ohnishi, Y., Nishiyama, S., Miki, S., Hirakawa, Y. & Mori, S. (2004) The application of three-dimensional DDA with a spherical rigid block to rockfall simulation. *International Journal of Rock Mechanics and Mining Sciences,* 41, 476–476.

Yeung, M.C.R. & Goodman, R.E. (1992) *Use of Shi's Discontinuous Deformation Analysis on Rock Slope Problems.* In: Seed, R.B. & Boulanger, R.W. (eds.) *Stability and Performance of Slopes and Embankments-II,* Vols 1 and 2.

Yeung, M.R. (1991) *Application of Shi's Discontinuous Deformation Analysis to the Study of Rock Behaviour.* PhD Thesis. Berkeley, CA, Civil Engineering, University of California.

Yeung, M.R. (1993) Analysis of a mine roof using the DDA method. *International Journal of Rock Mechanics and Mining Sciences,* 30, 1411–1417.

Yeung, M.R., Jiang, Q.H. & Sun, N. (2003) Validation of block theory and three-dimensional discontinuous deformation analysis as wedge stability analysis methods. *International Journal of Rock Mechanics and Mining Sciences,* 40 (2), 265–275.

Yeung, M.R., Jiang, Q.H. & Sun, N. (2007) A model of edge-to-edge contact for three-dimensional discontinuous deformation analysis. *Computers and Geotechnics*, 34 (3), 175–186.

Yeung, M.R., Sun, N., Jiang, Q.H. & Blair, S.C. (2004) Analysis of large block test data using three-dimensional discontinuous deformation analysis. *International Journal of Rock Mechanics and Mining Sciences*, 41, 521–526.

Zaslavsky, Y. & Shapira, A. (2000) Experimental study of topographic amplification using the Israel seismic network. *Journal of Earthquake Engineering*, 4, 43–65.

Zaslavsky, Y., Shapira, A. & Arzi, A.A. (2000) Amplification effects from earthquakes and ambient noise in the Dead Sea rift (Israel). *Soil Dynamics and Earthquake Engineering*, 20, 187–207.

Zaslavsky, Y., Shapira, A. & Leonoy, J. (1998) *Topography effects and seismic hazard assessment at Mt. Massada – Dead Sea*. The Geophysical Institute of Israel. Report no. 522/62/98.

Zelig, R., Hatzor, Y.H. & Feng, X.T. (2015) Rock burst simulations with 2D-DDA. In: *Proceedings of the 49th U.S. Rock Mechanics Symposium, June 28–July 1 2015. San Francisco.*

Zhang, C., Feng, X.-T., Zhou, H., Qiu, S. & Wu, W. (2013) Rockmass damage development following two extremely intense rockbursts in deep tunnels at Jinping II hydropower station, southwestern China. *Bulletin of Engineering Geology and the Environment*, 72, 237–247.

Zhang, H., Liu, S.G., Chen, G.Q., Zheng, L., Zhang, Y.B., Wu, Y.Q., Jing, P.D., Wang, W., Han, Z., Zhong, G.H. & Lou, S. (2016a) Extension of three-dimensional discontinuous deformation analysis to frictional-cohesive materials. *International Journal of Rock Mechanics and Mining Sciences*, 86, 65–79.

Zhang, H., Liu, S.G., Han, Z., Zheng, L., Zhang, Y.B., Wu, Y.Q., Li, Y.G. & Wang, W. (2016b) A new algorithm to identify contact types between arbitrarily shaped polyhedral blocks for three-dimensional discontinuous deformation analysis. *Computers and Geotechnics*, 80, 1–15.

Zhang, L. & Einstein, H.H. (1998) Estimating the mean trace length of rock discontinuities. *Rock Mechanics and Rock Engineering*, 31, 217–235.

Zhang, L., Einstein, H.H. & Dershowitz, W.S. (2002) Stereological relationship between trace length and size distribution of elliptical discontinuities. *Geotechnique*, 52, 419–433.

Zhang, Y., Wang, J., Xu, Q., Chen, G., Zhao, J.X., Zhenge, L., Han, Z. & Yu, P. (2015) DDA validation of the mobility of earthquake-induced landslides. *Engineering Geology*, 194, 38–51.

Zhao, G.F., Zhao, X.B. & Zhu, J.B. (2014) Application of the numerical manifold method for stress wave propagation across rock masses. *International Journal for Numerical and Analytical Methods in Geomechanics*, 38, 18.

Zhao, Z., Zhang, Y. & Wei, X. (2010) Discontinuous deformation analysis for parallel hole cut blasting in rock mass. In: Ma, G. & Zhou, Y. (eds.) *Analysis of Discontinuous Deformation: New Developments and Applications*. Singapore, Nanyang Technological University. pp. 169–176.

Zhao, Z.Y., Gu, J. & Bao, H. (2007) Understanding fracture patterns of rock mass due to blast load – A DDA approach. In: Ju, Y., Fang, X. & Bian, H. (eds.) *The Eighth International Symposium on Analysis of Discontinuous Deformation, Beijing, China*. pp. 150–156.

Zhao, Z.Y., Zhang, Y. & Bao, H.R. (2011) Tunnel blasting simulations by the discontinuous deformation analysis. *International Journal of Computational Methods*, 8, 277–292.

Zhao, Z.Y., An, X.M. & Zhou, Y.X. (2013) DDA/NMM developments and applications in Nanyang Technological University, Singapore. In: Chen, G., Ohnishi, Y., Zheng, L. & Sasaki, T. (eds.) *The 11th International Symposium on Analysis of Discontinuous Deformation*. Fukuoka, Taylor & Francis Group. pp. 67–80.

Zheng, H. & Li, X.K. (2015) Mixed linear complementarity formulation of discontinuous deformation analysis. *International Journal of Rock Mechanics and Mining Sciences*, 75 (1), 23–32.

Zhong, Z.H. & Nilsson, L. (1989) A contact searching algorithm for general contact problems. *Computers & Structures*, 33 (1), 197–209.

Zhu, H., Wang, S. & Cai, Y. (2007) A discontinuous sub-block meso-damage evolution model for rock mass. In: Ju, Y., Fang, X. & Bian, H. (eds.) *The Eighth International Symposium on Analysis of Discontinuous Deformation, Beijing, China*. pp. 162–167.

Zhu, H., Wu, W., Chen, J., Ma, G.W, Liu, X. & Zhuang, X. (2016) Integration of three dimensional discontinuous deformation analysis (DDA) with binocular photogrammetry for stability analysis of tunnels in blocky rockmass. *Tunnelling and Underground Space Technology*, 51, 30–40.

# Subject index

# ISRM Book Series

*Book Series Editor: Xia-Ting Feng*

ISSN: 2326-6872

Publisher: CRC Press/Balkema, Taylor & Francis Group

1. Rock Engineering Risk
   Authors: John A. Hudson & Xia-Ting Feng
   2015
   ISBN: 978-1-138-02701-5 (Hbk)

2. Time-Dependency in Rock Mechanics and Rock Engineering
   Author: Ömer Aydan
   2016
   ISBN: 978-1-138-02863-0 (Hbk)

3. Rock Dynamics
   Author: Ömer Aydan
   2017
   ISBN: 978-1-138-03228-6 (Hbk)

4. Back Analysis in Rock Engineering
   Author: Shunsuke Sakurai
   2017
   ISBN: 978-1-138-02862-3 (Hbk)

5. Discontinuous Deformation Analysis in Rock Mechanics Practice
   Author: Yossef H. Hatzor, Guowei Ma & Gen-hua Shi
   2017
   ISBN: 978-1-138-02768-8 (Hbk)

Printed in the United States
by Baker & Taylor Publisher Services